Sustainable Water Engineering

The authors dedicate the book to
their families and colleagues

Sustainable Water Engineering

Theory and Practice

Ramesha Chandrappa
Diganta B. Das

WILEY

This edition first published 2014

Registered office
John Wiley & Sons Ltd, The Atrium, Southern Gate, Chichester, West Sussex, PO19 8SQ, United Kingdom

For details of our global editorial offices, for customer services and for information about how to apply for permission to reuse the copyright material in this book please see our website at www.wiley.com.

Library of Congress Cataloging-in-Publication Data

Chandrappa, Ramesha.
Sustainable and water engineering : theory and practice / Ramesha Chandrappa, Diganta B. Das.
pages cm
Includes index.
ISBN 978-1-118-54104-3 (cloth)
1. Water quality. 2. Water–Purification. 3. Water-supply–Management. 4. Hydraulic engineering.
I. Das, D. B. (Diganta Bhusan), 1974– II. Title.
TD370.C484 2014
628.1028′6–dc23 2013050546

A catalogue record for this book is available from the British Library.

ISBN: 9781118541043

Set in 10/12pt Times by Aptara Inc., New Delhi, India.
Printed and bound in Singapore by Markono Print Media Pte Ltd

1 2014

Contents

Preface

Having spent two decades working in sustainable development we feel that it is not a bed of roses. Problems in the form of corruption, illiteracy, data inadequacy, skill deficiencies and the inaction of governments are keeping many people away from adequate quality water.

Many books have been written on water engineering but theories published four decades back often cannot be used today because the quantity and quality of the water that is available has changed and so has the world's population and its ways of living. Similarly the large dams built in the past have not proven to be environmentally friendly and have caused conflicts in many cases. Conserving flora and fauna has become more important all over the world and eating meat will just leave a large water footprint.

Sustainability does not happen overnight and needs investment in terms of money, honesty, knowledge, information and time. There have been substantial examples all over the world where the sustainable use of water has been practised, setting an example for others. This book is a package of theory and practice concerning sustainable and unsustainable water use picked from different parts of the world. Unlike books that only elaborate on theoretical knowledge or simply criticize, this book makes an effort to go beyond theories and to explain the practical world we are exposed to. All is not well all over the world but at the same time not all is wrong. The photographs in the book show both the interest taken in some parts of the world and the negligence in other parts.

We would like to thank Dr Vaman Acharya, D. R. Kumaraswamy and S. Nanda Kumar of Karnataka State Pollution Control Board, Bangalore, India, for their encouragement in our endeavour.

We would also like to thank S. Madhusudhan, Anil Kumar, Amar Yeshwanth, of Karnataka State Pollution Control Board and R. Savyasachi, K. Rahitha, N. Kamalamma and S. Rekha for their help towards the completion of book. We thank the British Council, the Swedish International Agency and the Swiss Development Cooperation for their financial aid extended to the first author during his career. The authors are extremely grateful to the Centre for Science and Environment, Delhi, India and the Swedish Environmental Agency, Stockholm, for arranging extensive international training to the first author in Sweden, which was helpful and gave an opportunity to take the photographs presented in this book.

We thank John Wiley publishers for their faith in us and for investing time and resources. We have worked hard to meet the expectation of the publishers and readers and look forward to any feedback.

Preface

Abbreviations

AC	alternating current
ADP	air-dried pulp
AFO	amorphous ferric oxide
AIDS	acquired immunodeficiency syndrome
ANN	artificial neural networks
AOP	advanced oxidation process
As	arsenic
ASP	activated sludge process
BOD	biochemical oxygen demand
BOF	basic oxygen furnace
Br–	bromide
BrO^{3-}	bromate ion
$CaCO_3$	calcium carbonate
$CaCl_2$	calcium chloride
C and D	construction and demolition
Cd	cadmium
Ce	cerium
CETP	common effluent treatment plant
CFL	compact fluorescent lamp
CN	cyanide
CN^-	cyanide ion
Co	cobalt
COD	chemical oxygen demand
CP	cleaner production
Cr	chromium
CTMP	chemithermal mechanical pulping
Cu	copper
DBP	disinfection byproduct control
DC	direct current
DCB	dichlorobenzine
DDD	dichlorodiphenyldichloroethane
DDT	dichlorodiphenyltrichloroethane
DMP	disaster management plan
Dy	dysprosium
EAF	electric arc furnace
ECF	elemental chlorine free

EIA	environment impact assessment
Er	erbium
FAO	Food and Agriculture Organization
FDI	foreign direct investment
FTW	floating treatment wetland
FDNPP	Fukushima Dai-ichi Nuclear Power Plant
Fe	iron
FOG	fat, oil, grease
Gd	gadolinium
GDP	gross domestic product
GFCI	ground-fault circuit-interrupters
GHG	greenhouse gas
GPP	green public procurement
GTZ	German technical cooperation
Hb	haemoglobin
HCB	Hexo Chloro Benezenes
HCl	hydrochloric acid
HEX-BCH	Hexachlorobicycloheptadiene, Bicyclo(2.2.1)hepta-2,5-diene
Hg	mercury
$HgCl_2$	mercury chloride
$HgSO_4$	mercury sulfate
HIV	human immunodeficiency virus
H_2SO_4	sulphuric acid
IARC	International Agency for Research on Cancer
ICLEI	International Council for Local Environmental Initiatives
IFC	International Finance Corporation
IFRC	International Federation of Red Cross and Red Crescent
IGES	Institute for Global Environmental Strategies
IUCN	International Union for the Conservation of Nature
IWRM	integrated water resource management
KCl	potassium chloride
$K_2Cr_2O_7$	potassium dichromate
KSPCB	Karnataka State Pollution Control Board
kVA	kilovolt-ampere
kWh	kilowatt hour
La	Lanthanum
LDC	least developed countries
LEED	leadership in energy and environmental design
LNWT	low or no waste technology
lpd	litres per day
Lu	lutetium
MCB	monochlorobenzene
MCM	million cubic metres
MDG	Millennium Development Goal
MED	multi-effect distillation
metHb	methomoglobin

MLD	million litres per day
MLSS	mixed liquor suspended solids
Mn	manganese
MSDS	material safety data sheet
MSEW	mechanically stabilized earth wall
MSF	multistage flash distillation
NaCl	sodium chloride
NaOH	sodium hydroxide
NAPL	nonaqueous phase liquid
Nb	niobium
Nd	neodymium
NDMA	*N*-nitrosodimethylamine
NF	nanofilter
NGO	nongovernment organization
Ni	nickel
$Ni(NO_3)_2$	nickel nitrate
NIOSH	National Institute for Occupational Safety and Health
NO_3	nitrate
NO_3^-	nitrate ion
NOx	nitrogen oxide
NTO	nanocrystalline titanium dioxide
NTUA	National Technical University of Athens
OF	overflow
OSHA	Occupational Safety and Health Administration
PAHs	polynuclear aromatic hydrocarbons
Pb	lead
PCP	pentachlorophenol
PIM	potentially infectious material
PO_4	phosphate
POTW	publicly owned treatment works
PPE	personal protective equipment
Pr	praseodymium
PRB	permeable reactive barriers
PVC	polyvinyl chloride
RBC	rotating biological contactors
RFB	river bank filtration
RI	rapid infiltration
RO	reverse osmosis
RWI	recreational water illnesses
SAT	soil-aquifer treatment systems
Sb	antimony
SBR	sequential batch reactors
SCE	snow cover extent
Se	selenium
SIDS	small island developing states
Sm	samarium

SMZ	surfactant modified zeolite
Sn	tin
SO_4	sulfate
SOC	synthetic organic compound
SR	slow rate
STP	sewage treatment plant
TA	technology assessment
Tb	terbium
Tc	technetium
TCF	total chlorine free
TCU	true colour units
Th	thorium
THMs	triholomethanes
Ti	titanium
TKN	total Kjedal nitrogen
Tm	thulium
TOC	total organic compound
U	uranium
UDDT	urine diversion dehydrating toilets
UFW	unaccounted-for water
UNECA	United Nations Economic Commission for Africa
UNEP	United National Environment Protection
UNESCO	United Nations Educational, Scientific and Cultural Organisation
UNICEF	United Nations Children Fund
UPS	uninterrupted power supply
USEPA	United States Environmental Protection Agency
VLH	volatile liquid hydrocarbons
VOC	volatile organic compounds
WCED	World Commission on Environment and Development
WHO	World Health Organization
WWF	World Wide Fund for Nature
WWTP	wastewater treatment plant
Y	yttrium
Yb	ytterbium
Zn	zinc
Zr	zirconium

Glossary

Acidity: The capacity of wastewater or water to neutralize bases.
Activated sludge: Sludge generated in wastewater by the growth of microbes in aeration tanks. In other words it is flocculated sludge of micro-organisms.
Advanced primary treatment: Primary treatment using additives before treatment to augment settling.
Aeration: The process of adding air to water.
Aerobic processes: Biological treatment processes in the presence of oxygen.
Aqua-privy: Watertight tank placed immediately below the latrine floor where excreta drop directly into the water tank through a pipe.
Algae: Variety of plant without distinct functional plant tissue.
Algal bloom: Increase in algae population in water.
Alkalinity: A measure of a substance's ability to neutralize acid.
Alumina: Synthetically produced aluminium oxide that is used as a starting material for the production of aluminium metal.
Anaerobic processes: Biological treatment processes that occur in the absence of oxygen.
Anoxic denitrification: This process is also known as anaerobic denitrification. In this process nitrate nitrogen is converted to nitrogen gas biologically in the absence of oxygen.
Aquifer: Water stored in the saturated zone below the water table.
Attached-growth processes: The biological treatment processes in which the microbes are attached to media.
Autotroph: Organism that uses carbon dioxide as the only carbon source.

Backflow prevention: Preventing the reverse flow of water in water supply system.
Backflush valve: three-way diaphragm valves used in filtration applications.
Backpressure: Pressure opposing the free flow of liquid/gas; it can suck foreign substances into the water-supply system.
Backsiphonage: Backflow due to a differential pressure that sucks foreign substances into the water-supply system.
Batch reactor: Reactors that are operated in batches.
Biochemical oxygen demand (BOD): Measure of the quantity of oxygen used by microbes to degrade organic matter.
Biodegradability: Capable of being decomposed by living things, especially micro-organisms.
Biodiversity: Overall diversity of organisms in the world.
Biogas: Mixture of gases released from anaerobic digestion.

Biological wastewater treatment: wastewater treatment using living organisms.

Biological nutrient removal: The term applied to the removal of nitrogen and phosphorus in biological treatment processes.

Biosolids: The nutrient-rich organic materials from the treatment of sludge.

Or

Organic, rich material left over from aerobic wastewater treatment.

Or

Treated sludge from wastewater treatment.

Blackwater: Wastewater with high organic and pathogen content, consisting of urine, faeces, flushing water, anal cleansing water and greywater.

Boiler feed water: Water fed to a boiler for the generation of steam.

Borehole latrine: The borehole latrine is an excreta disposal system where a borehole is combined with a slab as well as a superstructure.

Borewell: Wells made by drilling boreholes in the earth.

Bottle irrigation: The bottle is first filled with water and then placed in the ground next to the plant and water is made to trickle through it.

Brackish water: Water containing less salt than salt water and more salt than fresh water.

Brownwater: Water consists of faeces and flushwater.

Bund: Embankment constructed from soil.

Capnophilic: Organisms that require increased carbon dioxide.

Carbonaceous BOD: BOD exerted by carbon fraction of organic matter.

Carbon sequestration: The elimination of atmospheric carbon dioxide by biological or geological processes.

Chemical oxygen demand (COD): Standard technique to measure the amount of organic compounds that cannot be oxidized biologically in water.

Chlorination: A process in a water-treatment system where chlorine or a chlorine compound is added to kill harmful micro-organisms such as bacteria.

Clarifier: A tank used for reducing the concentration of suspended solids present in a liquid.

Cluster wastewater system: Wastewater collection and treatment system, which serves some of the dwellings in the community but less than the entire community.

Coagulation: A process of aggregation of colloidal suspended solids by floc-forming chemicals.

Combined sewer: Combining the storm drainage with municipal sewer systems.

Constructed wetlands: Wetlands designed and constructed to treat wastewater.

Cross-connection: The result of a connection between contaminated and noncontaminated water in a water network.

Dead zone: Low-oxygen (hypoxic) areas in the oceans.

Decentralized wastewater treatment: A system divided into groups or clusters where wastewater is treated independently instead of a centralized system.

Denitrification: Microbiological process where nitrities/nitrates are reduced to nitrogen gas, or, removing nitrate biologically and converting it to nitrogen gas.

Desalination: Process of removing salt from water.

Detention time: The time required for a liquid to pass through a tank at a given rate of flow.

Dewatering: Removing water from sludge for further handling and disposal.

Direct surface groundwater recharge: Groundwater recharge to the aquifer via soil percolation.

Disinfection byproduct: Chemical byproducts, formed after disinfection.

Downstream ecosystem: Ecosystem of a lower watercourse.

Drip irrigation: Irrigation in which plants are irrigated through special drip pipes.

Drying bed: Shallow ponds with drainage layers used for the separation of the liquid and solid fraction of sludge.

Dual flush toilet: Flush toilet designed with two handles/buttons to flush different levels of water to save water.

Economic instruments: Fiscal and other economic incentives along with disincentives to include environmental costs as well as benefits.

Ecosystem services: The services provided by ecosystem like habitat for flora and fauna, biological diversity, oxygen production, biogeochemical cycles and so forth.

End-of-pipe approach: Waste-treatment methods conducted at the end of the process stream.

Enteropathogenic serotypes: *E. coli* strains that can cause harmful effects to human beings when consumed in contaminated drinking water.

Eukaryotes: Organisms whose cells contain a nucleus as well as other organelles enclosed within membranes.

Eutrophication: A process of transformation from nutrient-deficit conditions to nutrient-rich conditions, leading to algal blooms in water bodies.

Factor of safety (safety factor): Capacity of a system beyond the expected loads.

Facultative processes: Biological treatment process in which the microbes can function in the absence or presence of oxygen.

Filamentous organism: Threadlike bacteria serving as the backbone of floc formation.

Floc: Particulate or bacterial clumps formed during wastewater treatment.

Flocculation: The process of forming flocs.

Fog harvesting: Collecting fog for anthropogenic activities.

Food to micro-organism ratio (F/M): Amount of food (BOD) available to micro-organisms per unit weight microbes (usually analysed for mixed liquor volatile suspended solids).

Free water surface wetland: A constructed wetland exposed directly to the air.

Green infrastructure: Also known as blue-green infrastructure which highlights the importance of natural environment when making decisions about planning the use of land.

Grey water: Wastewater from baths, sink and wash that can be recycled for *in situ* consumption.

Grit chamber: A chamber or tank in which primary influent is slowed down to remove inorganic solids.

Groundwater: Available natural water found underground in the soil or in between rocks.

Headworks: Structure at the head of a waterway. In the context of water/wastewater treatment, the commencement of the treatment.

Heavy metal: Heavy metals are relatively dense metals like cadmium, chromates, lead and mercury.

High-temperature short-time pasteurization: Passing the milk through heated as well as cooled plates or tubes.

Humus: A dark-brown or black material consisting chiefly of nonliving organic material derived from microbial degradation of plant and animal substances.

Hydrolysis: A decomposition process that breaks down a compound by reaction with water.

Hypernatraemia: A condition where blood sodium level is too high.

Imhoff tank: It is type of treatment in which solids settle in the upper settling compartments and sludge sinks to the bottom of the lower settling compartment where it is decomposed.

In conduit hydropower: Production of hydroelectric power in existing manmade water conveyances like canals, tunnels, pipelines, aqueducts, ditches and flumes.

Indicator organism: Organisms that serve as a measure of the environmental conditions.

Industrial ecology: Industrial ecology is concerned with the flow of dd and materials through systems.

Infiltration basins: Basins used for collecting water for surface groundwater by percolation.

Influent: Liquid that enters into a place/process. Wastewater entering treatment plant.

Ion exchange: Process in which ions of one substance are replaced by ions of another substance.

Lacustrine: Any living organisms growing along the edges of lakes.

Lamella clarifier: Primary clarification device composed of a rack of inclined metal plates to filter materials from water that flow across the plate.

Leachate: Wastewater that trickles in landfill or waste dumps.

Littoral/sublittoral: Any living organisms living along coastal areas.

Lockout: The placement of devices to separate energy to ensure that equipment to be serviced is operated till the lockout device is removed.

Macrophyte: Aquatic plant that grows near or in water.

Microaerophilic: Organisms that require decreased oxygen.

Mixed liquor: The combination of wastewater and return activated sludge in the aeration tank.

Mixed liquor suspended solids: Concentration of suspended solids comprising biomass in an aeration tank in the activated sludge process.

Mutagenic: Capable of inducing mutation and increasing the rate of growth.

Organic loading: Amount of additional organic materials or BOD applied to the filter per day per volume of filter media.

Oxidation pond: Lagoon designed to treat sewage wastewater biologically in secondary treatment with the aid of sunlight, microbes and algae.

Ozonation: A process that introduces ozone into water molecules.

Pathogenic organisms: Bacteria that can cause infectious diseases and harmful effects to humans when infected.

Percolation basins: Seepage of water through soil under gravity.

Permaculture: Branch of ecological design, ecological engineering and environmental design that develops sustainable architecture, human settlements and self-maintained agricultural systems.

Permeable reactive barrier: *In situ* treatment zone that passively captures a plume of contaminants and breaks down or removes the contaminants, releasing uncontaminated water.

Photochemical oxidants: Chemicals that can undergo oxidation reactions in the presence of light.
Photolysis: A process of decomposition of molecules by light.
Phytoplankton: The plant forms of plankton.
Plankton: Microscopic aquatic organisms that swim or drift weakly.
Pour-flush latrine: Latrines are fitted with a trap for providing water seal.
Primary wastewater treatment: The first process usually associated with municipal wastewater treatment to remove the large inorganic solids and settle out sand and grit.
Prokariotes: A group of organisms whose cells lack a membrane-bound nucleus.

Quenching: Rapid cooling of a substance to impart certain material properties.

Reggio Emilia: Approach to teaching young children to improve close relationships they share with their environment.

Salt-water intrusion: Displacement of fresh surface/groundwater by the movement of salt water.
Sequential batch reactor: Aerobic wastewater treatment process that combines reaction and settling in one unit, thereby decreasing foot space.
Sludge: Solid matter generated from wastewater treatment.
Substrate: Organic matter converted during biological treatment.
Suspended-growth processes: The biological treatment process in which the microbes responsible for the changing of the organic matter to biomass.
Swale: Grassed area of depression.

Tagout: Placement of a tagout device on an energy-isolating device to indicate that the energy-isolating device and equipment are being controlled and should not be operated until the tagout device is removed.
Thermotolerant coliforms: Group of bacteria that can withstand and grow at elevated temperatures.
Total Kjeldahl nitrogen: An analysis to find out both the ammonia nitrogen and the organic nitrogen content of organic substances.
Toxoplasmosis: An infectious disease caused by *T. gondii* harmful to human beings. Symptoms include lesions of the central nervous system that can cause brain damage and blindness.
Turbidity: The capacity of suspended solids in water to scatter/absorb light.

Ultrafiltration: A kind of membrane filtration.
Ultrasonic: Ultrasonic is adjective referring to ultrasound (sound with a frequency more than the higher limit of human hearing (20 kHz).
Ultraviolet disinfection: Disinfection using UV rays.
Ultraviolet radiation: Radiation with wavelengths from about 10 nm to 400 nm.
Unconfined aquifers: Saturated permeable soil not capped by impermeable layer.
Urban heat island: Phenomenon where central urban locations will be hotter than nearby rural areas.
UV-A Radiation: UV radiation with wavelength in the range of 315 and 400 nm.
UV-B Radiation: UV radiation with wavelength in the range of 280 and 315 nm.
UV-C Radiation: UV radiation with wavelength between 100 and 280 nm.

Valency: The valency of an atom or group is number of hydrogen atoms of that can displace it or combine with it in forming compounds.
Vat pasteurization: Heating a material for a long period in a vat followed by cooling.
Vector (in the context of epidemiology): Any agent (micro-organism, person or animal) that carries and transmits a pathogen into another living organism.

Water seal: The trap that retains a small quantity of water after the fixture's use.
Watershed: Area of land that contributes rainwater to a water body or stream.
Water table: Top level of the groundwater.
Well casing: Tubular material that gives support to the walls of the borehole.
Well development: Development procedures designed to restore or improve the performance of the borehole.
Well rehabilitation: Cleaning and disinfection of the well and well development procedures to obtain quality water.
Well remediation: Cleaning of oil wells to improve performance.
Well screen: Filtering device that permit groundwater to enter the well.
Wet well: Underground pit used to store wastewater.
Windrow composting: Composting process in which the material is piled up in elongated heaps called windrows.

Yellow water: Urine mixed with flushing water.

Zoonosis: Diseases that occur normally in animals and that are transmitted to people.
Zooplankton: The animal forms of plankton.

1

Water Crisis

Water is essential for life; our food cannot grow without water and millions of plants and animals live in it. Despite this, it is taken for granted in many parts of the world. At times it may feel as though there is an infinite stock of freshwater but available freshwater in the world is less than 1% of all the water on earth. The human population has increased enormously and data show that freshwater species are threatened by human activities. The average population of freshwater species fell by around 47% between 1970 and 2000 (UNESCO, 2006). The problems we face today are numerous but we experience only some of them directly. For example, while many people and animals have died due to water scarcity in various parts of the world, excess nitrate runoff is responsible for dead zones (low-oxygen areas in the oceans) in other parts of the world.

Drinking water that is clean and safe is one of the basic needs for the survival of human beings and other species. It has a large effect on our daily lives and therefore civilizations are concentrated around water bodies (Figure 1.1). We may have to pay a certain amount of money to water suppliers to access drinking water, or we may receive the water supply as an amenity from governments.

Although our planet has a large amount of water, estimated at 1.4 billion km^3, only 2.8% consists of freshwater. Moreover, most of this freshwater is contained in polar glaciers, which dramatically reduces the amount of water available to human beings. Renewable water resources decreased from 17 000 m^3 per inhabitant per year in 1950, to 7500 m^3 in 1995 (UNESCO, 1996), and they are continuing to decrease. Water resource distribution is not uniform on the planet and some countries suffer from natural disasters, such as floods or earthquakes. In such cases, the shortage of drinking water becomes a major problem. Water quality can be dramatically reduced, as was the case after the tsunami in Indonesia in 2004 (Barbot *et al.*, 2009).

Statistically there are many problems associated with a lack of a clean freshwater supply. Diseases and contamination are spread through unsafe water and many people become sick as a result. Problems with water are expected to grow worse in the coming decades, with water scarcity occurring globally. In regions currently considered water rich, primary

Sustainable Water Engineering: Theory and Practice, First Edition. Ramesha Chandrappa and Diganta B. Das.

Figure 1.1 *Civilization has been mainly concentrated adjacent to water bodies.*

water treatment may not be accessible when natural disasters occur (Shannon *et al.*, 2008). Problems with drinking water in the event of natural disasters often concern microbial pollutants, although organic and inorganic chemical pollutants can also play a role (Ashbolt, 2004). Access, to potable clean and safe drinking water has been reported as a major problem faced by the people affected by natural disasters.

Virtually all business decisions will affect natural resources. Of these natural resources, water is the most affected by business decisions all over the world. As other resources have been extracted, the water fit for direct human consumption diminished; often it is not even directly suitable for other purposes, for example industrial and agricultural uses.

Water stress can be defined as a situation where there is insufficient water for all uses. It results from an increase in population, invention of new uses for water and the use of water bodies as disposal points for wastes. Technology has also made it easy to extract water from the groundwater table, divert surface water flows and transport the water to water-scarce locations. Intense urbanization and industrialization have resulted in climate change, thereby enhancing water scarcity and reducing the sustainable supply. Changing climate has increased water shortages due to variation in precipitation patterns and intensity. The subtropics and mid-latitudes, where most of the world's poorest people live, are likely to become substantially drier (Chandrappa *et al.*, 2011). An increase in the temperature has been linked to glacier/snow-cap melting. This water will ultimately reach the sea, so that it will no longer be useful unless it is treated in costly desalination plants. Extreme weather patterns may result in disasters, affecting the quality of water.

Groundwater-dependent areas (where open wells were once sunk) have now adopted drilling technology to extract ground water through bore wells. This technology was attractive as it reduced the time for sinking a well from 3 months to a day. Failure at one spot does not discourage people from sinking another bore well a few metres away at a greater depth than the earlier one. Competition amongst neighbours resulted in emptying ground water, within a decade, which had accumulated over thousands of years.

As the perception of water as an infinite resource is diminishing, many attempts have been made around the world to adapt to the situation using wisdom within the community. Some ideas were successful over time; others failed. While the people in Greenland used melted snow to meet their water needs, the people in the Sahara settled around oases. While people in dry areas of India took a bath once a week or once a month, others in the same country tried to build huge dams across rivers and diverted the water course through a system of canals. While the urban agglomeration grew, these approaches could not be sustained. The wisdom of engineers four decades back is no longer meeting the needs of present population. Systems designed half a century ago have placed environmental and economic burdens on countries and communities alike.

Many of the solutions have now become problems. Examples include huge wastewater treatment plants that are not adequate to cater for today's sewage generation. The entrepreneurs who built industries in the past did not bother to construct sound waste-treatment plants. As a result, mankind depends on technology that requires large amounts of energy and chemicals, resulting in high carbon emissions and large ecological footprints.

Negligence and lack of consideration by government (legislative, executive and judiciary) as well as inadequate investment in public drinking water supplies led to adaptive measures like selling water in sachets in some parts of the world. While pollution has encouraged the bottled water industry, water scarcity has adversely affected food security. Irrigation has helped to improve agricultural yields in semi-arid and arid environments (Hanjra *et al.*, 2009a, 2009b) but 40% of the world's food is produced by 19% of the irrigated agricultural land (Molden *et al.*, 2010). Continued demand for water for urban and industrial use has put irrigation water under greater stress.

Figure 1.2 shows the availability of water per person in different regional of the world based on the information available in Ramirez *et al.* (2011). These figures lead to the conclusion that fresh rain water is more available for a person in America than for one in Asia. This is true because Asia has historically more populous countries. Asia also experiences a lower amount of rain due to its geographical location. Some of the largest deserts are in this continent.

Not all of the 112 100 km^3 of water on the surface of the earth is available to humans. It flows and reaches the sea, making less than 3% of the world's water fresh, of which 2.5% is frozen, locked up in the Arctic, on Antarctica as well as in glaciers. Thus, humanity and terrestrial ecosystems have to rely on the 0.5% of global water. But global freshwater distribution is not equal. The following countries possess nearly 60% of the world's freshwater resources: Brazil, Russia, China, Canada, Indonesia, the United States, India, Columbia and the Democratic Republic of Congo. But, it does not mean that all the people in these countries have sufficient water to fulfil their needs. Local variations within these countries are highly significant.

Given that 120 l/person/day is just sufficient to fulfil the water needs of one person, precipitation across the globe is sufficient to meet requirements. Unfortunately, not all the water is shared equally amongst the people across the globe. As shown in Figure 1.3, only 8% of the water used in the world (not water received by the world through precipitation) is supplied to the public by governments across the globe (www.climate.org/topics/water.html, accessed 13 December 2013) and not all people are fortunate enough to have water supplied to their

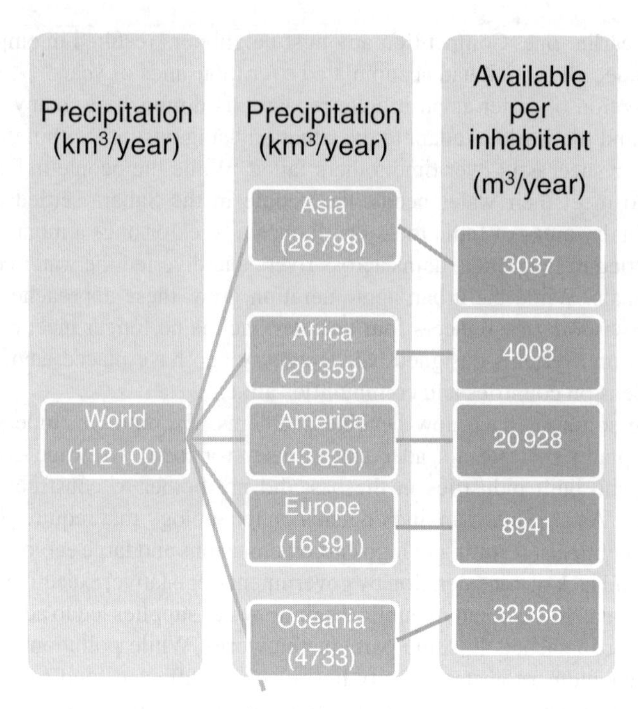

Figure 1.2 Availability of water per person in different regional of the world (based on the information available in Ramirez et al.*, 2011).*

home. Apart from human domestic consumption (drinking, cooking, bathing/sanitation and washing) there has been shift in water consumption by industry since the Industrial Revolution. Industrial activity currently consumes 25% of water and agriculture consumes around 67%, leaving behind the rest of the water for other purposes.

Apart from the discrepancies in water availability, discrepancies in purchasing power due to differential financial distribution have created artificial water scarcity. Some of the pets in rich people's houses will have easier access to water than poor and marginal people (Figure 1.4). While the rich and elites enjoy the water (swimming, car washing, long showers in bath tubs, etc.), poor and marginal people may have to satisfy their needs with less than 10 l/person/day.

The Industrial Revolution and associated poor practices and waste management have resulted in pollution and resource degradation. The cost-cutting principles of industrialists lead to poor treatment of wastewater generated. Many entrepreneurs discovered that

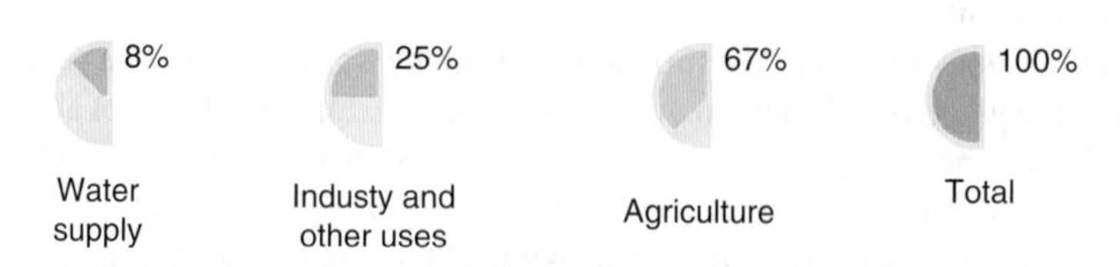

Figure 1.3 Global water use pattern.

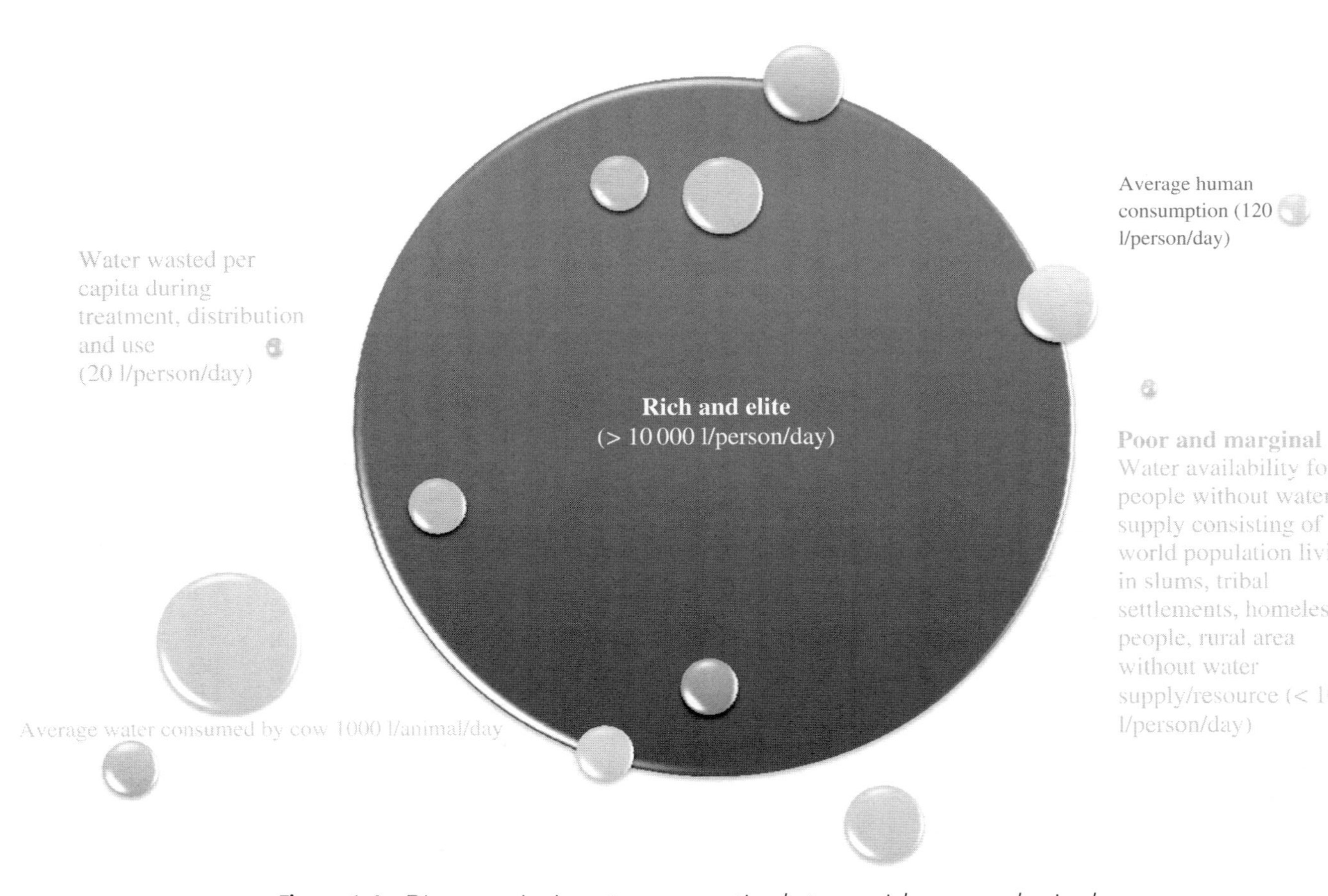

Figure 1.4 Discrepancies in water consumption between rich, poor and animals.

'corruption is cheaper than correction' and discharged wastewater without treatment all over the world. But the enactment and enforcement of stringent laws in developed countries make it possible to regain the quality to greater extent. As a result, some of the developed countries lost manufacturing business to other countries like China and India. The textile mills that were landmarks of Manchester in the United Kingdom and Norrkoping in Sweden are no longer manufacturing textiles but India and China, which export garments to Europe and the United States, have added pollution to water bodies.

Agriculture requires more than 60% of global water use and 90% of the use in the developing countries. Global freshwater consumption has more than doubled after World War II and is likely to rise another 25% by 2030.

Asia has 32% of global total freshwater resources but Asia is home to about 60% of the global population. It is projected that 2.4 billion people in Asia will suffer from water stress by 2025 (IGES, 2005). Developing countries have invested in water infrastructure but not in sustainable infrastructure.

Economic development has made countries thirsty. The situations in Europe during the Industrial Revolution made the countries thirsty during the late eighteenth century. China, with an economic growth rate of 10% per annum since the late 1970s, currently has 20% of the global population and has only 7% of the global freshwater to quench its thirst.

On average, the people of southern China have four times more water than the people in the north whereas people in northern India have more water than their counterparts in the south, the reason being that the Himalayan mountain range, with glaciers that feed perennial rivers, is located towards the north in the case of India and the south in the case of China.

Population projections by the UN in 1996 revealed that world population growth is slowing more than previously thought. The UN projections prove that even slight variations in population growth rates can have affect the quantity and quality of water available to each person. Slower population growth has resulted from the desire of millions of people to have fewer children, which is a welcome development for the future.

1.1 Water Resource Issues

Water is used much faster than nature can replace it. Water is a finite resource circulating between the atmosphere and the earth. Long-term water security cannot be guaranteed if rainwater accumulated in aquifers is mined and overused.

Water stress is caused by (i) excessive withdrawal from surface water and groundwater; (ii) water pollution and (iii) inefficient use of water.

Despite water stress, people stay and face water crises for many reasons, some of which are:

1. Inheritance of property/business in the locality.
2. Absence or lack of skills to move to new place.
3. Lack of confidence to live in new place.
4. Resistance from other region or countries to accepting people from some other countries/religion/region.
5. Cultural, linguistic and financial issues.
6. Attachment to land and people.
7. Dependents like children and old people who cannot move to new place independently.

Migration of people to water-abundant areas is not possible in the current context of the political fragmentation of the globe, thereby ruling out this solution. Countries just cannot accept environmental refugees as it will put burden on their citizens and in time may cause poverty among their original citizens.

Sharing water with other countries located far away is not considered for financial reasons. Sharing of water by neighbouring states might be a solution but there are numerous examples where there has been conflict between such states. States/countries/regions release water when there is abundance and hold water when there is water scarcity, thereby causing floods and droughts respectively downstream.

As a result, people are left only with combinations of the following choices:

1. Reduce the population.
2. Reduce consumption of water.
3. Reduce wastage of water.
4. Reduce/avoid water pollution.
5. Reuse/recycle water.

Discussing how to reduce the human population is beyond scope of this book and there have been many attempts across the globe using legislation, increasing awareness and providing incentives in this regard. A reduction in consumption could be done by avoiding water-intensive crops but, people just refuse to switch over from foods with a higher water footprint to those with a lower water footprint. People do not switch over to vegetarian food instead of meat and dairy products to save water, even though the water footprint of vegetarian food is far smaller than that of food derived from animals. Hence, the only choices people prefer to make is (i) reducing wastage, (ii) reducing/avoiding pollution and (iii) reusing/recycling water. This book elaborates on various methodologies, strategies, issues and challenges in achieving these three objectives.

Manmade and natural disasters are often followed by considerable loss of life and temporary disruption of normal life, which may result in suffering and substantial damage to infrastructure, society and the economy. It has been reported that more than 90% of all disasters occur naturally and 95% of disaster-related casualties occur in developing countries (Thuy, 2010). It has also been reported that Asia and the Pacific are the regions that are particularly affected by disasters (Thuy, 2010).

The earthquake in the Republic of China in 1976 was ranked as one of the most devastating events in terms of the number of people killed and economic damage. Asia, with its geographic position and topographic conditions, has special climatic characteristics, resulting in serious disasters such as floods, typhoons, tornados, tsunamis, earthquakes and droughts.

Meanwhile manmade disasters such as war and political violence, apart from death and destruction, also cause disruption to economic networks and contribute to environmental degradation, which in turn jeopardizes food production, water quality and living conditions (Thuy, 2010). It was reported that, in 2008, about 5600 people lost their lives because of human-made disasters such as shipping disasters, mining accidents, stampedes and terrorism (Thuy, 2010).

As well as food, shelter and medical aid, providing clean water is usually one of the highest priorities in the event of an emergency (Reed, 1995). During emergency situations, the shortage of drinking water is not only an inconvenience but its availability and use under such conditions is also associated with risks that threaten human lives (Thuy, 2010). Effective primary water treatment may not be available to a huge percentage of undeveloped countries. It is therefore essential to have a fully functional portable water purification device in order to live when natural disaster happens. Failure to provide safe water can often be fatal in the wake of natural disasters.

The most popular water treatment methods nowadays include filtration such as sand filtration (Thuy, 2010), bio-sand filtration (Elliott *et al.*, 2008) or membrane filtration (Butler, 2009; McBean, 2009; Park *et al.*, 2009), and coagulation (Garsadi *et al.*, 2008). Normally, these processes do not ensure the disinfection of treated water so a chlorination process is necessary. However, due to the adverse effects from disasters, there is limited access to chemicals such as chlorine or iodine for disinfection and aluminium sulfate for coagulation and also electricity to supply power in order to operate systems. This is the main reason why chemical and electricity requirements are the most important factors that restrict the use of these methods in emergencies (Thuy, 2010).

Membrane technology has emerged and has proven to be an advanced technology for water treatment to produce safe drinking water. Its application is increasing day by day. As compared with conventional treatment methods, water treatment using membrane technology produces a better water quality, uses a much more compact system, is easier to control in terms of operation and maintenance, requires fewer chemicals and produces less sludge (Nakatsuka and Nakate, 1996). The methods to create the driving force for this filtration are more flexible and less dependent on electrical energy; they include use of gravitational force (Butler, 2009), bicycle-powered filtration (McBean, 2009) or wind-powered renewable energy (Park *et al.*, 2009). Research will focus on developing a membrane-based portable water purification system that could be deployed to countries in the wake of natural disasters or for emergencies.

1.1.1 Water Footprint

The water footprint is an indicator of water use with respect to consumer goods (Hoekstra *et al.*, 2011). The water footprint of a product/service is the quantity of freshwater used/evaporated/polluted to produce the product/service. A water footprint has three components: blue, green, and grey (Figure 1.5). The quantity of freshwater evaporated from the surface/groundwater is considered to be the blue footprint. The green water footprint is the quantity of water evaporated from rainwater stored in the soil. The grey water footprint is the quantity of water required so that the quality of the ambient water will be above

Figure 1.5 Definition of blue, green and grey water footprint. (For a colour version of this figure, see the colour plate section.)

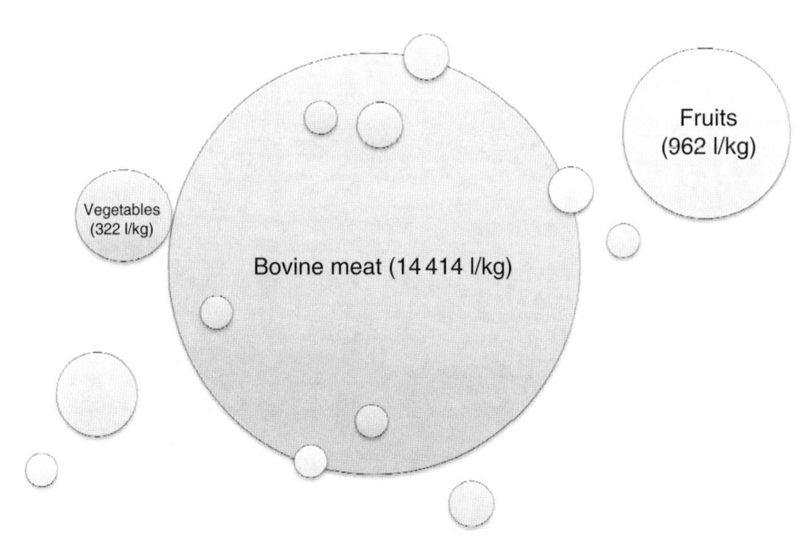

Figure 1.6 *Average footprint for production of vegetables, bovine meat and fruits. (Source: based on data in Mekonnen and Hoekstra, 2010.)*

water quality standards (Hoekstra and Chapagain, 2008). Figure 1.6 shows the average footprint for the production of vegetables, bovine meat and fruits. Twenty-seven per cent of the global water footprint is due to the production of animal products (Mekonnen and Hoekstra, 2010).

Apart from water consumption for food and drinking, the world has witnessed an increase in the use of goods and services, which has left a greater footprint than food production. Hydropower, which accounts for nearly 16% of the global electricity supply, has a blue water footprint of around 90 Gm^3/year which is equivalent to 10% of the blue water footprint of worldwide crop production in the year 2000 (Mekonnen and Hoekstra, 2012). The increase in the number of cars in the world has also placed a high demand on petroleum-based fuel, which is a nonrenewable resource. Hence, the blending of ethanol has been considered as a sustainable solution in many parts of the world and many governments are encouraging production of sugar beet and sugar cane to enhance production of ethanol. The demand for and subsequent diversion of water to grow raw material for ethanol has also resulted in competition for water with conventional uses.

1.2 Climate Change and Its Influence on Global Water Resources

During the Palaeolithic period (before 10 000 BCE) people lived as nomads and there were no permanent settlements and hence no stress on water resources. Humans developed the first stone tools but did not put any stress on water bodies. This was followed by the Neolithic period (or New Stone Age), which started in about 9500 BCE in the Middle East. The Neolithic period was followed by the terminal Holocene Epipalaeolithic period, when farming was started. The Mesolithic period, which occurred 10 000–5000 years ago, saw a growth in population, which started using water resources as well as other natural resources.

Figure 1.7 Raw material storage in cement plant.

The era was characterized by widely dispersed, small, semi-permanent settlements and nomads. The Bronze and Iron Ages occurred 5000 years ago, resulting in villages resulting in early forms of human settlement that covered a few acres and supported a population of several thousand. These villages further developed into permanent settlement in dense aggregations.

The Industrial Revolution, from the 1760s to the mid-1800s in Western Europe and North America, witnessed improvements in industrial machinery, iron smelting, cement manufacture (Figure 1.7), thermal power generation, as well as specialization and division of labour. The era witnessed a decline in mortality, an increase in population and an increase in carbon emissions resulting in global warming due to the greenhouse effect leading to change in active layers, the ice cap, ice flow, ice sheets and ice shelves.

A few centuries ago the rivers of the world had sufficient water to meet the needs of humans and animals. Groundwater was manually extracted from wells and crops were usually grown with rainwater. Even today, rainwater-fed agriculture is practised across the globe (Figure 1.8) both for the market and for home consumption. Water pollution was

Figure 1.8 Preparation of land for rain-fed agriculture.

Figure 1.9 *Freshwater sources, once abundant, are declining due to climate change, destruction of watersheds and distortion of natural streams.*

caused mainly due to small quantity of human sewage. The world's population was less than 25% of today's level.

Global warming has changed the global water cycle due to rising temperatures and hydrological processes. Changing climatic patterns are causing extended periods of drought or flood, a change in rain patterns, an increase in summer, an increase in average temperature, changes in seasons, changes in flowering patterns in plants, an increase in the sea level and melting of ice in glaciers in mountainous and polar regions. Climate change may increase freshwater availability in some places but it causes stress in most places. Freshwater sources that were once abundant are declining due to climate change, destruction of watersheds and distortion of natural streams (Figure 1.9). The decline in the annual snow-cover extent (SCE) has been about 10% in the Northern Hemisphere since 1966 (Robinson, 1997). Global land precipitation has increased by nearly 2% since the beginning of the twentieth century (Jones and Hulme, 1996; Hulme *et al.*, 1998) but the rise is neither spatially nor temporally uniform (Karl and Knight, 1998; Doherty *et al.*, 1999).

In the mid-twentieth century nearly half a million hectares of land irrigated in Pakistan was going out of production every year due to a buildup of salt as well as an increasingly saline groundwater table (Postel, 1999). Another nearly five million hectares of agricultural land was vulnerable with reduced productivity because of high salt concentrations (UNEP, 2007).

About 3 billion humans will join the current of 6.5 billion in the next 100 years, increasing the scarcity created by climate change. Groundwater reserves that were sufficient to meet needs have already exhausted or are on the verge of exhaustion. Continuation of current trends in global greenhouse gas (GHG) emissions will lead to at least a 25% greater water shortage in the next century (Ackerman and Stanton, 2011).

Climate change in sub-Saharan Africa will have consequences that go beyond agriculture. Studies with respect to Northern Kodofan in Sudan explain that temperatures will increase by 1.5 °C between 2030 and 2060, with rainfall declining by 5% (UNDP, 2007).

1.3 Protection and Enhancement of Natural Watershed and Aquifer Environments

Despite abundant rainfall, Africa is facing water scarcity and 25 African countries are likely to face water scarcity by the year 2025 due to increasing demand for water as a result of population growth, economic development, watershed degradation, pollution and inefficient water use (UNECA, 2008). As shown in Figure 1.10 many watersheds now have urban settlement and waste dumping yards, reducing infiltration and enhancing evaporation as well as pollution. The momentum of urbanization throughout the world due to migration was fuelled by failure of agriculture and loss in crops in rural areas. Migration has led to destruction of watersheds and natural drains. In a hurry to accommodate the incoming population, urban planners neglected the protection of natural and aquifer environments. In some cases the population just did what it felt like for survival and settled within the water drains by diverting and obstructing the natural water flow.

1.4 Water Engineering for Sustainable Coastal and Offshore Environments

The quality of coastal and offshore environment is degrading rapidly due to excess human activities. Growing coastal populations as well as overuse of resources from oceans are the major source of the problem.

Figure 1.10 *Many watersheds have been converted into urban settlement and waste-dumping yards.*

Figure 1.11 *Activities in coastal areas have been a source of pollution in the fragile ecosystem of estuaries, deltas and the marine environment.*

Activities in coastal area have been a source of pollution in the fragile ecosystems of estuaries, deltas and marine environments (Figure 1.11). Marine pollution is caused by (i) discharge of untreated or poorly treated sewage from human settlements, (ii) pollutants from industries, (iii) runoff from urban area, (iv) solid waste entering due to offshore and onshore activities.

Waste from ships, like oily sludge, food packaging and food waste, finds its way into water despite the discharge of solid waste from ships being forbidden by the International Convention for the Prevention of Pollution from Ships 1973, modified by its Protocol in 1978.

Ports and harbours generate a range of wastes, which include minerals and damaged products (Figure 1.12).

1.5 Endangering World Peace and Security

There are more than 263 international river basins all over the world and more than 145 countries are riparian to these basins (MacQuarrie *et al.*, 2008). Conflicts over water consist of three key spheres: economic, hydrosphere, and political (Cosgrove, 2003). Water scarcity, speeded up by climate change, may threaten peace and security (Tignino, 2010). While some predict wars (Starr, 1991; Bulloch and Darwish, 1993) others point out that no 'water war' has occurred for nearly 45 centuries (Wolf, 1998). Many regions in the Middle East and North Africa are likely to face further decreases in water availability. Reduced agricultural production due to a decline in rainfall is a factor in conflict in Darfur. During the 1950s, the UN Security Council adopted two resolutions concerning development projects in the

Figure 1.12 *Ports and harbours generate a range of wastes, which include minerals and damaged products.*

Jordan River. Water issues were incorporated in the mandate of the Commission in 1980 by the Security Council (Tignino, 2010).

A decline in rainfall and in water resources has increased damage to fragile ecosystems (Figure 1.13) and has caused forest fires. Agriculture has failed (Figure 1.14) in many parts of globe, leading to migration and endangering peace due to changes in demography and the creation of conflicts.

Figure 1.13 *The drop in water resources has increased damage to fragile ecosystem as well as forest fires.*

Figure 1.14 *Failure of agriculture due to drop in water resources.*

1.6 Awareness among Decision Makers and the Public across the World

Awareness among decision makers and the public varies from place to place. Stringent laws in Europe and the United States generally ensure the quality of water supplied meets prescribed standards. In developed countries, cases are disposed of speedily, there are effective redress mechanisms and the public knows its rights. On the other hand slow dispute-resolution and complaint-redress mechanisms in developing countries often discourage the public from approaching the civic authorities and using the judicial system. In such situations, the public often does not bother to know its rights and accept whatever is made available, thinking 'beggars can't be choosers'. Nobody from the public participated in the decision-making process prior to the incorporation of public consultation mechanisms with respect to mega projects around the world. But, since such involvement leads to delay in project and hindrance from people with vested interests, some projects have been exempted from the public consultation process.

Decision makers are sets of people who can withstand the stress, controversy and opposition with respect to decisions taken by them. They are not subject experts and are usually advised by groups of advisors who are, in turn, supplied with data and information about projects or issues by supporting staff. On the other hand the general public will not have advisors and information providers and awareness comes from their educational background, access to information, interest in particular issues, and the time that is available for them to spare dealing with the issues. A poor person who is bothered about wages for survival usually does not bother about arsenic and fluoride content in the groundwater where he works. There have been many instances where people hardly have choice.

The increase in the consumption of packaged water has been attributed to awareness amongst some people. Despite the cost of packaged water being 500 to 1000 times greater than tap water, people choose packaged water in many parts of the world as they have lost faith in quality of tap water.

Since the Stockholm Convention in 1972, many international laws have been passed. In response to international conventions, much environmental legislation has been passed at national level all over the world but the degradation of the environment continues, throwing the effectiveness of this legislation into doubt. This has attracted environmentalist to produce a variety of recommendations like environmental education at school, awareness and effective communication but the key issue of lack of professionalism in environmental law enforcement and policy-framing organization has stayed in the background. Environmental organization is part of government. The people working in these institutes have either deputed from other government organizations or have recruited in-house people who take motivation from other organizations. The lack of professionalism in government organization ranges from coming into the office late to using office space and time for private purposes. Poor time management coupled with lack of motivation, poor emotional intelligence and lack of respect for the profession add to the inefficiency of the system. It is common to observe many professionals reading newspapers at their desks instead of attending to office work. It is common in some government offices to see people pealing vegetables and tying garlands during office hours for private use. Many government employees use their paid time for worshipping at offices. The late arrival of electricians and hardware/software professionals often hinders the work of others.

Other unprofessional and unethical practices may include use of office staff/vehicles for private purposes. In many cases, the loyalty of government staff/officials to the spouse/children of the head of the office may outweigh their commitment to the responsibilities they are paid for (a practice deriving from monarchy and colonial rule in many developing countries). Official vehicles may be used for personal purposes while personnel struggle to travel for official purposes. Unprofessional record management also often leads to time wasting as the records are searched for quite a long period while decisions are required. Absence of procedural knowledge often results in professionals not attending to the work. Many citizens avoid approaching government offices as the staff in offices beg or demand money even to attend to simple queries. The misplacement of records, corruption, lack of motivation/ethics/ punctuality/loyalty/values and common sense often leads to delay in work, which results in degradation of the environment.

1.7 Criteria for Sustainable Water Management

Sustainable water use and management should provide guidance for the individuals and institutions that use, manage and resolve conflicts. As per, sustainable use of water requires the maintenance of a preferred flow of benefits to a particular place or group, undiminished over time (Gleick, 1998). The World Commission on Environment and Development has indicated that sustainable development is meeting current needs without compromising the position of future generations (World Commission on Environment and Development, 1987).

The standard industrialized world model is not affordable in the developing world. The public drinking water supply is unreliable in Nigeria (Egwari and Aboaba, 2002) and, therefore, drinking water is sold in polythene sachets. In Lagos state, up to 70% of the population is meeting its daily water requirements from sources other than the state municipal supply (Coker, 2004).

Water has historically been managed by governments at local, regional and national level. Decisions are usually taken by a group of people without involving the public (Figure 1.15). Decisions about different development activities occurred discretely without being integrated into water management. As a result, surface and ground water was affected

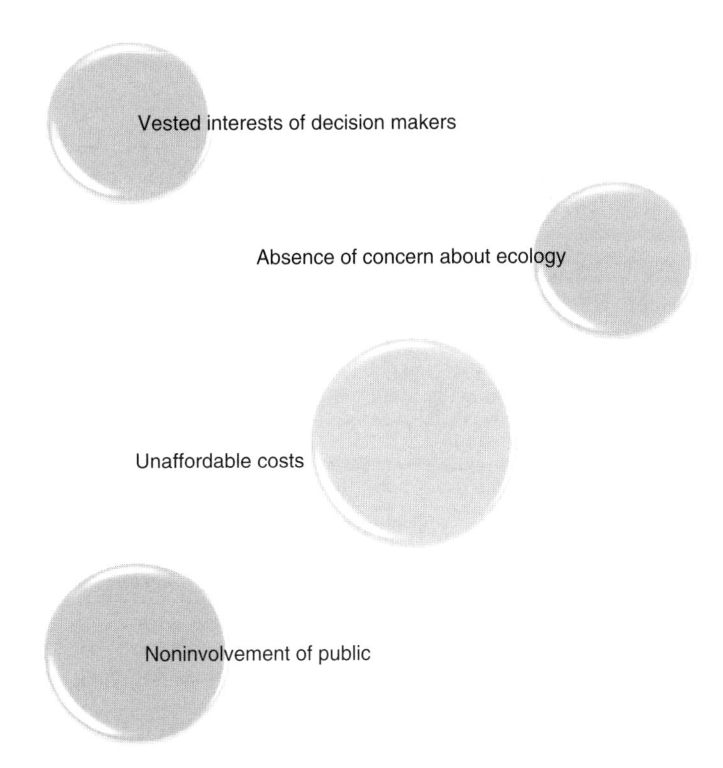

Figure 1.15 *Noninteraction of components in unsustainable water management.*

in various regions. Development activities like the expansion of cities and transportation networks often lead to the fragmentation of watersheds, thereby changing the water flow pattern. Changes in the permeability of soil lead to stagnation and evaporation, resulting in poor ground water recharge capability. The construction of massive dams and water distribution systems without considering their future impact has resulted in disasters as well as ecological imbalance. Such unscientific decisions usually involve corruption and absence of transparency. Where procedures are confidential, without transparency or the involvement of the community, the decisions may be made so as to favour decision makers rather than society or the environment.

Sustainable water management is the current management of water without compromising the needs of future generations. Some of the major recommendations by Postel *et al.* (1996) can be used as criteria for sustainable water management: (i) allocate water to protect the health of ecosystems in the water basin, (ii) eliminate long-term groundwater overdraft, (iii) restructure water institutions to encourage planning for sustainable water use, (iv) link land-use planning and water planning and (v) fill data and information gaps with respect to water. Other criteria for sustainable water management are: (i) maintain a constant water balance within the region, (ii) avoid degradation of the quality of water in the region, (iii) restore water quality within the shortest possible time after disasters, (iv) dissipate information about water quality and quantity continuously to the public, (v) avoid a sudden rise in population density in a few cities due to concentrated development, (vi) predict the impact of all developmental activity on water quality and quantity in the region.

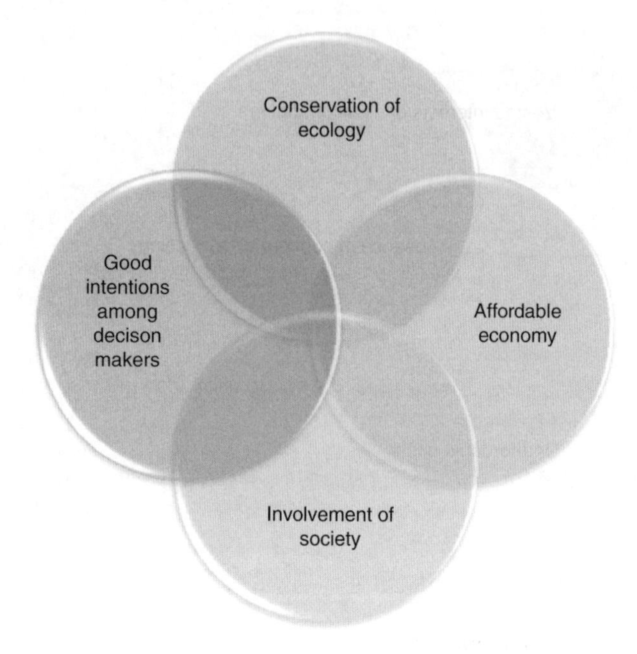

Figure 1.16 *Interaction of components in sustainable water management.*

In a nutshell, unsustainable water management is mainly due to noninteraction of components required for sustainable water management (Figure 1.16): (i) conservation of ecology, (ii) good intentions amongst decision makers, (iii) affordable economy and (iv) the involvement of society.

1.8 Water Scarcity and Millennium Development Goals

The Millennium Development Goals (MDGs) are international development goals that all United Nations member states have agreed to achieve by the year 2015. These goals are:

- eradicating extreme poverty and hunger;
- achieving universal primary education;
- promoting gender equality and empowering women;
- reducing child mortality rates;
- improving maternal health;
- combating HIV/AIDS, malaria and other diseases;
- ensuring environmental sustainability; and
- developing a global partnership for development.

The limits for sustainable water resources have exceeded the threshold of 75% in western Asia as well as northern Africa (UN, 2011). Southern Asia, the Caucasus and Central Asia are approaching the threshold of 60% (UN, 2011).

Domestic budget allocations for drinking water and sanitation have been increasing in some countries but other countries still report insufficient financial resources for them to meet their targets (WHO, 2012). About 1.1 billion urban people and 723 million rural inhabitants have gained access to an improved drinking water source in the period 1990–2008 across the world. Access to water supply and sanitation (WSS) is very low in Africa and nearly half of the population in Africa suffers from one of six major water-related diseases. Improvement in eastern Asia increased from 69% in 1990 to 86% in 2008 and in sub-Saharan Africa access to improved drinking water nearly doubled from 252 million in 1990 to 492 million in 2008 (UN, 2011). Despite increased access to drinking water, the poorest of the poor as well as those disadvantaged because of their sex, ethnicity, age or disability may not receive any benefit.

External assistance in the provision of drinking water and sanitation is provided by multilateral organizations, countries, NGOs and private foundations. Out of the amount invested, some of the infrastructure will be unutilized or underutilized after the project period due to absence of capacity, commitment or funds. Most of the time funding agencies will not look back at funded projects after the project period is over. Even though investments in drinking water and sanitation are increasing, accountability for results achieved remains weak (WHO, 2012).

1.9 Lack of Access to Clean Drinking Water and Sanitation

According to water.org (2009; accessed 13 December 2013), 3.575 million people die every year from water-related disease and 98% of these deaths occur in the developing world. About 1.2 billion people across the world have no access to sanitation at all. These figures clearly indicate issues about historic settlement patterns across the globe. Some countries have made good progress in drinking water coverage. Tanzania and Namibia increased drinking-water coverage from 38% and 58% in 1990 to 73% and 80% respectively in 2002 (WHO and UNICEF, 2004).

Lack of access to clean drinking water could be due to one or many of the following reasons:

1. Financial inability to dig wells or drill bore wells.
2. With the evolution of mankind, nomads started settling where there is an abundance of water but with time people might have moved away from water due to clashes or in search of land. People who inherited land would have lost access to water resources.
3. Rural people who migrated to urban areas in search of jobs ended up in slums and places where water is not available.
4. Earlier generations could have overused water sources beyond their recharge capacity.
5. Water demands due to urbanization and migration might exceed the water available to fulfil the demands.
6. Pollution / contamination due to human/natural causes.
7. Improper use (using ground water in water deficient region to grow crops like sugar cane or rice, which need abundant water).
8. Wastage of water (water loss due to seepage and leakage) and the use of furrow agriculture instead of drip irrigation.

In the past unsustainability of clean drinking water supplies and sanitation occurred mainly due to a poor understanding of the subject and the economics of treatment but the problem has extended into the present despite advances in science and technology. This could be mainly due to professionals in the field making it uneconomical. Hence this book not only looks at various options but also makes an attempt to reduce expenditure by providing orientation to curb expenditure during the establishment and operational stages of water infrastructure.

Water conservation and efficient use and reuse of water have become the need of the day as the planet is facing reduced groundwater and surface-water levels. The water scarcity issue is complicated by drought and changing climate patterns. Water demand management is a useful tool for decreasing luxury water demand, enhancing efficiency and raising awareness about water scarcity. The importance of reducing water demand is highlighted in many parts of the world. One such example is the water-management strategy implemented in Namibia. Due to severe drought, a water-demand-management strategy was implemented in 1994 in Windhoek, Namibia. The key objective was to eliminate luxury water demand and reduce the pressure and the reliance on primary water sources. The strategy was further refined and consists of information campaigns, policy issues, legislation and technical measures.

Another example of water scarcity is that of the Arabian peninsula. Urban water demand has increased rapidly since the 1990s in the Arabian peninsula due to high population growth, better living standards and rapid urbanization. The urban population in the peninsula increased from 6.08 million in 1970 to around 23.12 million in 1995, and is likely to rise from 33.38 million in 2000 to 65 million in 2025. Demand for drinking water is expected to rise from 2269 MCM in 1990 to around 4264 MCM in 2000 and to around 10 580 MCM in 2025 (Abderrahman, 2000). Due to the arid nature of the Arabian Gulf, water is playing a dominant role in human settlements as well as socio-economic interactions (Arab Water Council, 2009). Even though the area was originally dependent upon agriculture, oil and gas have played a major role in the economies of this region since the early twentieth century (Beaumont, 2002). Rapid population growth and urbanization during the first decade of the twenty-first century in the region have created challenges in the provision of infrastructure and public services, including clean water (Kajenthira *et al.*, 2011). Even though agriculture uses the highest percentage of water withdrawal in the Arabian Gulf as well as worldwide, the present economic growth in the region is likely to result in increased quantity of wastewater in the municipal/industrial sectors (Jimenez and Asano, 2008), which should be effectively reclaimed and reused to avoid water scarcity. At present only 65% of the wastewater in the Arabian Peninsula is treated (Kajenthira *et al.*, 2011) and an increase in population and wastewater quantity has resulted in overloaded treatment plants (Qadir *et al.*, 2010).

1.10 Fragmentation of Water Management

Law, responsibility and business bind different authorities and agencies with different territories. The major types of fragmentation in water management include jurisdictional, temporal, territorial, financial, cultural and biophysical (Figure 1.17). Jurisdictional fragmentation occurs when a number of agencies have varied responsibilities over a spatial territory. Jurisdictional fragmentation is created by the political and legal institutions that hold authority in a territory. Water management is shared amongst numerous government actors and can generate a governance gap when jurisdictional responsibility is uncertain.

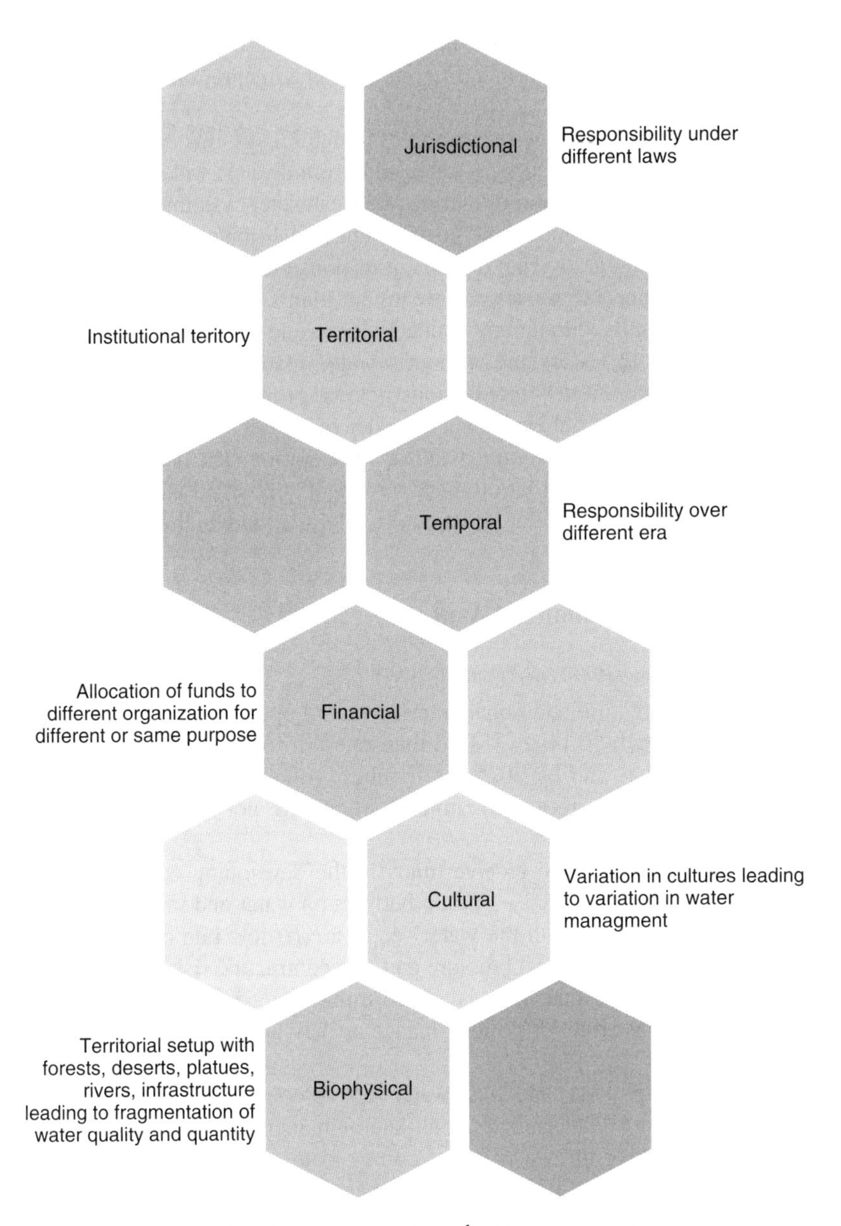

Figure 1.17 *Fragmentation of water management.*

Allocation of responsibility with respect to water governance amongst multiple actors/agencies may lead to relatively little or no coordination.

The agency responsible for water distribution hardly has any jurisdiction to curb pollution by polluters; similarly the agency that is responsible for safeguarding quality does not bother about a reduction in the quantity of water resources. Similarly, the irrigation department usually does not bother about the quality of water or the shortage of water faced by industries.

Temporal fragmentation exists when the effectiveness of one agency is obstructed by the existence of another agency from a different era. The integration of fragmentation is a challenge all over the world.

Biophysical fragmentation of water resources in natural systems can result in fragmentation between land and water; surface water and groundwater; water quantity and water quality; upstream and downstream; or marine and freshwater systems.

Financial fragmentation compels the government to distribute funds amongst different agencies thereby creating an overlap or gaps in responsible spending. The funds released in some places for water and wastewater treatment plants will therefore only ensure the construction of these units without any connectivity to end users. It is also quite common to see urban bodies with sewers and without sewage treatment plants (STPs). There are numerous examples where water-intense industries and projects are cleared by government agencies without looking into their impact on water resources.

Another example of fragmentation is wildlife conservation. The department responsible for conservation of wildlife will hardly have any legal support to initiate legal action on those who deplete water resource or affect quality of water on which wildlife depends.

1.11 Economics and Financial Aspects

The quality of water used in various industrial processes is critical and therefore polluted water resources require additional investment and operational costs. Reduced water availability is affecting food prices. More than one-third of the global population lives in water-stressed nations and by 2025 the number is likely to rise to two-thirds (FAO, 2007). When alternative sources or treatment options are not feasible, industry will be disrupted.

Water and sanitation tussle to receive funds in the developing countries. Developing countries spend between 1% and 3% of their budgets on water and sanitation.

The main risks associated with the water sector are (i) low rate of return, (ii) foreign exchange risk, (iii) risk of political pressure and (iv) contractual risk. Water projects have the additional disadvantage that there is a high minimum size of project finance. Many water projects are not feasible for project finance as they are below the minimum size for financing.

Major funding for the water and wastewater sectors comes from governments, NGOs, banks and private investors, aid agencies, multilateral financial institutions and international organizations. Finance for the water sector comes from (i) state and federal grants, (ii) international donor agencies, (iii) lending from financial institutions and (iv) infrastructure bonds.

The water sector attracted private participation in 68 developing countries between 1990–2005, with most of the investment concentrated in China (56%) and Algeria (34%) (GTZ, 2006).

Some financial approaches for water problems are given in Box 1.1 and case studies of sustainable water financing are given in Box 1.2. The water-financing systems should be coherent, sustainable and adequate. The ideal commercial water finance is low-interest, long-term loans available for subsovereign borrowers. Such loans will usually not have any surety as in the case of private loans.

Box 1.1 Some financial approaches for water problems

Loans with subsidies: Practised in many parts of world, especially with respect to bore wells drilled by farmers. Farmers will pay only part of the amount they borrowed while the rest is paid by government.

Loans without subsidy: Loans borrowed for residences, industries and commercial activity will use part of the expenditure from the loan for drilling bore wells or water connections. Loans from international donors are common practice at a national level.

Loan waive off: Practised in many parts of world by government especially to lure 'vote banks' by waiving off loans for the farming community at the taxpayers' expense.

Equities: Mobilization of funds by the public equity participation of water companies has been the approach in many countries. An investor shares the risk and the profit of a company or project by the purchase of equity shares.

Micro credit: A type of loan extended to poor communities. This approach is commonly used for sanitation.

Micro savings: Saving small amounts for use at a later date.

Lifeline approach: A state/government/donor will provide water free of cost. Donors can be anthropogenic agencies, industries (under corporate social responsibility), temples and so on.

Reimbursement: Reimbursement can be for the urban poor (as in case of Santiago de Chile where municipality reimburses 20% of the poorest urban population) or for those who can afford to pay (as in military settlements and some select government officials as well as people's representatives).

Bonds: A public corporation or municipality may issue bonds that are purchased by private financial institutions. Similarly, private company bonds may be purchased by the public. Usually income-tax benefits would be given to such investments (that is, instead of paying tax the amount can be invested in bonds or the taxpayer will be given a tax rebate).

Infrastructure cess: Cess is nothing but a subtax for a specific purpose. It is usually paid by those who have an income and can afford to pay. In such a scenario a small additional tax is imposed, which can be a small percentage of the total amount of income tax being paid. Alternatively bulk consumers like industries may pay cess calculated according to the amount of water consumed.

Stealing: Even though it is unethical, stealing is carried out by industries, commercial establishments and households by bypassing a connection around the water metre or by providing a faulty water metre. Some users may obtain an illegal connection by paying a bribe to officials. Such stolen water is usually considered as a loss to avoid further inquiries and action against the water-supplying organization.

Box 1.2 Case studies of sustainable water financing

Case 1: France

Water policy preparation, legislation and regulation are funded by national budgets. Regional water agencies (*agences de l'eau*), pertaining to the major river basins, are responsible for managing water resources, which includes abstraction and discharges. Each agency has a council of consumers and other stakeholders to review and vote on expenditure. Based on abstraction and pollution, taxes are raised from water users. Revenues are distributed through *agences de bassin* for environmental improvement and water-management measures. Local authorities provide the services directly, or through companies. Investments are financed and carried out by local bodies, which can draw soft loans from central government. Water consumers pay for water through tariffs (Cap Net, 2008).

Case 2: Netherlands

Policy, administration and supervision in the water sector is the responsibility of the central government at national level while the provinces are responsible for strategic policy, management and operation, as well as supervision of the water boards and local bodies. VEWIN (the water-planning agency) charts out ten-year plans. The Dutch Water Bank is the main source of investment in the country and it will lend only to the public sector. The water boards raise revenues by collecting property taxes. Drinking water companies treat and distribute drinking water whereas municipalities are responsible for sewerage and wastewater treatment.

Globally, some 200 billion litres of bottled water are consumed, generating around 200 billion bottles of waste water. Table 1.1 shows the cost of water in some countries. Owing to the impact on the environment, the US town of Concord in Massachusetts banned the sale of bottled water in bottles containing less than 1 litre.

1.11.1 Water Treatment and Distribution

People in many parts of the world go to river banks to carry out daily activities like washing clothes (Figure 1.18), washing animals, and taking baths, exposing themselves as well as others to risk. Population growth and economic development are causing increases in water demand in agriculture, homes and industries. Semiconductor firms need huge quantity of

Table 1.1 Cost of water.

Sl. no.	Description	India	United States	Switzerland	Sweden	United Kingdom
1	Rainwater	Free	Free	Free	Free	Free
2	Bottled water (per litre)	Rs. 20	$ 0.5–2.0	1–4.5 sfr	14–18 Kr	40p to £3

Figure 1.18 *Village dwellers on a river bank washing cloth.*

ultraclean water and 11 out of world's 14 largest semiconductor industries are in the Asia-Pacific region, where water quality is already severely stressed. Beverage bottlers lost their operating licenses in some parts of India due to water shortages. Freshwater consumption has more than doubled globally after World War II and is likely to rise by 25% by 2030 (Wild *et al.*, 2007).

The informal sector not only supplies water through tankers to slums but also to high-rise apartments – for example, in Bangalore, India, where the sudden population explosion has increased demand for water for construction activities as well as domestic activates in the city. About 60% of Delhi's people are not served by domestic connections and receive water from tube wells with hand pumps as well as water tankers.

Depending on the nature of the project, adequate financing need not necessarily guarantee sustainability. Southern Nevada Water Authority (SNWA) intends to supply 11 billion gallons water to urban dwellers from rural north-eastern Nevada, about 300 miles away despite opposition from Clark County farmers and conservationists. The water pumped to Las Vegas under the proposal does not sustain the city's annual growth and 40 million annual visitors (Sweet, 2008). The Southern Nevada Water Authority plans to finance project with tax-exempt bonds but the bonds may present long-term risks due to (i) significant environmental concerns, (ii) Clark County farmers and conservationists feel that there are high energy costs in withdrawing the water and pumping and (iii) water pumped will not sustain the city's annual growth and visitors (Morrison *et al.*, 2009).

In general, a reverse osmosis (RO)-based water-treatment system is most commercially available especially in the developed countries. It consists of a multistage process that involves both pretreatment and post-treatment stages in addition to an RO spiral-wound membrane module. Typical pretreatment stages involve sediment filters or microfilters and activated carbon. Meanwhile, post-treatment stages will include activated carbon filters. Such systems are placed on a table top or under a sink system. They are normally installed in order to purify tap water coming from water supply. The system can work perfectly without the use of electricity; the pressure will be provided by feed tap water in the system. The maintenance in most cases will be the replacement of pre- and postfilters once in 16–18 months. The membrane lifetime is expected to be 2–3 years, depending on the quality of tap water. The price for such system varies according to the flow rate in the range from US$ 200 to 700. Furthermore, their annual operating costs are approximately US$ 85–135.

In general, such RO-based water treatment systems with multistage pre- and post-treatment required expensive installations that need service and parts replacement and a good-quality water source. Due to these factors this system may not be applicable in poor countries even if it is widely used in developed countries.

The 'ROSI' system was developed by Schafer and Richards (2005), which is an RO-based system designed independently without the use of an energy source. This system treats water from different types of water sources, ranging from highly turbid surface waters to highly saline brackish. The process involves two stages: a pretreatment stage using ultrafiltration (UF) membrane followed by a desalination stage, which uses either an RO or nanofilter membrane. The function of a UF membrane is to remove most pathogens, such as bacteria, particles and some colloidal material; in this way the RO/NF membrane is protected from excessive biofouling and the cleaning frequency of the modules is reduced (Schafer and Richards, 2005). It is equipped with photovoltaic or solar modules and the ROSI system may be used independently of any energy source in regions with high sunshine intensity. This system has been tested in remote rural areas in Australia. However, there are no published data on the costs of this system. The equipment is relatively complex, which includes UF, RO/NF and photovoltaic modules. Hence the investment costs are high and maintenance would require skilled personnel.

Most water-quality problems are due to the presence of pathogens, which are completely removed by ultrafiltration membranes. Reverse osmosis membranes require much higher pressures than UF membranes due to higher resistance and also because RO generates osmotic pressure, which counteracts the water transported through the membrane (Peter-Varbanets *et al.*, 2009).

Many commercially produced microfiltration filters are for travellers' use.

Another newly implemented microfiltration application is the 'filter pen', operated by sucking raw water through a strawlike device, which is actually a microfiltration membrane. The membrane has an average pore size of 0.15 μm and a surface area of 0.02 m^2. According to the manufacturer's data, initial clean water flow rates are about 0.1 l/min at a pressure difference of 0.1 bar. Depending on the quality of the feed water, the service life of the filter pen is approximately 4 weeks or 100 litres of treated water, which is equal to a water production of about 3.5 l/day.

The use of membrane systems has increased significantly, especially for water and wastewater treatment (Peter-Varbanets *et al.*, 2009). Although this technology has become more efficient and the costs of the membranes have decreased significantly (Churchouse and Wildgoose, 1999), it remains unaffordable for the world's poorest. Previous research has shown that research into, and development of, membrane systems aimed specifically for the developed countries remains limited to isolated cases (Wessels, 2000; Pillay and Buckley, 2003; Goldie *et al.*, 2004) and is often not published in the available literature (Peter-Varbanets *et al.*, 2009), although this is important to ensure that the knowledge/skills in developing these systems is not restricted to a few individuals or groups or companies.

1.11.2 Wastewater Treatment, Collection and Disposal

Despite the Millennium Development Goal (MDG) targets and the increase in awareness about sanitation, financial commitment is lacking. Despite the understanding that an adequate and reliable sanitation service is essential to improve health, urban and rural settlements are largely unsuccessful in meeting basic needs. Each sanitation facility will cost nearly the price of a cow (or ten piglets or four sheep), which is not attractive to a villager (Chandrappa and von Munch, 2008). Hence, poor people in rural settlements would opt for open defecation, thereby spreading pathogens. Total bilateral aid for water as well as sanitation from the major OECD donors was 25% lower in 2001 and 2002 compared to 1998 and 1999 (GTZ, 2006). Due to the failure to prioritize the water supply and sanitation in national development perceptions, the sector has to compete with other sectors and it does so with only limited success due to low recognition of its importance. The sector is therefore frequently underfunded. In sub-Saharan Africa, East Asia, South Asia and the Pacific, less than 50% of the population has access to sanitation. Central Europe, Eastern Europe, Latin America and Caribbean have the maximum rates of access to sanitation.

Apart from pollution due to poor sanitation, pollution from industries is posing a major challenge especially that from small- and medium-scale industry. In many parts of the world, sourcing goods and services, small- and medium-scale industries are economically attractive as they are operated and owned by entrepreneurs, thereby reducing expenditure on manpower and overheads. Such industries also contribute to water contamination as they try to cut down the costs of pollution-control activities as well (Figures 1.19 and 1.20).

On the other hand, the organic compounds produced by the petroleum and other chemical industries can contaminate sites and leach to groundwater. These chemicals are persistent and have low solubility; light ones float on the water table and heavy ones will move down to bedrock and form a no aqueous phase liquid (NAPL). BTEX, composed of benzene, toluene, ethyl benzene and xylene, is a basic petroleum derivative that can be found as a groundwater contaminant.

Figure 1.19 Wastewater is being drained out without treatment from a small-scale industry.

Figure 1.20 Oil spillage in a waste-oil storage yard.

1.12 Legal Aspects

Legislation related to the environment and the conservation of natural water quality was enacted for the first time in many countries all over the world after United Nations Conference on the Human Environment held in Stockholm in 1972. Even though the implementation of these laws has not been as thorough in the developing countries, the situation is far better than in 1972, which has been a cause for hope in every country. Even though laws are flouted in many countries, activism by nongovernmental organizations (NGOs) and public-interest litigation in many parts of the world have restored confidence among citizens. Legislation related to water conservation will typically have the following: (i) objectives, (ii) institutional arrangements, (iii) institutional responsibilities, (iv) enforcement mechanisms, (v) standards for effluent and natural water bodies.

Responsibilities are often twofold: (i) administrative and (ii) enforcement. Administrative aspects include recruitment, the academic qualification of officers and staff, salary structure, budgeting responsibility, delegation of financial powers and so on. Enforcement usually involves filing cases, issuing closure orders, issuing permits and cutting down essential services like power/fuel/water.

Legal enforcement can be proactive or relative. Proactive approaches include prohibition of certain activities in certain areas, protecting sensitive areas/water bodies, protecting catchment areas and so on. Reactive approaches include lodging complaints/bringing cases in courts of law, giving direction to polluting units, giving direction to organizations that are providing essential services to stop services. Essential services include water/fuel/electric supplies, securities, loans and so on.

In many instances, failures of environmental laws (Figure 1.21) occur due to ignorance by the enforcing agency or too much knowledge amongst the polluters about loopholes in the law. Usually the government will protect the interest of polluters as they are a source of funds in democratic countries. Officers of the enforcing agency will often be used for purposes other than pollution control, like election duty, work related to national festivals and personal work by people's representatives and people in power. Too much time on paperwork, low work ethics, poor record-keeping practice and lack of decision-making ability in top

Figure 1.21 *Reasons for failure of environmental legislation.*

management also sometimes lead to the weakening of organizations' ability to deliver their objectives. Many of the developing countries will have poor manpower due to the inability of organizations to recruit eligible candidates or because of pressure from politicians. Some environmental protection organizations are bogged down with indiscipline and disputes among the staff.

External interference still dominates in many environmental protection organizations and staff are unable to update their knowledge about court verdicts, new policies due to international law and local legislation. It is often considered that family members of higher officials are more important than people who work in the organization. Use of vehicles, infrastructure and funds by political leaders, higher officials and their family members for private purposes has been a cause of a decline in morals amongst the staff in public organizations.

Holidays and nights are most often used by polluters and officers are often distracted by false complaints. The media attach greater importance to sports and glamorous events than to environmental issues and this has also often distracted ruling representatives. In many cases political leaders will be unable to see beyond the next election and hence they prioritize local issues that can bring them more votes than long-term environmental protection. Free electricity and loan waivers have resulted in the depletion of groundwater due to excessive drilling of bore wells and siphoning of ground water. The use of excessive agrochemicals and encouragement of industries has also caused environmental deterioration in many developing countries.

As per Shapiro *et al.* (2009), the five 'protector agencies' in the United States – Food and Drug Administration, the Occupational Safety and Health Administration, the Consumer Product Safety Commission, the National Highway Traffic Safety Administration and the Environmental Protection Agency – grapple with their responsibility to protect the American public from hazards. The agencies' inability to act swiftly and decisively in recent decades is largely the result of severe shortfalls in funding, political interference, outdated authorizing statutes and an ageing, demoralized civil service (Shapiro *et al.*, 2009).

Rapid urbanization due to poor migration policies and job creation has led to poor urban planning. In many instances, industries established within the city in previous decades continue to operate while new residential apartments are established, leading to clashes in the community. Too much emphasis by financial planners on increasing gross domestic product (GDP) has led to urbanization being prioritized over improvements in the rural economy. Establishment of industrial areas is often considered the only way to improve the economy rather than preserving the environment and water quality. As a result deteriorating heath due to a poor environment and waterborne diseases has led to an increase in GDP due to burgeoning medical expenditure and strengthening of pharmaceutical sector. The deteriorating water quality has also led to an increase in sales of bottled water across the globe, thereby contributing to GDP.

Unsafe working conditions due to terrorism or workplace harassment have also resulted in poor performance by environmental protection organizations in many countries. The absence of transparency where it is required (as in decision making) and the absence of confidentiality in certain situations (like drafting legal cases against defaulters) have also been reasons for failure of enforcement in many situations.

According to Pallangyo (2007) there are over 50 principal laws in Tanzania related to environmental issues and many of these laws are outdated, not understood and overlap in terms of functional authority. Most of these laws fail to induce compliance since the *ex ante* value of the penalties is far below the compliance cost. As per Pistone (2010) at least 103 children in Toms River, located in Dover Township, New Jersey in the United States, had been diagnosed with cancer and problem is that the loopholes in the laws concerning how remediation is carried out rely greatly on agencies' discretion.

References

Abderrahman, W.A. (2000) Urban water management in developing arid countries. *Water Resources Development* **16** (1), 7–20.

Ackerman, F. and Stanton, E.A. (2011) *The Last Drop: Climate Change and the Southwest Water Crisis*, Stockholm Environment Institute, Stockholm.

Arab Water Council (2009) Arab Countries Regional Report, Arab Water Council, Cairo.

Ashbolt, N. (2004) Microbial contamination of drinking water and disease outcomes in developing regions. *Toxicology* **198**, 229–238.

Barbot, E., Carretier, E., Wyart, Y. *et al.* (2009). Transportable membrane process to produce drinking water. *Desalination* **248**, 58–63.

Beaumont, P. (2002) Water policies for the Middle East in the 21st century: the new economic realities. *International Journal of Water Resources Development* **18**, 315–334.

Bulloch, J. and Darwish, A. (eds) (1993) *Water Wars: Coming Conflicts in the Middle East*, Victor Gollancz, London.

Butler, R. (2009) Skyjuice technology impact on the UN. MDG outcomes for safe affordable potable water. *Desalination* **248**, 622–628.

Cap Net (2008) Training Manual and Facilitators' Guide, UNDP, New York.

Chandrappa, R., Gupta, S. and Kilshrestha, U.C. (2011) *Coping With Climate Change*, Springer-Verlag, Berlin.

Chandrappa, R. and von Munch, E. (2008) Urban and Rural Sanitation Problem in India. World Water Week, Stockholm (August 17–23).

Churchouse, S. and Wildgoose, D. (1999) Membrane bioreactors progress from the laboratory to full-scale use. *Membrane Technology* **111**, 4–8.

Coker, O.O. (2004) Reforming the water sector in Lagos state: the Lagos Model city development strategies. Paper presented at the Development Strategies from Vision to Growth and Poverty Reduction conference, 24–26 November 2004, Hanoi, Vietnam, 24–26 November.

Cosgrove, W.J. (2003) Water Security and Peace, UNESCO, Paris.

Doherty, R.M., Hulme, M., and Jones, C.G. (1999) A gridded reconstruction of land and ocean precipitation for the extended tropics from 1974–1994. *International Journal of Climatology* **19**, 119–142.

Egwari, L. and Aboaba, O. (2002) Environmental impact on the bacteriological quality of domestic water supplies in Lagos, Nigeria. *Revista de Saude Publica* **36** (4), 513–520.

Elliott, M., Stauber, C., Koksal, F. *et al.* (2008) Reduction of *E. coli*, echovirus type 12 and bacteriophage in an intermittently operated household-scale slow sand filter. *Water Research* **42** (10–11), 2662–2670.

FAO (2007) Making Every Drop Count. UN-FAO press release, 14 February.

Garsadi, R., Salim, H., Soekarno, J. *et al.* (2008) Operational experience with a micro hydraulic mobile water treatment plant in Indonesia after the 'Tsunami of 2004'. *Desalination* **246**, 91–98.

Gleick, P. H. (1998) Water in in crisis: paths to sustainable water use. *Ecological Applications* **8**(3), 571–579.

Goldie, I., Sanderson, R., Seconna, J. *et al.* (2004) A Guidebook on Household Water Supply for Rural Areas with Saline Groundwater. WRC Report TT 221/04, South African Water Research Commission, Pretoria.

GTZ (2006) Financing the Water and Waste Water Sector Financial Instruments to Meet Water Sector Challenges, Deutsche Gesellschaft fur Technische Zusammenarbeit, Frankfurt.

Hanjra, M.A., Ferede, T. and Gutta, D.G. (2009a) Pathways to breaking the poverty trap in Ethiopia: investments in agricultural water, education, and markets. *Agricultural Water Management* **96** (11), 2–11.

Hanjra, M.A., Ferede, T. and Gutta, D.G. (2009b) Reducing poverty in sub-Saharan Africa through investments in water and other priorities. *Agricultural Water Management* **96** (7), 1062–1070.

Hoekstra, A.Y. and Chapagain, A.K. (2008) *Globalization of Water: Sharing the Planet's Freshwater Resources*, Blackwell Publishing, Oxford.

Hoekstra, A.Y., Chapagain, A.K., Aldaya, M.M. and Mekonnen, M.M. (2011) *The Water Footprint Footprint Assessment Manual: Setting the Global Standard*, Earthscan, London.

Hulme, M., Osborn, T.J. and Johns, T.C. (1998) Precipitation sensitivity to global warming: comparison of observations with HadCM2 simulations. *Geophysical Research Letters* **25**, 3379–3382.

IGES (2005) Sustainable Asia 2005 and Beyond: In the Pursuit of Innovative Policies, IGES White Paper, Institute for Global Environmental Strategies, Kanagawa.

Jimenez, B. and Asano, T. (eds) (2008) *Water Reuse: An International Survey of Current Practice, Issues and Needs* (Scientific and Technical Reports 20), IWA Publishing, London.

Jones, P.D. and Hulme, M. (1996) Calculating regional climatic time series for temperature and precipitation: methods and illustrations. *International Journal of Climatology* **16**, 361–377.

Kajenthira, A., Diaz, A. and Siddiqi, A. (2011) The Case for Cross-Sectoral Water Reuse in Saudi Arabia: Bringing Energy into the Water Equation. Dubai Initiative Policy Brief. Energy Technology Innovation Policy Program, Belfer Center for Science and International Affairs, Harvard Kennedy School, Harvard University, Cambridge, MA.

Karl, T.R. and Knight, R.W. (1998) Secular trends of precipitation amount, frequency, and intensity in the USA. *Bulletin of the American Meteorological Society* **79**, 231–241.

MacQuarrie, P.R., Viriyasakultorn, V. and Wolf, A.T. (2008) Promoting cooperation in the Mekong region through water conflict conflict management, regional collaboration, and capacity building. *GMSARN International Journal* **2**, 175–184.

McBean, E. (2009) Evaluation of a bicycle-powered filtration system for removing 'clumped' coliform bacteria as a low-tech option for water treatment. *Desalination* **248**, 91–98.

Mekonnen, M.M. and Hoekstra, A.Y. (2010) The Green, Blue and Grey Water Footprint of Farm Animals and Animal Products. Value of WaterRes. Rep. Ser. No. 48.UNESCO-IHE, Delft, the Netherlands.

Mekonnen, M.M. and Hoekstra, A.Y. (2012) The blue water footprint of of electricity from hydropower. *Hydrology and Earth System Sciences* **16**, 179–187. www.hydrol-earth-syst-sci.net/16/179/2012/ (accessed 7 December 2013). doi: 10.5194/hess-16-179-2012.

Molden, D., Oweis, T., Steduto, P. *et al.* (2010) Improving agricultural water productivity between optimism and caution. *Agricultural Water Management* **97** (4), 528–535.

Morrison, J., Morikawa, M., Murphy, M. and Schulte, P. (2009) Water Scarcity and Climate Change: Growing Risks for Businesses and Investors, Pacific Institute, Oakland CA.

Nakatsuka, S. and Nakate, L. (1996) Drinking water treatment by using ultrafiltration hollow fibre membranes. *Desalination* **106**, 55–61.

Pallangyo, D.M. (2007) Environmental law in Tanzania; how far have we gone? *Law, Environment and Development Journal* **3** (1), 26, available at http://www.lead-journal.org/content/07026.pdf (accessed 7 December 2013).

Park, G., Schafer, A. and Richards, B. (2009) Potential of wind powered renewable energy membrane system for Ghana. *Desalination* **248**, 169–176.

Peter-Varbanets, M., Zurbrugg, C. and Pronk, C.S. (2009) Review: decentralized systems for potable water and the potential of membrane technology. *Water Research* **43** (2), 245–265.

Pillay, V.I. and Buckley, V.P. (2003) Evaluation and optimization of a crossflow microfilter for the production of potable water in rural and peri-urban areas. WRC Report 662/1/03, South African Water Research Commission, Pretoria.

Pistone, R.A. (2010) Loopholes in environmental laws allow for incomplete remediation thwarting environmental sustainability. *Journal of Sustainable Development* **3** (2), 35–39.

Postel, S. (1999) *Pillar of Sand: Can the Irrigation Miracle Last*? W.W. Norton & Company, New York.

Postel, S., Gleick, P. and Morrison, J. (1996) The Sustainable Use of Water in the Lower Colorado River Basin, Pacific Institute for Studies in Development, Environment, and Security, Oakland, CA.

Qadir, M., Bahri, A., Sato, T. and Al-Karadsheh, E. (2010) Waste water production, treatment, and irrigation in Middle East and North Africa. *Irrigation and Drainage Systems* **24** (1–2), 37–51.

Ramirez, M.C., Maldonado, A., Calvo, D. *et al.* (2011) Water engineering: a challenge for sustainable development for vulnerable communities – case Colombia, in *Water Engineering* (ed. D.P. Torres), Nova Science Publishers, Inc., New York.

Reed, B. (1995) *Emergency Water Supply*, http://cidbimena.desastres.hn/docum/crid/CD_Agua/pdf/eng/doc14624/doc14624-contenido.pdf (accessed 7 December 2013).

Robinson, D.A. (1997) Hemispheric snow cover and surface albedo for model validation. *Annals of Glaciology* **25**, 241–245.

Schafer, A. and Richards, B.S. (2005) Testing of a hybrid membrane system for groundwater desalination in an Australian national park. *Desalination* **183**, 55–62.

Shannon, M.A., Bohn, P.W., Elimelech, M. *et al.* (2008). Science and technology for water purification in the coming decades. *Nature* **452** (7185), 301–310.

Shapiro, S., Steinzor, R. and Shudtz, M. (2009) Regulatory Dysfunction: How Insufficient Resources, Outdated Laws, and Political Interference Cripple the 'Protector Agencies', Center for Progressive Reform, Washington DC.

Starr, J.R. (1991) Waterwars. *Foreign Policy* **82** (spring), 17–36.

Sweet, P. (2008) Gibbons Takes Another Whack at Pipeline Plan, Las Vegas Sun (21 February). http://wwwlasvegassun.com/news/2008/feb/21/gibbons-takes-another-whack-pipeline-plan/ (accessed 6 December 2008).

Thuy, N.T. (2010) Development of a water treatment system for emergency situations, doctoral thesis, Asian Institute of Technology, Vietnam.

Tignino, M. (2010) Water, international peace, and security. *International Review of Red Cross* **92** (879), 647–674.

UN (2011) The Millennium Development Goals Report 2011, United Nations, New York.

UNDP (2007) *Human Development Report 2007/2008, Fighting Climate Change: Human Solidarity in a Divided World*, Oxford University Press, Oxford.

UNECA (United Nations Economic Commission for Africa) (2008) Sustainable Development Report on Africa: Five-Year Review of the Implementation of the World Summit on

Sustainable Development Outcomes in Africa, United Nations Economic Commission for Africa, Addis Ababa.

UNEP (2007) Post-Conflict Environmental Assessment, United Nations Environment Programme, Nairobi.

UNESCO (1996) The makings of a water crisis. *UNESCO Surveys* **84**, 12–13.

UNESCO (2006) Water: A Shared Responsibility, United Nations World Water Development Report – 2. http://ww2.unhabitat.org/programmes/water/documents/water report2.pdf (31 December 2013).

Wessels, A. (2000). *Development of a Solar Powered Reverse Osmosis Plant for the Treatment of Borehole Water.* WRC Report 1042/1/00, South African Water Research Commission, Pretoria.

WHO (2012) UN-Water Global Annual Assessment of Sanitation and Drinking Water (GlAAS) 2012 report: The Challenge of Extending and Sustaining Services, World Health Organization, Geneva.

WHO and UNICEF (2004) Meeting the MDG Drinking Water and Sanitation Target, A Mid-Term Assessment of Progress, WHO, Geneva.

Wild, D., Francke, C.-J., Menzli, P. and Schon, U. (2007) Water: A Market of the Future – Global Trends Open Up New Investment Opportunities, Sustainability Asset Management (SAM), Zurich.

Wolf, A.T. (1998) Conflict and cooperation along international waterways. *WaterPolicy* **1**(2), 251–265.

World Commission on Environment and Development (1987) *Our Common Future*, Oxford University Press, New York.

2

Requirements for the Sustainability of Water Systems

Water is essential for civilization. Water-related disasters include the 2004 Indian Ocean tsunami, the 2004 and 2005 hurricanes in the west Pacific, in the Caribbean and the United States, the 2005 floods in Central/Eastern Europe and in many other regions, the extensive droughts in Mali, Niger, Spain and Portugal, the 2010 cloudburst in Ladhak in India and the 2010 floods in Pakistan and India.

While the developed countries enjoy stable populations, the developing countries are experiencing rapid population growth/shifts in urban areas. In many rapidly growing cities it is difficult to build the required infrastructure to meet growing demand. In many parts of the world, internally displaced and international refugees, migrants and tourists are exerting a stress on water supplies.

Water is a global issue. Countries engage in trading products rather than the physical transportation of water itself. As a result, billions of tonnes of water will remain where it is abundant. Some water-scarce nations, which include countries in the Middle East, are net importers of water to meet the needs of their growing populations. This is also true of many European nations, due to consumer tastes for particular foods and products (Hoekstra, 2011).

The top 11 global water consumers are India – 13%; China – 12%; US – 9%; Russia – 4%; Indonesia – 4%; Nigeria – 3%; Brazil – 3%; Pakistan 2%; Japan 2%; Mexico 2% and Thailand 2%. Nations with the biggest per capita water footprints tend to have large gross domestic products and huge durable consumption rates. The United States has the largest water footprint of any country but the United States ranks third globally with respect to water consumption. India and China, which have smaller per capita water footprints, are the largest water consumers.

Sustainable water management is not an end itself; it is only part of sustainable development. Without proper design of water distribution and wastewater collection, one cannot attain sustainability with respect to water. Improper water distribution will use energy

Sustainable Water Engineering: Theory and Practice, First Edition. Ramesha Chandrappa and Diganta B. Das.

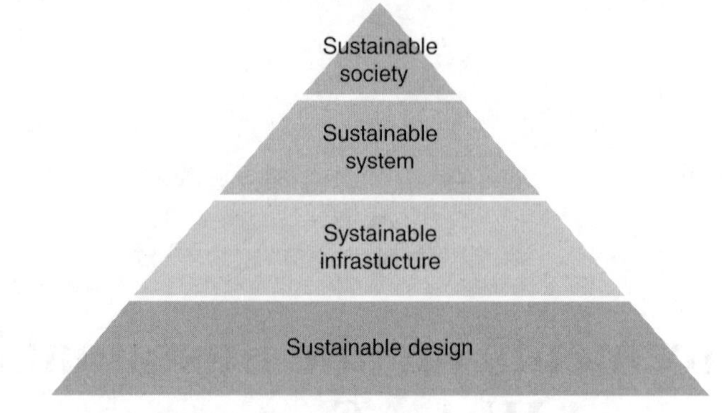

Figure 2.1 *The relation between sustainable design and sustainable society.*

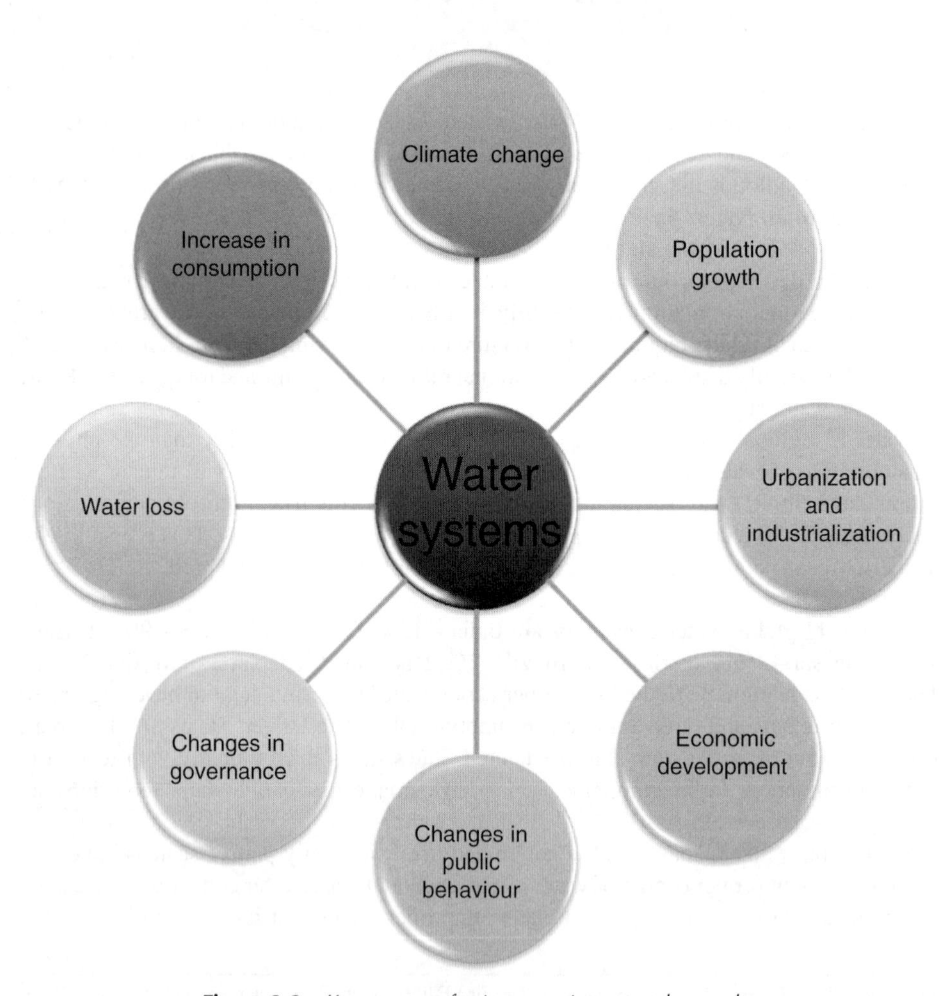

Figure 2.2 *Key reasons for increase in water demands.*

unnecessarily and cause a loss of water. Improper collection of waste water will lead to leakage of waste water, leading to water pollution. Inadequate corrosion protection, old pipelines, poorly maintained valves as well as mechanical damage contribute to leakage. Many leaks remain undetected. Water lost after treatment and pressurization is unsustainable as water, money and energy are wasted. In the absence of knowledge about the fundamentals of water systems, the water engineers will end up with costly solutions that will not serve the citizens. The main requirements for the sustainability of water systems are (i) an asset-management framework that ensures the right investments at the right time, (ii) water and energy efficiency ensuring sustainable practices and technologies to improve their efficiency, (iii) infrastructure financing and the price of water services to pay for water infrastructure needs and (iv) alternative technologies and assessment.

Many cities and towns around the world are supplied with high-quality water but large infrastructure systems are needed to meet these requirements. The rivers and aquifers that provide water to urban areas are highly stressed and degraded, increasing conflicts amongst the urban and rural consumers. Figure 2.1 shows the relationship between sustainable design and sustainable society. The key reasons for increase in water demand are shown in Figure 2.2. They are: (i) climate change, (ii) population growth, (iii) urbanization and industrialization, (iv) economic development, (v) changes in public behaviour, (vi) changes in governance, (vii) water loss and (viii) increase in consumption.

Historically, projects involving large infrastructure developments resulted in improvements in living conditions. Ecological impacts due to these projects may, however, be ignored or downplayed (Johnson and Rix, 1993). This is now evident and, as a consequence, opposition to new projects, for example, construction of new dams, is observed. The recognition that centralized water supply/disposal solutions are reaching their limits has resulted in calls for new solutions. The major challenges in providing safe water and sanitation are: (i) water contamination in distribution systems, (ii) rising water scarcity, (iii) implementing innovative cost-effective sanitation systems, (iv) reducing global and regional disparities with respect to water and sanitation and (v) providing sustainable water supply and sanitation for megacities (Moe and Rheingans, 2008).

Sustainable water systems apart from distribution should also address the issue of conservation of quality. Table 2.1 gives the comparison of water-quality threats in the developed and developing countries. Box 2.1 gives a glimpse of efforts in Copenhagen about good practice in sustainable water management.

Box 2.1 Learning from Copenhagen, Denmark

Half of the human population now lives in cities. Within two decades, about 60% of the world's population will be urban dwellers. Water management in cities should therefore be sustainable. Compared to other countries, Denmark is good at handling water, with one of the lowest loss rates within the distribution system, very low per capita loss and a high level of security in the water supply system. For more than 60 years, Denmark has been developing physical planning. Planning has considered how to preserve groundwater. Through the modernization of the sewage system, the public harbour baths were officially opened in 2002 (Kristiansen, 2012).

Table 2.1 Examples of general water-quality threats and comparison for the developed and developing countries.

Sl. no.	Threat	Developing countries	Developed countries
1	Awareness regarding environmental conservation	Very low	High–very high
2	Chewing and spitting in public place	Common	Negligible
3	Stray dogs	Present	Absent or rare
4	Defecation of pet dogs	Open defecation	Will be collected and placed in solid waste bins
5	Solid waste management	Improper and inefficient	Proper and efficient
6	Slaughtering	Carried out inside the city and individual shops	Centralized slaughter house with wastewater treatment
7	Illegal waste disposal through vehicles	Common	Uncommon/absent
8	Number of staff in enforcement of environmental law	Inadequate	Adequate
9	Political priority to conserve environment	Very low	Very high
10	Allocation of funding to conserve environment in national budget	Inadequate	Adequate
11	Corruption	High	Low
13	Nosocomial infection	High	Low
14	Cattle rearing in urban areas	Yes	No
15	Religious rituals that lead to pollution	Common	Uncommon/absent
16	Air pollution	High	Low
17	Open defecation and urination	Common	Uncommon/absent
18	Slums	Present	Absent
19	Industries in residential area	Present	Absent

2.1 History of Water Distribution and Wastewater Collection

Ancient civilizations settled around water sources but it was common to settle in water-scarce places like deserts. Drinking water quality was mainly restricted to aesthetics. Figure 2.3 shows water storage for drinking, agriculture, animal rearing and temperature maintenance of a building from ancient India.

The first water-distribution pipes were used in Crete in 1500 BC during the Minoan civilization and the city of Knossos developed tubular conduits to convey water. In 1455 the first cast-iron pipe was manufactured in Siegerland, Germany and was installed at Dillenburg Castle. Residential wastewater in seventeenth-century colonial America consisted primarily of a privy with the outlet at ground level discharging into the yard, gutter, street,

Figure 2.3 *Water storage for drinking, agriculture, animal rearing and temperature maintenance of a building in ancient India.*

or an open channel (Duffy 1968). As the population density increased, wastewater was discharged until it was soaked into the ground or removed manually and disposed far away from the residence. Another alternative form of sanitation practised in the United States in the nineteenth century was the placing of containers below the seats of privies and disposing the human excrement in dry condition (Folwell, 1916). This was followed by construction of sewers in the United States. The sewers constructed prior to the 1850s did not involve trained engineers because sewers were not perceived as technically complex systems requiring the services of an engineer (Burian *et al.*, 2000). Groundwater contamination in Philadelphia in 1798 and Baltimore in 1879 was the greatest source of disease, requiring the lining of privy vaults and cesspools.

Water-supply systems were in place in most of the major US cities by the early to mid-nineteenth century (Armstrong, 1976; Dworsky and Berger, 1979; Fair and Geyer, 1954).

The late eighteenth and early nineteenth centuries saw major changes in the agriculture, manufacturing, mining, production and transportation sectors. The Industrial Revolution became the turning point for pollution. The Industrial Revolution had an effect on socioeconomic and cultural conditions in the United Kingdom, followed by Europe, North America, and ultimately the world. In the latter part of the eighteenth century, Great Britain's manual labour and animal-based economy transformed to machine-based manufacturing processes. Mechanization of the textile industry was followed by the development of iron making, which led to increased use of water. The developments of machine tools in the first decades of the nineteenth century led to the use of more machines for other industries. Due to rapid urbanization during the Industrial Revolution that occurred during second half of the nineteenth century, there was a rise in the urban population in Europe, which resulted in several epidemics in the cities, resulting in numerous deaths. While cholera spread over the whole of Germany, typhoid had a permanent presence in many cities due to bad sanitary conditions. Such conditions still exist today in many poor countries across the world.

London witnessed dramatic changes with respect to water problems during the Industrial Revolution. In 1848, there were 50 water closets with privies for 82 000 people in the Bethnal Green area of London, (Eveleigh, 2008). In 1900, about 75% of the London dwellers did not have a bath at their home (Eveleigh, 2002). In 1845, only six of the 50 towns in Britain had adequate water supplies adequate and in 31 they were bad (Falkus, 1977).

A sewer system was constructed in 1842 in Hamburg, Germany and a sewer system was constructed in Frankfurt/Main in 1867 after one of the biggest cholera epidemics (Seeger, 1999). The discharge of urban wastewater into the surface water bodies through centralized sewer networks resulted in pollution, requiring wastewater treatment.

Poor water quality throughout the world has been one of the factors causing people to turn to bottled water. The bottled water market increased rapidly with a global consumption of more than 200 000 million litres per annum. The United States is the largest consumer of bottled water and China has increased use by more than 15% since 2003. The production of the bottles requires 17 million barrels of oil per annum in the United States (Corcoran *et al.*, 2010).

2.2 Integrated Water Management

One of the main objectives of integrated water management is to deliver effective water resource services with minimum risks and improved sustainability of the system. Figure 2.4 shows the major themes that need to be integrated for a sustainable water management system. Planning for individual goals will affect the other themes shown in the figure. This is because integrated management cannot be achieved by engineering and technical solutions alone. As Figure 2.5 shows, it is important to have various instruments for integrated water management. It requires appropriate financial support, a legal framework, engineering solutions and education for all stakeholders. Integrated water management can happen at international, national, watershed, river basin, or city level. The boundary of the project depends on people implementing it. Twenty countries had plans/strategies in place or a process underway, 50 countries were in the process of preparing national strategies and 25 countries had taken only initial steps towards integrated water resource management (IWRM) at the end of 2005. Studies based on a UN-Water survey sent to governments of all UN member countries in 2011 showed that 64% of countries had developed IWRM plans and 34% were in an advanced stage of implementation (UNEP, 2012).

Agenda 21 of the UN Conference on Environment and Development in 1992 called for 'the application of integrated approaches to the development, management and use of water resources.'

Since 1992, the population increased from around 5.3 billion in 1992 to around 7 billion, resulting in increased demand for food, energy and water. Competition between uses has resulted in difficult situations. Climate change has prompted concern about existing resources and an interest in adapting them and making them more resilient. Mobile phones and the Internet have improved and accelerated the sharing of knowledge.

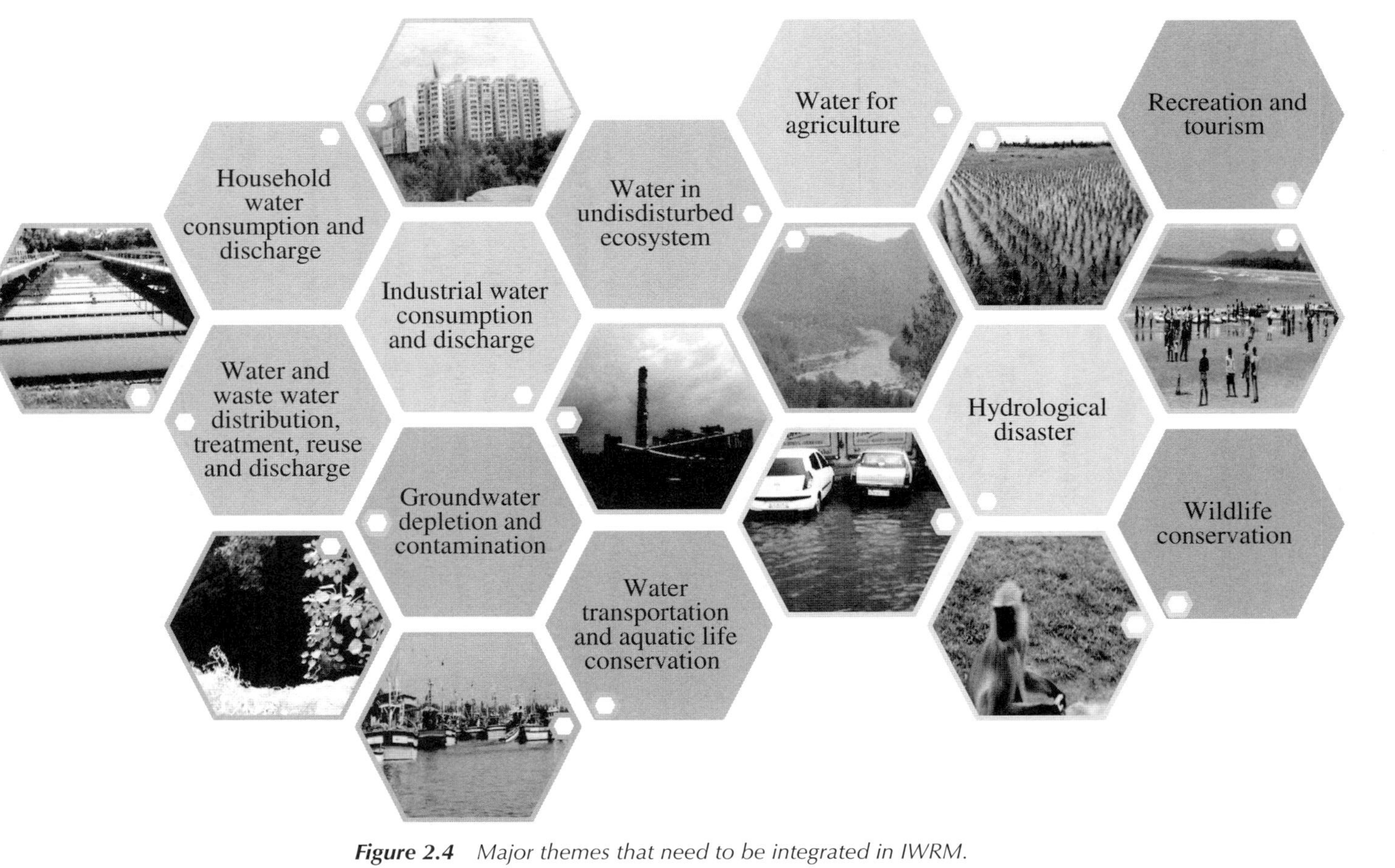

Figure 2.4 Major themes that need to be integrated in IWRM.

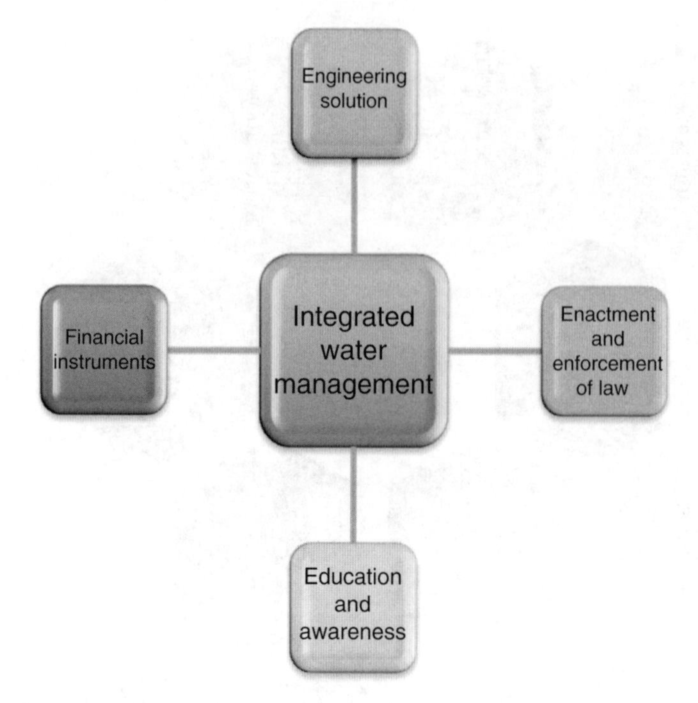

Figure 2.5 *Supporting instruments of integrated water management.*

Liao River Basin management resulted in a reduction in conflicts and pollution and improvement in river water quality. Conserving La Cocha Lagoon in Columbia resulted in increasing the income of 387 poor families, a reduction in the threat to La Cocha Lagoon and surrounding wetlands and 'shelving' (gradual sloping) of land. Management of scarce water resources in Morocco resulted in a reduction in watershed soil loss significantly. Improving water accessibility through IWRM in Fergana Valley helped Uzbekistan, Kyrgyzstan and Tajikistan. Integrated water resource management and the use of a water-efficiency plan in Sri Lanka benefitted a considerable number of people there. New York City Water Supply acting as a partner in watershed management projects resulted in improvements in water quality in the area.

Private agents only pay the *private costs* of economic activity but there is no market price that reflects the costs to society (*social cost*) in terms of damage from pollution. Too much pollution is generated, which causes costs to society (*social costs*) in terms of damage. Economic instruments aim to put a price on the use of natural resources like air, water and soil. The economic instruments adopted by Sweden are (i) pollution taxes, (ii) marketable permits, (iii) deposit refund systems, (iv) charges, (v) subsidies.

Today about 50 environmentally motivated economic instruments are in use in Sweden to achieve environmental objectives. About ten are taxes/charges: mainly on energy and transport. Total revenue from environmentally related taxes about 7 billion EUR per year, which represents about 3% of GDP. Examples of economic instruments include: (i) a carbon dioxide (CO_2) tax, (ii) a nitrogen oxide (NO_X) charge, (iii) congestion tax, (iv) car-scrapping premiums and environmental car premiums, (v) water and waste-management fees.

Box 2.2 IWRM in Sustainable Housing Estate Eva-Lanxme, Culemborg in the Netherlands

EVA-Lanxmeer contains about 250 houses, several offices and a 'city farm'. Precipitation on roofs is collected in ponds. Precipitation falling on the street infiltrates into infiltration trenches. Restored old riverbeds provide further water-storage capacity. The design of the district and the building is based on Reggio Emilia, permaculture and organic design principles. Permaculture is a branch of ecological engineering, ecological design and environmental design, which develops human settlements and sustainable architecture as well as self-maintained agricultural systems modelled from natural ecosystems. The Reggio Emilia is an approach to teaching young children that emphasizes the natural development of children and the close relationships they share with their environment. The project has mixture of 'red and green' development with integrated decentralized technologies for wastewater as well as organic waste treatment with carbon, nutrients and energy recovery. The concept is called 'sustainable implant' (SI). The project has a small-scale biogas installation for the treatment of black water as well as organic waste/garden and park waste. The project also has combined heat power (CHP) as well as an accompanying closed greenhouse with integrated wastewater treatment based on the 'Living Machine®' concept (van Timmeren *et al.*, 2007). Living Machine uses plants and micro-organisms to treat wastewater into clean water. The purification of kitchen and laundry effluent takes place on the project terrain containing reed beds. All houses are thoroughly insulated to avoid energy loss. All houses have been provided with systems for solar energy, Sun boilers preheat the tap water. Solar cells supply the houses with electricity. The local water board supplies heat from a central heating system using the warmth from subsoil water. The project also integrates small-scale sustainable food production by cultivating fruits and vegetables.

Voluntary instruments do not put any regulatory obligation on any person/organization. The voluntary instruments used by industries/government in European countries are: (i) lifecycle approach/assessment, (ii) adopting the ISO 14 001-environment management system, (iii) green public procurement (GPP), (iv) eco-labelling, and (v) adopting ISO 26 000-guidance on social responsibility.

Box 2.3 Sustainable Urban Water Management – Swedish Experience

Swedish people consume around 200 lpd of water per person. The country has shifted the method of treatment in urban bodies from no treatment in 1965 to special nitrogen removal. As a result Swedish people enjoy a high-quality environment and uncontaminated food that is produced in Sweden. The biogas generated is being used to run buses in the public transportation system. The regulation of wastewater from dwellings is mainly the responsibility of local bodies. The environmental board at the local council is responsible for permits. Each plant must apply for a permit for establishment and operation. In Sweden, 850 000 properties in Sweden have an on-site sewage treatment system. Requirements for specific treatment plants vary from location to location. Treatment plants vary from soil-based systems to source-separation systems with the recycling of nutrients.

Sweden has GPP in place so that government agencies will buy eco-friendly goods/services. It started in the 1990s by engagement with NGOs, local municipalities and so on followed by a committee for ecologically sustainable public procurement. This was followed by the development of the Swedish Instrument for Ecologically Sustainable Procurement and the Swedish Environmental Protection Agency guidelines, and surveys on GPP. Strong support was provided for GPP by the Prime Minister in a government declaration in 2002. As a result of legislation, a minimum of 75% of all cars purchased/leased by public authorities in Sweden after 1 January 2006 must be green cars.

2.3 Sewerage Treatment and Urban Pollution Management

The majority of urban and rural settlements in the developing world are not successful in meeting the sanitation requirements of the poor. Unplanned slum and squatter settlements often lead to problems associated with unsafe hygiene. Without a safe water supply poor people seem to collect whatever they can find (Chandrappa and von Münch, 2008). Figure 2.6 shows some examples of sustainable and unsustainable plumbing. The maximum reuse within the household can save water instead of using public water supply of potable

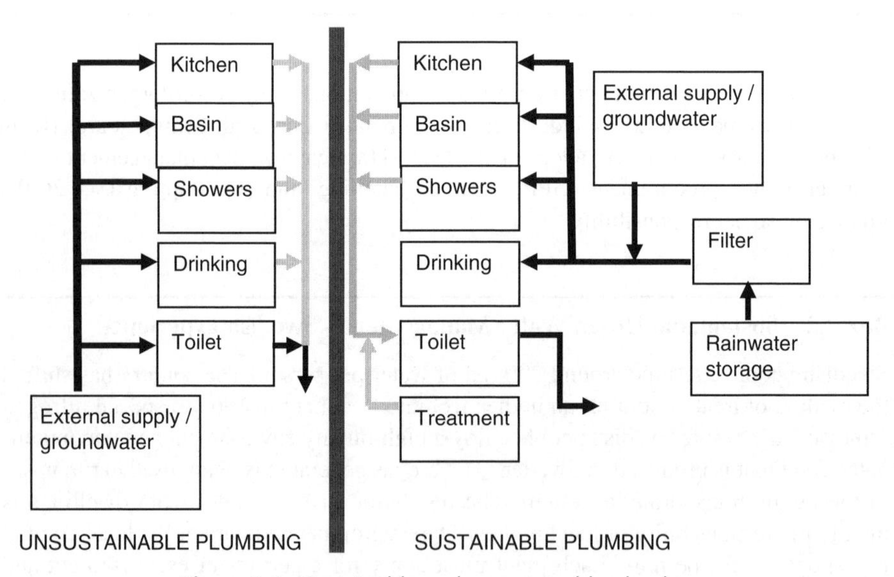

Figure 2.6 *Sustainable and unsustainable plumbing.*

water quality for flushing, gardening and other uses that do not require water that is of drinking quality.

Box 2.4 Source Separating Wastewater System in Ekoporten, Norrkoping, Sweden

Ekoporten in Norrkoping, Sweden, is a house that was originally constructed in 1967. It has three floors with 18 apartments. The house is heated with local district heating in the cold season and during the warmer season sun panels heat the tap water. The tenants are educated in type of detergents to be used, in how to sort out the garbage and how to use the toilets.

The toilets are urine separating and urine is used by ecologically cultivating farmers. The faeces are composted together with paper, kitchen as well as garden waste for the cultivation of vegetables and flowers. The flush water was originally treated with ultraviolet radiation and, together with the grey water from kitchens, it is treated in a reed bed where the plants absorb nutrients. Since the reed bed did not match the treatment effect in waste water treatment plant in Norrköping, sewage was connected to the regional sewage water system. Ekopotern uses a computer for controlling and registering flows.

The water consumption in Ekoporten was 44% lower than in conventional houses in 1996–1997 (Cardiff, 2012).

The burden on sewerage treatment plants can be greatly reduced with proper urban planning, which is deficient in the developing countries. The people with influence and money have developed many projects in cities in the developing world despite the lack of treatment plants. Many of these new buildings have their own treatment plants but an absence of skilled labour has resulted in the release of untreated wastewater, resulting in the pollution of lakes and groundwater.

2.4 Conventional Water Supply

A water-supply system involves the transmission of water from the source to point of consumption through channels or conduits (Figure 2.7). The source could be tube wells, lakes, rivers or desalination plants. Depending on local conditions, the water is either pumped or supplied under gravity.

Gravity supply is used where the source of supply is at an elevation sufficiently above the consumer so that the required pressure can be maintained. Wherever consumers are located at an elevation above the water source, a pumped water supply is used – water is pumped to the tank/reservoir to provide the necessary pressure. The tank can be either overhead or on ground located at an elevation above the consumer point.

Disruption in Hassan city, India, due to a break in the pipeline that brings water from Gorur to Hassan resulted in cancellation of many scheduled operations in the Hassan Institute of Medical Science and attendants of patients were asked to bring water from

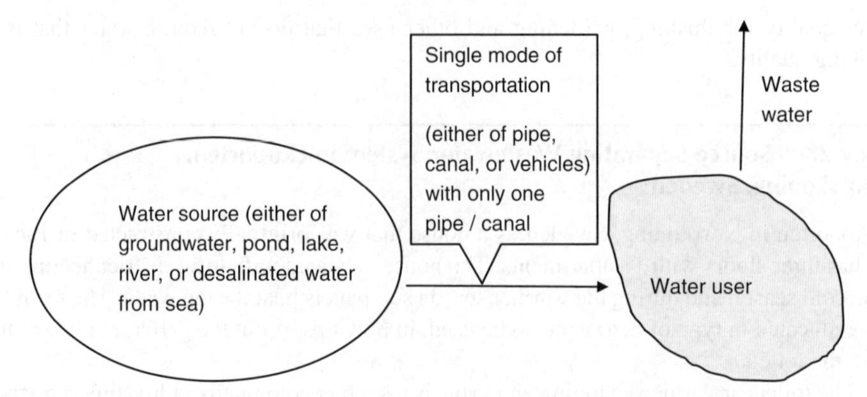

Figure 2.7 Single source and single mode of transportation leads to unsustainable water supply in case of failure of water source or mode of transportation.

outside (The Hindu, 2012). Such a scenario occurs due to dependence on a single source for water without foreseeing failures of the water source and supply systems (Figure 2.7). Multiple sources and multiple supply systems will ensure sustainability for many years (Figure 2.8). Water supply engineers usually do not carry out risk assessments to systems, as chemical engineers do when designing chemical manufacturing plants, but multiple treatment and nontreatment barriers have been set up to safeguard the supply of water in Windhoek, Namibia.

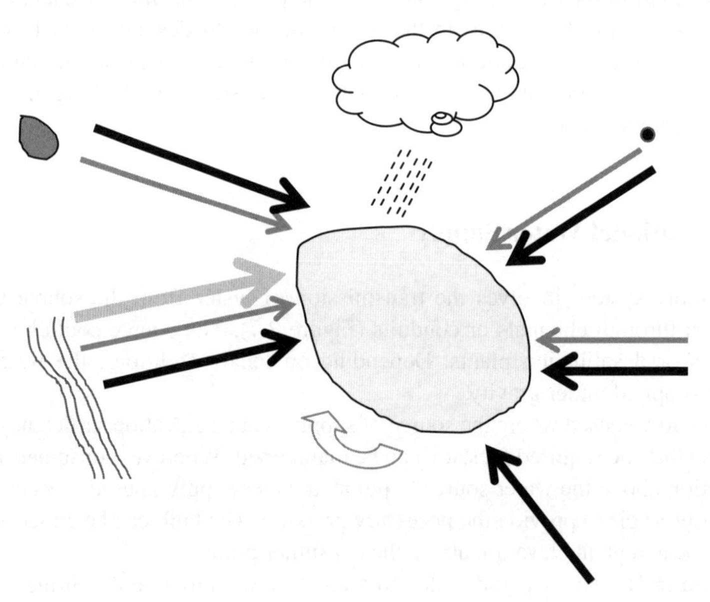

Figure 2.8 Multiple options for source and transportation would ensure sustainability in water supply.

Box 2.5 Reclamation in Windhoek, Namibia

Windhoek, in Namibia, has a population of 0.25 million. The average rainfall in the city is 360 mm whereas annual evaporation in 3400 mm. Three dams at a distance of 70 to 160 km from each other were built from1978 to 1993 on rivers that run only during the few days following a heavy rainfall. The first municipal bore well was made in 1912 and about 60 municipal borewells were developed by 2006, with an assured yield of 1.73 Mm^3/y. After independence the population grew at a rapid rate. Hence, a new 21 000 m^3/d reclamation plant was built with a loan from European financial institutions.

The nontreatment barriers in the reclamation plant are: (i) diversion of industrial wastewater to another treatment process, (ii) monitoring at inlet and outlet or sewage treatment plant, (iii) monitoring of drinking water quality, (iv) blending about 35% of water from the reclamation plant with other sources of drinking water. The reclamation plant comprises coagulation/flocculation, dissolved air floatation, rapid sand filtration, ozonation, biological and granular activated carbon filtration, ultrafiltration, chlorination/stabilization, and distribution.

Calculation of consumption patterns depends on: (i) the size of the city, (ii) the culture of the city, (iii) climatic variation, (iv) pricing of water, (v) assumed population growth and (vi) the income of the consumer. Designers often overdesign the system considering the future demands, safety factors and losses in the system. Further, the assumptions made when forecasting population growth during system design will not be met in reality (Figure 2.9). This was true especially with respect to those cities located in countries in transition or cities where the economy was booming. Unsustainability with respect to water supply to a city occurs also due to absence of policy regarding the population growth in the city. It is usual practice in many parts of the world to encourage investments without foreseeing the environmental impacts, including the impact on water resources. Such unscientific decisions would add to the burden on both the original and the new population.

Some cities in India, which grew rapidly due to the software boom, witnessed a sudden increase in water demand for construction and domestic use, resulting in high withdrawal of groundwater for construction and domestic purposes. Bangalore, which grew horizontally until the 1990s, started growing vertically to accommodate population coming and settling in the city. Many of the plots allotted to single individual were used to build eight to ten flats to make profit due to demand. Due to the rapid increase in demand, apartments withdrew water from bore wells, which exhausted within couple of years. As a result, apartment dwellers started buying water from agencies through tankers mounted on trucks.

The various types of pipes used in a conventional water supply are given in Table 2.2.

Figure 2.10 shows a typical water-supply system using gravity and pumping. For the water supply to be sustainable in these cases the pipeline materials used should have: (i) hydraulic smoothness (ii) structural strength for external loads, (iii) strength to sustain internal pressure, (iv) ease of handling, transportation and storage, (v) the capacity to withstand damage in handling and maintenance, (vi) resistance to internal corrosion, (vii) resistance to external corrosion, (viii) resistance to heat/sunlight, (ix) they should be resistant to rodent attack, (x) they should be sustainable in black cotton soil, (xi) they should

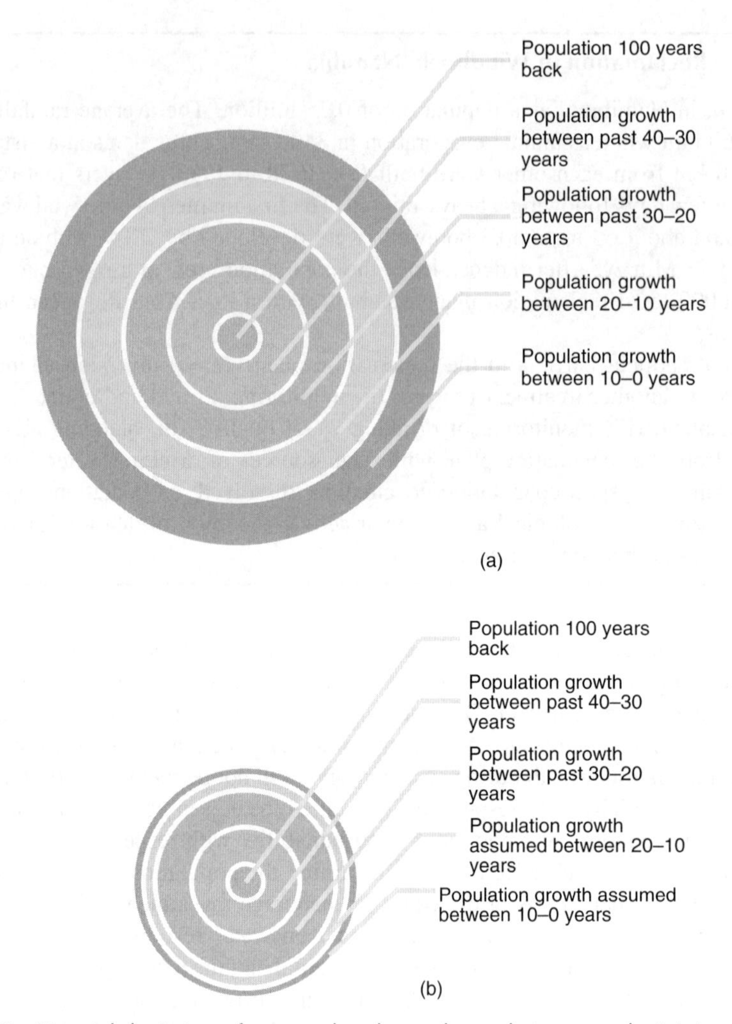

Figure 2.9 *Pictorial depiction of assumed and actual population growth. (a) Actual population growth; (b) forecast population growth.*

Table 2.2 *Types of pipes used in conventional water supply.*

Type	Subtype	Example
Metallic	Unlined	Cast iron, ductile iron, mild steel, galvanized iron
	Lined	Metallic pipes lined with cement mortar, epoxy lining
Nonmetallic	Reinforced concrete, pressurized concrete, bar wrapped steel cylinder concrete, asbestos cement	
	Plastic pipes	Poly vinyl chloride (PVC), Polyethylene

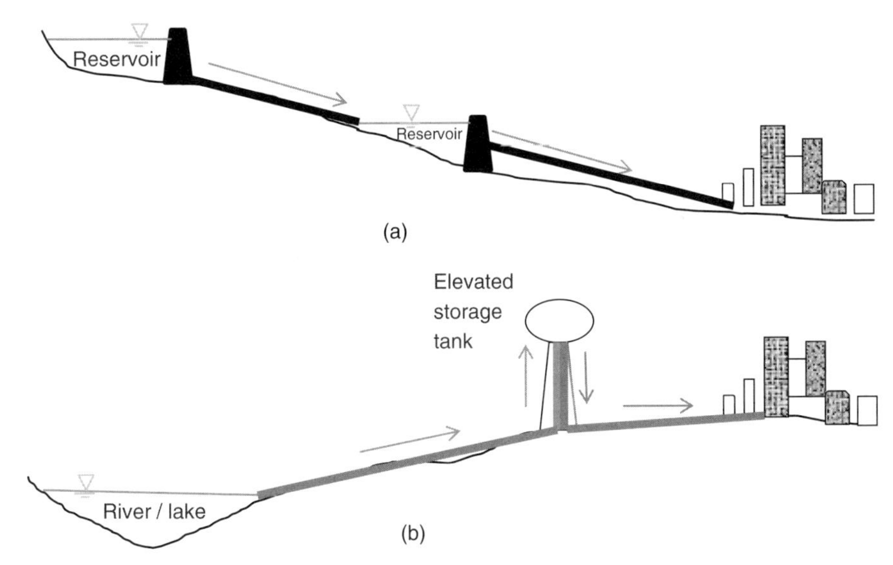

Figure 2.10 Typical water supply system: (a) gravity, (b) pumped.

have reliable and effective joints, (xii) they should be capable of absorbing surge pressure, (xiii) they should be easy to maintain and repair, (xiv) they should be durable.

2.4.1 Features

Conventional water distribution is characterized by: (i) huge investment, (ii) energy consumption, (iii) operation and maintenance, (iv) a water source with an assured supply throughout the year, (v) skilled manpower, (vi) resource to sustain the system, (vii) regular analysis and quality assurance.

2.4.1.1 Distribution Systems

A system of interconnected pipes to supply water from the point of source is known as a distribution system. Different types of water distribution systems are shown in Figure 2.11.

Types of water distribution system usually adopted are (i) a branching system, (ii) a grid arrangement, (iii) a grid arrangement with loops, and (iv) a grid arrangement with dual mains.

A branching system is typically used in old cities or rural areas where planning has not been carried out properly. The early cities and current villages that developed haphazardly without considering water supply and sanitation as well as transportation often have dead ends. Such scenarios force water supply organizations to adopt a branching system. The grid arrangement will have the advantage of water flow from other loops in case any of the pipes fail or have to be repaired.

Figure 2.12 shows the layout of a pipe in progress in Europe along a road. Such sophistication and planning in terms of safety precautions, leakage prevention and insulation (to avoid freezing during winter) is often missing in the developing world, which is often stressed by waterborne diseases.

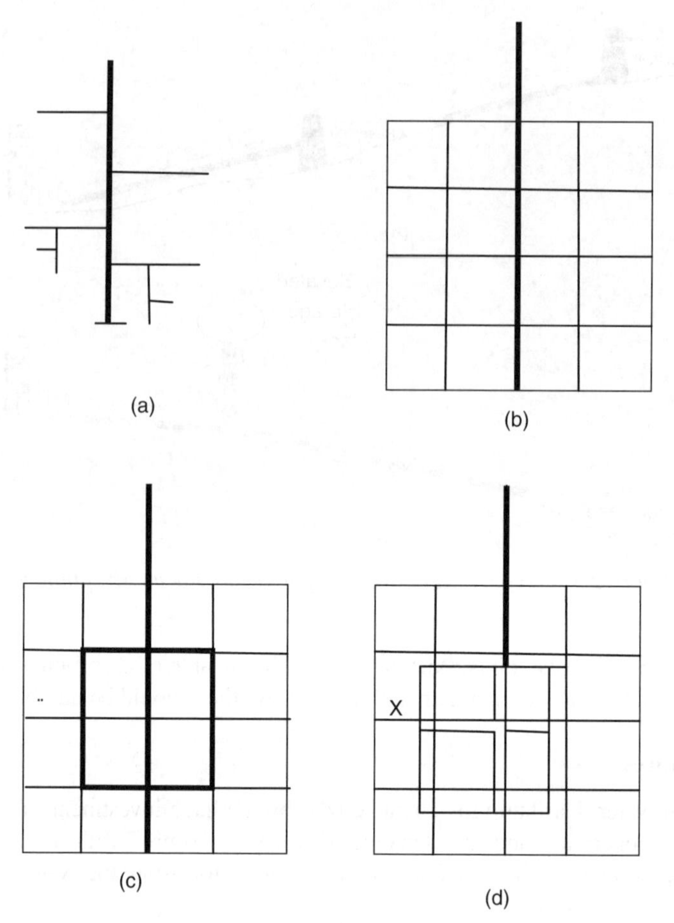

Figure 2.11 Types of water distribution system: (a) branching system, (b) grid arrangement, (c) grid arrangement with loops, (d) grid arrangement with dual mains.

2.4.2 Capacity and Pressure Requirements

Depending on topography, water can be transported by canals, aqueducts, flumes, tunnels and pipelines. Gravity, pumping or both may be used to supply water to the consumer with adequate pressure.

Gravity supply is used where one or more sources are at an elevation above the consumer so that the required pressure is maintained. In a pumped water supply, pumps are used to develop the required pressure. In a pumped storage supply, storage tanks or a reservoir are constructed to maintain pressure.

It is desirable to maintain the maximum flow and pressure to avoid damage to the water supply system. Further, a minimum pressure and velocity should be ensured to meet consumer demand. For a typical residential area, normal maximum water pressure should be 275 kPa and minimum pressure should be 140 kPa. Pressure at fire hydrants should be between 300 to 400 kPa. These values have been derived empirically and by experience over

Figure 2.12 *Laying of pipe in progress in Europe.*

a long period of time. The pipes and pipe fixtures are normally designed to withstand these pressures. Hence, any pressure above these values in a distribution system is not advisable and should be avoided. Typical dimensions adopted in the distribution component are given in Table 2.3. A pressure beyond the permissible water pressure will lead to damage of faucet and pipes. The velocity of 1.5 m/s would be necessary at the point of consumption to ensure adequate water flow whereas a velocity of 3 m/s is required at fire hydrants to ensure

Table 2.3 *Dimension of distribution system components.*

Item	Value
Pipes	
Smallest pipe in grid	150 mm
Smallest branching pipe	200 mm
Largest spacing of 150 mm grid	180 m
Largest spacing of supply mains or feeders	600 m
Valves	
In single and dual main system	Three at crosses, two at tees
Largest spacing on long branches	250 m

adequate water flow to extinguish fire. Excessive velocity damages the system and lower velocity will lead to settling of solids in pipe and dissatisfaction amongst consumers. Due to water scarcity, many countries are not installing fire hydrants and are not considering fire safety requirements during design.

Due to water scarcity people attach pumps to water supply systems to draw more water from the system. Such practices in commercial establishments and apartments will lead to scarcity in other parts of the city.

2.4.3 Design and Hydraulic Analysis of Distribution System

The laws of conservation of mass and energy are used for network design and analysis.

2.4.3.1 Pipes in series

The total head loss in case of pipes in series (Figure 2.13) is the sum of the head loss in individual pipes:

$$\text{total head loss} = \text{head loss in pipe 1} + \text{head loss in pipe 2} + \ldots$$

$$\mathbf{H_L} = \sum_{i=1}^{i=n} H_{L,i}$$

where

H_L = total head loss
$H_{L,i}$ = head loss in i^{th} pipe

$$\text{flow in pipe 1} = \text{flow in Pipe 2} = \text{flow in pipe 3}$$

2.4.3.2 Pipes in parallel

In case of pipes in parallel (Figure 2.14) the flow is split amount different pipes whereas pressure across the pipe will remain constant.

Figure 2.13 *Pipes in series.*

Pipes in parallel

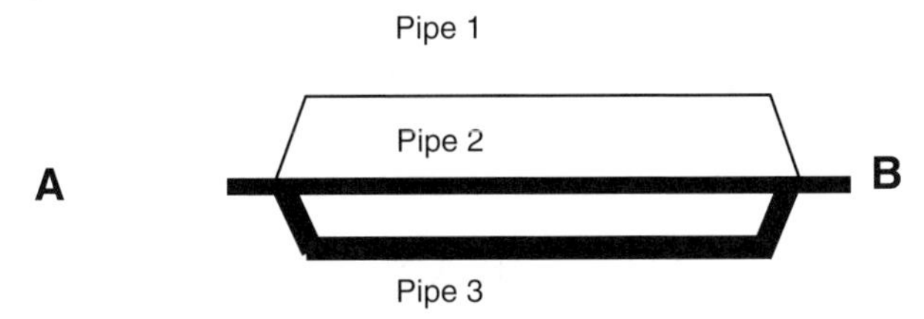

Figure 2.14 *Pipes in parallel.*

head loss between each pipe between A and B = head loss between all pipes between A and B flow between A and B = flow in pipe 1 + flow in Pipe 2 + flow in pipe 3

$$Q = \sum_{i=1}^{i=n} Q_i$$

where

Q = total flow rate
Q_i = flow rate in i^{th} pipe

Distribution systems (Figure 2.15) are usually analysed by computer. The following equations should be satisfied throughout the network to ensure the desired flow without damage to the network.

At every junction

$$\sum Q_{inflow} = \sum Q_{outflow}$$

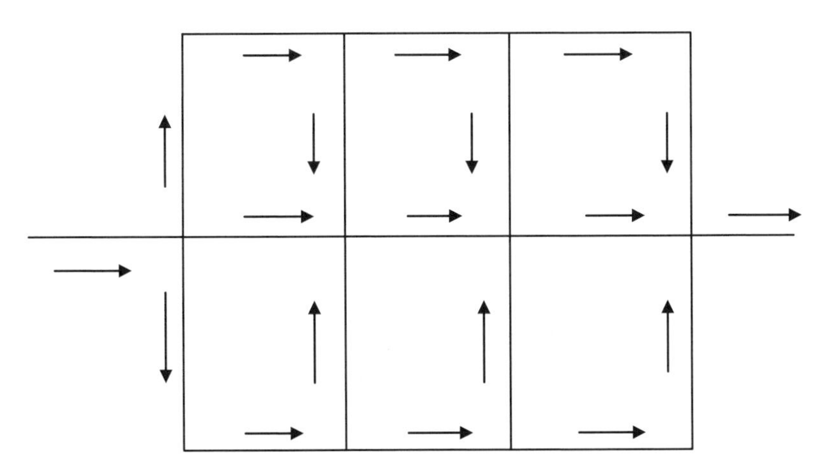

Figure 2.15 *Typical distribution system.*

For the complete circuit

$$\sum H = 0$$

For an individual pipe

$$H = kQ^n$$

where, k and n values have to be defined empirically depending on pipe material.

When a new system is first filled, all valves are opened so that air can escape; excess pressure is created if air is not evacuated.

The distribution system is usually designed to cater for a combination of peak firefighting and peak consumer demand but, due to water scarcity, many cities will supply water once or twice in a day for a limited period. It is common, in some parts of the world, to distribute water once in a few days.

2.4.3.3 *Cross Connection in the Distribution System*

Cross connection in the distribution system is an actual or potential connection linking the public water supply with a source of contamination/pollution. The epidemic that occurred in Chicago in 1933 was due to defective and improperly designed plumbing and fixtures; it resulted in the contamination of water, as a result of which 1409 people contracted amoebic dysentery, which ultimately ended in 98 deaths (USEPA, 2003). In an another episode, creosote entered the water distribution system of a south-eastern county in Georgia, in November 1984, due to a cross-connection between a hose that was used as a priming line between a fire service connection and the suction side of a creosote pump (USEPA, 2003). Cross connection can occur due to backflow, back pressure and back syphonage. Backflow can be prevented by (i) air gaps, (ii) barometric loops, (iii) vacuum breakers, (iv) double check valves and (v) reduced pressure principle devices

An *air gap* (Figure 2.16) is separation between a water outlet and a contamination source. *Reduced pressure principle backflow preventer* (RPBP) devices will have two spring check valves with a pressure-relief valve at the centre of them vented to the atmosphere. *Vacuum breakers* have a check valve that seals in the uppermost position from the thrust of water pressure. *The double-check valve* will have two single-check valves joined within one body with test cocks to determine leakage and two closing gate valves. A *barometric loop* is inverted U-tube as shown in Figure 2.17.

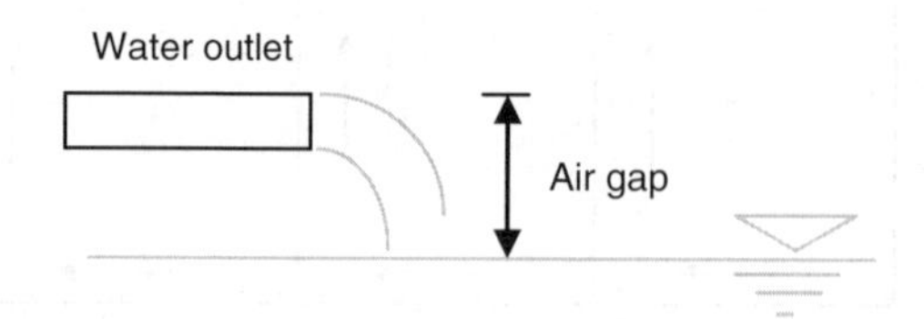

Figure 2.16 *Air gap.*

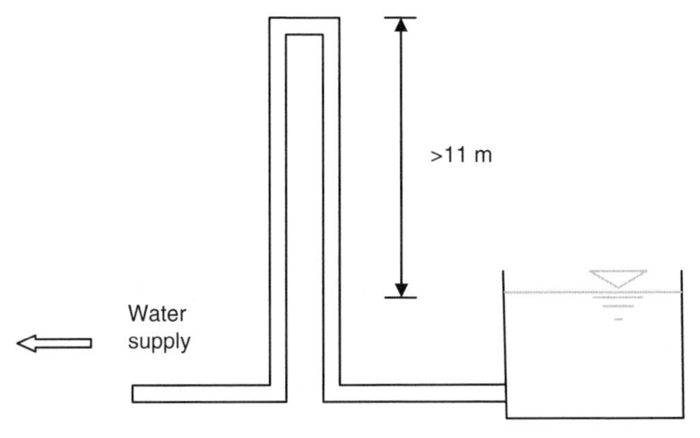

Figure 2.17 *Barometric loop.*

2.4.3.4 Pumping Requirements

Pumping requirements depend on (i) the quantity of water supplied, (ii) the duration of water supply, (iii) pressure requirements in the system. In the past, water pressure in urban areas was calculated based on pressure required for firefighting but the higher pressure and flow results in higher losses. Lowering the pressure in the pipes reduces the flow leaking water and the holes do not enlarge. Distribution through lower pressure results in less leakage and therefore infrastructure costs will reduce. Further, due to rapid urbanization, many cities in the developing world neglect water requirements for firefighting despite repeated fire hazards.

2.4.4 Unsustainable Characteristics

A sustainable approach includes not just treating polluted water before use and discharge but also avoiding pollution of water itself. The major water pollution sources are (i) sewage, (ii) storm water, (iii) urban and rural runoff, (iv) industrial water, (v) use of agrochemicals in agriculture, (vi) storage of fuel, raw material and products (Figure 2.18), (vii) solid waste (Figures 2.19 and 2.20) and (viii) air pollution (Figures 2.21–2.24).

Sustainability should aim to reduce water pollution and consumption. Solid-waste management and air pollution control are often not considered as approaches to solve the water pollution problem. The population explosion and concentration in few urban locations has been major source of merging chemical pollutants as these urban bodies use and discard a variety of chemicals compared to rural setups. The pollution from leaking petroleum storage and the absorption of toxic chemicals used in paints, varnishes and other surface coating material is often not addressed in national laws and policy.

The absence of proper solid-waste management, the presence of slums with poor sanitation, indiscriminate unplanned industries within the residential/commercial area, open defecation by humans/animals and discarded household hazardous and biomedical waste contribute to water pollution as the chemicals from these sources dissolve in rainwater and percolate into groundwater or reach surface-water bodies in developing countries.

Figure 2.18 *Storage of raw material in an industry open to the sky.*

Water distribution systems are vulnerable to contamination resulting in endemic and epidemic waterborne disease. Contamination due to cross connections, back syphonage and biofilms present in water distribution systems may provide favourable conditions for some bacterial pathogens and these problems are serious in middle-income and developing countries where resources are inadequate to maintain the distribution system (Moe and Rheingans, 2008). It is necessary to disinfect the system after repairs or additions by

Figure 2.19 *Indiscriminate solid waste disposal.*

Figure 2.20 *Solid waste stored in open area before disposal in a solid-waste-disposal facility.*

introducing water with residual chlorine of 50 mg/L and permit it to remain in the system for at least 12 to 24 hours before it is drained out.

Air pollution from vehicles, industries, agriculture sprays, dust, volcanic eruptions, solid waste burning, fuel combustion and biodegradation dissolve in rain water and groundwater. Seepage of contaminants into groundwater from waste dump sites and rupture of liners in landfill is also a major challenge for sustainable water management.

Rapid urbanization in the developing world is often accompanied by illegal connections. Leakage of 5 to 25 L/mm per km is common for many existing systems. Many systems will have cracks and high leakage with a loss of 25% to 45% in the developing countries, 15% to 24% in middle-income countries and 8% to 24% in the developed countries (WHO, 2001).

Figure 2.21 *Open kitchen using wood in roadside eatery.*

Figure 2.22 *Air pollution due to vehicular emission.*

Frequent power cuts contribute to low/negative pressure in the pipes and contaminated water surrounding the pipes is drawn in through cracks. Hence, there is an increase in household treatment by portable treatment systems that include filtration, activated carbon filtration, UV disinfection, ion exchange and reverses osmosis. Alternatively consumers are switching over to packaged water purchased from nearby shops or delivered to the door on call.

Minimizing the leakage in water systems has many benefits: (i) reduced water system operational costs, (ii) enhanced operational efficiency, (iii) reduced potential for contamination, (iv) reduced water system liability, (v) extended life of facilities, (vi) improved public relations and (vii) reduced water outage events.

Figure 2.23 *Industrial air pollution.*

Figure 2.24 *Unplanned urbanization.*

The hydraulic efficiency of pipes will diminish with time due to tuberculation, encrustation and sedimentation. Unaccounted-for water is normally discussed as:

$$\text{unaccounted-for water } (\%) = [(\text{production-use})/(\text{production})] \times 100\%$$

Water supplies are becoming scarcer in many cities, especially those using a single source of water that depends on the climate. A diversified range of water sources is needed to ensure economic and environmental sustainability. Some of the options to augment drinking-water supplies are: (i) recycling after advanced treatment, (ii) rain water harvesting, and (iii) supply through tankers. Corruption plays a major role in maintaining unsustainable practices (Figure 2.25). Corruption in government, private and nongovernment sectors often hinders new, sustainable ideas as individual interests often prevail over the interests of the state or society. Apart from corruption, negligence (Figure 2.26) and lethargy in saving water are also key issues for unsustainable water use.

Sustainability cannot happen just by advances in science and technology. Countries need to upgrade the honesty and emotional intelligence of public servants and to enhance water quality. In the absence of these qualities all the investment will be wasted within a short time and the impact of this could spread across the world.

Corrupt agencies can do a lot to make quick money. The magnitude of the problem and the approach taken varies from country to country, department to department and person to person but corrupt officers usually compare themselves with other departments, officers and countries to justify corrupt practices. Often, each officer and department would be given a 'bribe target' to achieve based on the amount that can be generated in the particular department and jurisdiction. Failing to achieve the target often results in transfer to problematic areas with poor infrastructure and facilities, which includes poor water supply, educational institutions for children and healthcare institutions. The honest people

Why do people give bribes

Cheaper than complying with the law
Changing / complicated legal procedures
Unfriendly officials
Too much competition
Failure of governance

Burden of corruption on environment, climate change and afterwards

Water pollution
Air pollution
Noise pollution
Thermal pollution
Radioactive pollution
Improper solid waste-management disposal
Climate change
Disasters
Burden of diseases

Possible remedies

Replace paper currency and coins with e-currency
Introduce corruption and its effect in school curriculum
Shift to e-governance
Provide CCTV in exam halls, public places, offices
Record job interviews
Include anticorruption in syllabus of professional courses
Include anticorruption in corporate and social responsibility
Ban expensive election campaign / political gatherings
Political parties to run own businesses
Avoid tips in restaurants, taxis and other public places

Why people are corrupt?

Responsibility without resources
Family background
Influence of society
Lack of ethics and values
Pressure from boss, colleagues and subordinates
Social and financial insecurity
Financial needs
Mismatching inflation and income growth
Saving for future
To give bribe to others
Absence of education on ill effects of corruption
Failure of governance

Modes of corruption

Currency and coins
Gifts
Tickets / passes to concerts and sport events
Hospitality, massage, hospital bills
Concessions in transactions / goods / commodities
Travel expenses/arrangements
Entertainment
Scholarships, fee concessions / waiving membership of elite clubs

Figure 2.25 *Fundamentals of corruption.*

may also be given responsibility for routine work like record keeping or interdepartmental communication, which do not require much skill or application of mind.

If an honest citizen/officer refuses to pay a bribe an adverse report may be generated to make honest people appear dishonest. An adverse report can arise during police verification of a passport, inspection of an industry, emission testing and so on. Overcoming these adverse remarks often becomes costly if citizens have to hire an expert legal advisors or

Figure 2.26 *Water being wasted in one of the railway stations in India.*

Table 2.4 *Sustainable water supply approaches for different type of settlement.*

Sl. no.	Type of settlement	Population	Source	Quality of water	Sustainable water supply approaches
1	Individual house	1–10	Groundwater	Poor	Rainwater harvesting, transporting water through vehicle, reuse wastewater
				Average	Rainwater harvesting, transporting water through vehicle from reliable source of good water, reuse, mix treated water with rainwater
				Good	Use without treatment
			Lake	Poor	Rainwater harvesting, transporting water through vehicle from reliable source of good water
				Average	Rainwater harvesting, transporting water through vehicle from reliable source of good water, reuse, mix treated water with rainwater
				Good	Use without treatment
			River	Poor	Rainwater harvesting, transporting water through vehicle from reliable source of good water, reuse, mix treated water with rainwater
				Average	Rainwater harvesting, transporting water through vehicle from reliable source of good water, treat and use, reuse wastewater
				Good	Use without treatment
2	Village	Up to 1000	Groundwater	Poor	Transporting water through vehicle/pipe, reuse, mix treated water with rainwater, water purifying and vending machines
				Average	Rainwater harvesting, Transporting water through vehicle from reliable source of good water, reuse wastewater, install water purifying and vending machines
				Good	Use without treatment
			Lake	Poor	Transporting water through vehicle/pipe, reuse, mix treated water with rainwater, water purifying and vending machines
				Average	Rainwater harvesting, transporting water through vehicle from reliable source of good water, reuse wastewater, install water purifying and vending machines
				Good	Use after disinfection
			River	Poor	Transporting water through vehicle/pipe, reuse, mix treated water with rainwater
				Average	Rainwater harvesting, Transporting water through vehicle from reliable source of good water, reuse wastewater
				Good	Use after disinfection

(continued)

Table 2.4 (Continued)

Sl. no.	Type of settlement	Population	Source	Quality of water	Sustainable water supply approaches
3	Temporary Camp	Up to 1000	Groundwater	Poor	Transporting water through vehicle/pipe, reuse, mix treated water with rainwater
				Average	Rainwater harvesting, transporting water through vehicle from reliable source of good water, reuse wastewater
				Good	Use without treatment
			Lake	Poor	Transporting water through vehicle/pipe, reuse, mix treated water with rainwater
				Average	Rainwater harvesting, transporting water through vehicle from reliable source of good water, reuse wastewater
				Good	Use after disinfection
			River	Poor	Transporting water through vehicle/pipe, reuse, mix treated water with rainwater
				Average	Rainwater harvesting, transporting water through vehicle from reliable source of good water, reuse wastewater
				Good	Use after disinfection
4	Linear settlement	-	Groundwater	Poor	People settle next to roads in leaner pattern even when the groundwater is poor quality due to adopted livelihood and business opportunity. Such pattern may not be served by local bodies in many cases and individual owners have to make own arrangements by transporting water by vehicles and rainwater harvesting. Individual bore well shall be discouraged.
				Average	Such pattern may not be served by local bodies in many cases and individual owners may have to make own arrangements by transporting water by vehicles and rainwater harvesting. Individual bore well shall be discouraged.
				Good	Use without treatment
			Lake	Poor	Transporting water through vehicle/pipe, reuse, mix treated water with rainwater
				Average	Rainwater harvesting, transporting water through vehicle from reliable source of good water, reuse wastewater
				Good	Use after disinfection
			River	Poor	Transporting water through vehicle/pipe, reuse, mix treated water with rainwater
				Average	Rainwater harvesting, transporting water through vehicle from reliable source of good water, reuse wastewater
				Good	Use after disinfection

5	Township	1000-20000	Groundwater	Poor	Piped water supply from reliable source of good quality water, rainwater harvesting, and reuse after treatment. Discourage individual bore wells.
				Average	Piped water supply from reliable source of good quality water, rainwater harvesting, and reuse after treatment. Discourage individual bore wells.
				Good	Adopt piped water supply. Discourage individual bore wells.
			Lake	Poor	Piped water supply from reliable source of good quality water, rainwater harvesting, and reuse after treatment. Discourage individual bore wells.
				Average	Piped water supply from reliable source of good quality water, rainwater harvesting, and reuse after treatment. Discourage individual bore wells.
				Good	Adopt piped water supply. Discourage individual bore wells.
			River	Poor	Piped water supply from reliable source of good quality water, rainwater harvesting, and reuse after treatment. Discourage individual bore wells.
				Average	Piped water supply from reliable source of good quality water, rainwater harvesting, and reuse after treatment. Discourage individual bore wells.
				Good	Adopt piped water supply. Discourage individual bore wells.
6	City/ metropolitan city	1000 and above	Groundwater	Poor	Piped water supply from reliable source of good quality water, rainwater harvesting, and reuse after treatment. Discourage individual bore wells.
				Average	Piped water supply from reliable source of good quality water, rainwater harvesting, and reuse after treatment. Discourage individual bore wells.
				Good	Adopt piped water supply. Discourage individual bore wells.
			Lake	Poor	Piped water supply from reliable source of good quality water, rainwater harvesting, and reuse after treatment. Discourage individual bore wells.
				Average	Piped water supply from reliable source of good quality water, rainwater harvesting, and reuse after treatment. Discourage individual bore wells.
				Good	Adopt piped water supply. Discourage individual bore wells.
			River	Poor	Piped water supply from reliable source of good quality water, rainwater harvesting, and reuse after treatment. Discourage individual bore wells.
				Average	Piped water supply from reliable source of good quality water, rainwater harvesting, and reuse after treatment. Discourage individual bore wells.
				Good	Adopt piped water supply. Discourage individual bore wells.
7	Slum				Install water purifying and vending machines

political leaders. Hence, it is often cheap to pay bribe to overcome an 'adverse' report as such reports can end someone's career or the life of a person or an enterprise. Since people live with limited resource the cost of enforcing the law would be 'cut' by paying a bribe.

Many private organizations that offer quality services do not participate in the tendering process of developmental projects, to avoid harassment, delays and difficulties in obtaining payment. Many government agencies would not release payment and would generate an adverse report unless a bribe is given. This may be one of the main reasons why quality infrastructure cannot be seen in many countries. On the other hand, corrupt agencies that have a 'good' relationship with government agencies would create poor infrastructure or obtain payment without providing a service by generating fake completion certificates.

Indirect potable reuse (IPR) is a water-recycling application developed as a result of advances in treatment technology. In IPR, municipal wastewater is externally treated and discharged into groundwater or surface-water sources (NRMCEPHC and NHMRC, 2008). It has been successfully implemented in the United States, Europe and Singapore and 3300 indirect reuse project nonpotable applications were implemented worldwide in 2005 (Bixio *et al.*, 2005). Recycled water is blended in ratios from 1% to 100% with water sources (Rodriguez *et al.*, 2009).

2.4.5 Sustainable Approach

There are many sustainable approaches and there have been many successful practices, which have not been replicated due to absence of knowledge and to some extent poor business models within trade and industry related to sustainable water practices.

Decisions about sustainable water supplies depend on climatic conditions, geographical location and settlement patterns. Settlement patterns can be categorized as dispersed, linear and nucleated. Dispersed settlement consists of houses far apart, including areas with low economic development as in cases where there is nomadic herding or smallholdings. Linear settlement or ribbon development occurs besides roads and rivers due to the advantages associated with the location. Nucleated settlement shows concentrations of buildings or houses, normally around centres of economic activity.

Apart from the economic advantages, the climate of the region plays a major role in the settlement pattern. The major Canadian population centres are within around 200 km of the United States. The southern region of Sweden is more populated compared to northern region, which is nearer to the Pole. Table 2.4 shows sustainable water supply approaches for different type of settlement.

Box 2.6 Learning from Denmark

The groundwater quality in Denmark in deeper aquifers is generally good and, hence, complex and expensive water purification is not necessary. Except for Copenhagen, the quality of tap water is better than that of bottled water but many shallow aquifers suffer due to groundwater pollution from nitrates and pesticides. Development during the twentieth century led to the depletion of water resources in Denmark and, in 1987, an Action Plan for the Aquatic Environment was approved by the Danish parliament with the objective of reducing nitrate leaching into aquatic environments by 50% and phosphate leaching by 80% within five years. While the phosphate goal was accomplished, the nitrate goal has not been fully achieved.

The first Pesticide Action Plan was introduced in 1986, followed by two further action plans to reduce the use of pesticides to avoid the negative effects on health as well as the environment, including protection of the groundwater.

In the period 1991–2005, 1306 wells were closed as water supply abstraction wells due to their pesticides or metabolite content and about 100 wells are closed every year due to pesticide contamination.

By the end of 1997, mapping was done according to the interest in drinking water.

Water-supply wells in Denmark have been protected at point sources, using two-level protection zones: a 10 m diameter protection zone round the well by a fence and a 300 m protection zone and site-specific protection zones may encompass the whole recharge area (Danish Ministry of the Environment, 2013).

Proper solid-waste storage on site with protection from rain as shown in Figure 2.27, photographed in Stockholm, can be a good example for countries that aim to provide safe water sources for their people.

Breathing spaces shown in Figure 2.28 are often missing in the urban landscape of developing countries as such spaces are often grabbed by influential people who convert them into slums overnight and rent them to poor immigrants. Such open spaces will turn out to be filthy settlements that will not have proper sanitation.

Identifying resources in waste contributes to sustainable development. In the example photographed in Figure 2.29, food waste is collected and sent to biodigesters to convert the waste to biogas, which ultimately fuels urban transportation.

A relatively new practice of waste disposal (Figure 2.30), where residents of a city haul the waste from their homes to waste-collection centres, has yet to gain popularity in developing countries. Rag pickers do pick the recyclable waste but the rag pickers often lead to scattering of waste that has questionable resale value.

Figure 2.27 *Proper onsite solid waste bins.*

Figure 2.28 *Open spaces within in the city.*

Figure 2.29 *Collection of food waste for biogas generation.*

Figure 2.30 *Wastewater collection centre where urban dwellers dispose of solid waste.*

Figure 2.31 *Waste-dropping place; pneumatic solid waste conveying system.*

Figure 2.31 shows a pneumatic dropping point for a solid-waste conveying system where the waste deposited will be conveyed by underground chute. This automated vacuum waste collection system, or pneumatic refuse collection, also known as an automated vacuum collection (AVAC) system, transports waste through underground tubes to a collection station from where it is sent for processing/disposal.

The total waste generated globally is more than 4 billion tons per annum. About 45% of this is municipal solid waste and the rest is industrial waste (Kaliampakos and Benardos, 2013). The use of subsurface space is a key issue towards achieving sustainable development in urban areas. Activities/infrastructures that are difficult, environmentally undesirable or less profitable for installation on the ground can be relocated underground. The underground systems with respect to solid waste collection can operate as stand-alone (Figure 2.32) collection points or AVAC. The management of waste below ground can therefore be a sustainable approach.

Sustainable transportation with renewable nonpolluting fuel and solid waste-treatment storage, treatment and disposal should not be seen in isolation with water problems. In fact curbing air pollution and solid waste problems are part of the solution to the water issues. Figure 2.33 depicts the electric points in one of the parking areas in Sweden, which encourages the sale and use of electric-powered cars, thus contributing to curbing air pollution, which in turn assures safe and clean water for citizens. The encouragement of green taxis (Figure 2.34) in Stockholm to control air pollution has also contributed to maintaining the air pollution and water quality.

As discussed in previousdraft section, storage of raw material and waste under the open sky often causes water pollution. The thought of costly storage facilities is the reason industries and waste mangers often avoid storing waste under cover. Figure 2.34 provides a simple solution to such situations so that water pollution due to waste/material can be avoided with simple and cheap protection from rain.

There are many solutions to the problem of water conservation. These include faucet aerators (fittings to be affixed to faucets), which add air to the water flow, reducing water flow by 25–50% and thereby reducing the water consumption to that extent. However, these are not used worldwide due to lack of availability. Similarly, integration of geographic

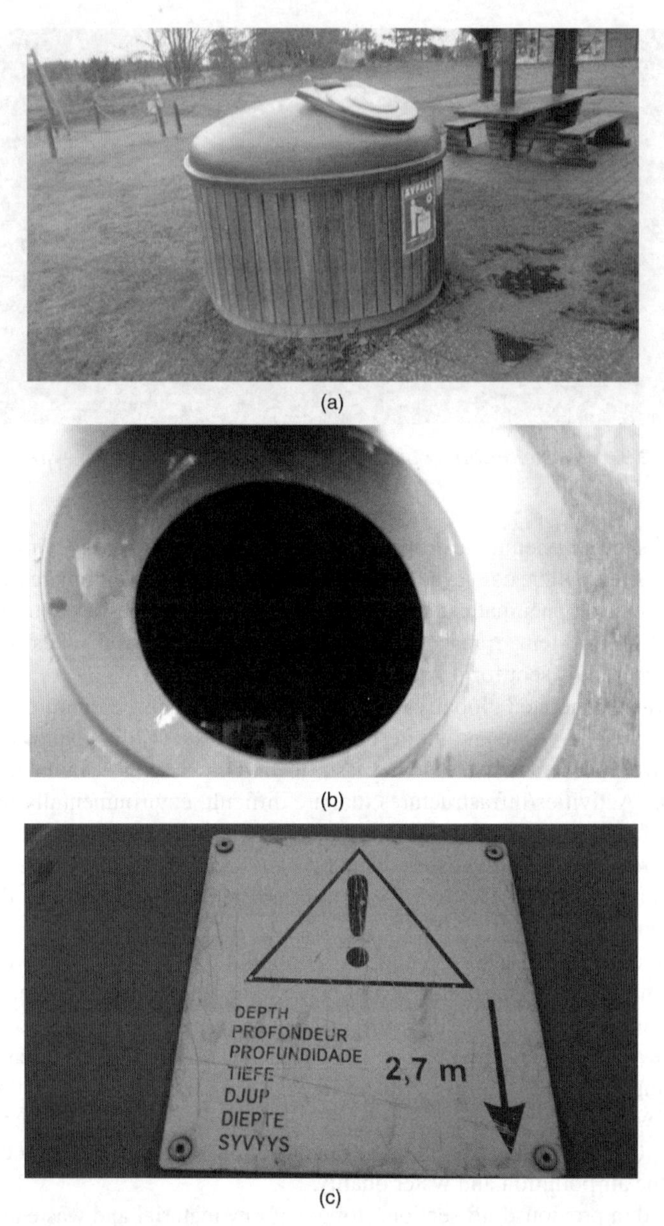

Figure 2.32 Stand along collection point: (a) front view; (b) top view; (c) depth of collection bin.

information system (GIS) and leak detection with water distribution have also not been adopted in many parts of the world due to poor education and poor capabilities of the people responsible for the water distribution. There are numerous ways to save water in houses, at workplaces and at other places. Simple precautions like washing dishes by hand and not letting the water run while rinsing can save water. Another method could be filling

Figure 2.33 *Electric driven cars being charged. Use of such cars solves air and water pollution problems simultaneously.*

one sink with wash water and the other with rinse water. Upgrading water-cooled appliance with air-cooled appliances can achieve significant water savings.

Proper solid waste storage facilities (Figure 2.35) prevent wastage of water through contamination. Operating dishwashers and clothes washers full will make optimum use of water as operating them when they are partially full will contribute to wastage of water. Monitoring water bills constantly can save water. Watering gardens in the morning or evening minimizes evaporation losses. Washing fruit and vegetables consumes a lot of water in houses and restaurants. Washing in a pan of water instead of using running water can save a considerable amount of water.

Erosion adds sediment and organic matter to water bodies. In many places human activities have changed natural erosion rates and have altered the rate, volume and timing of the flow of water bodies, affecting the adaptation of species and their physical/chemical processes in response to existing sediment regimes. Increased sedimentation decreases primary productivity, impairs spawning habitats and harms fish, plants and benthic (bottom-dwelling) invertebrates. Fine sediments attract nutrients like phosphorus and toxic contaminants like pesticides, altering water chemistry (Carr and Neary, 2008). Avoiding soil erosion by planting vegetation (Figure 2.36) on exposed soil and by covering soil with gravel (Figure 2.37) can greatly enhance the quality of water.

Figure 2.34 *Green taxis in Stockholm. (For a colour version of this figure, see the colour plate section.)*

Figure 2.35 *Proper solid waste-storage facility at one of the hazardous waste treatment and disposal facilities.*

Figure 2.36 *Avoiding soil erosion in rural and urban area can greatly enhance the quality of water.*

Figure 2.37 *Integration of urban landscape design with water-quality objectives – gravel provided on soil to avoid soil erosion and dust.*

The rinsed water can be used for watering plants or floor washing. Using a broom instead of a hose to clean a driveway and sidewalk can save water. Replacement of inefficient showerheads has saved water in many households and hotels. Outdoor spigots can be winterized prior to temperatures falling below freezing to prevent bursting/leaking of pipes.

2.5 Conventional Wastewater Collection Systems

Wastewater collection system involve transmission of wastewater from the source to point of treatment/disposal through channels or conduits. Sources could be household, industries, or commercial establishments. Wastewater is collected under gravity and pumped from intermittent storage tanks only if the point of treatment/disposal is at a higher elevation. Pumping is done only in exceptional cases as it involves energy and precautionary measures during power failures or disasters.

Quantity and flow in a collection system depends on: (i) the size of the city, (ii) the culture of the city, (iii) climatic variation, (iv) the pricing of water, (v) the income of the consumer, (vi) the type of industrial and commercial establishments and (vii) leakages in the collection system. Designers often overdesign the system considering the future demand, safety and losses in the system. The assumptions made during system design do not always come true, especially with respect to those cities or countries in transition or cities where the economy is booming. Some cities that grew rapidly witnessed a sudden increase in wastewater generation.

2.5.1 Features

Conventional wastewater collection with respect to sewage consists of a series of pipes connected with joints. It is characterized by (i) huge investment, (ii) energy consumption, (iii) operation and maintenance and (iv) expenditure to sustain the system.

Sewerage is, in many cases, considered as indispensable as it conveniently and hygienically transports out wastewater and prevents flooding. But, sewerage is expensive. Sewerage systems cost approximately 70% of the wastewater transport and treatment costs in densely populated rural/periurban areas in German (Otterpohl, 2000). During heavy rains the diluted wastewater combined with sewers is discharged directly from a pipeline or comes out of manholes. Hence, many advanced countries have adopted sewerage and stormwater drainage systems. In rural and periurban areas storm water can be infiltrated or directly discharged into surface water. In an improved separated system, the first flush of the stormwater with pollutants is directed to the wastewater treatment plants while the remaining stormwater needs to be directed to surface water. Another option is to make provision for extra storage capacity in the sewers/treatment plant to store wastewater during heavy rains.

In cases where wastewater generators cannot let the wastewater into sewerage due to absence of sewers or regulatory requirements, treated or untreated wastewater can be transported to a treatment plant or disposal site. Further sludge or urine can be transported for reuse by truck.

Box 2.7 Transportation of waste water from electroplating units in Bangalore, India

Bangalore (now Bengaluru), India, developed rapidly since the 1970s. As a result the industrial area once located on the outskirts of the city is now located at the heart of city as more residential dwellings were added to the city to accommodate people. Many electroplating units inside the city cannot discharge wastewater into sewers to comply with new legislation, so the wastewater from electroplating units is transported to common effluent treatment plants located in the outskirts of the city and discharged after treatment.

A vacuum system of four to five toilets can be feasible when situated under the water level or when water conservation is essential or space is major issue. The most frequently used system for approximately 100 people requires a 4 kW pump.

2.5.1.1 Types of Sewers

Sewers are made up of many materials. The commonly used materials are listed in Table 2.5. The following paragraphs briefly explain different types of sewers:

Building: Building sewers are used to connect building plumbing to lateral sewers.

Lateral or branch: They collect sewage from one or more building sewers and convey it to mains.

Main: Main sewers are used to connect lateral sewers to trunk sewers or intercepting sewers

Trunk: These convey wastewater from main sewers to a treatment/disposal facility or intercepting sewers. For example, an injunction was obtained in 1869, in Leeds, which prevented the discharge of sewage until it had been purified and deodorized sufficiently (Sellers, 1997). This resulted in the construction of trunk sewers to intercept the early outfalls and convey sewage to a point for disposal.

Intercepting: These are used to intercept a number of main or trunk sewers to carry sewage to a treatment/disposal facility.

Table 2.5 *Material used for sewers.*

Sl. no.	Material	Size range in mm
1	Asbestos cement	100–900
2	Ductile iron	100–1350
3	Reinforced concrete	100–3600
4	Prestressed concrete	400–3600
5	Polyvinyl chloride	100–375
6	Vitrified clay	100–900

2.5.1.2 Collection System Appurtenances

The major appurtenances of the sewage system are manholes, drop inlets to manholes, building connections and junction chambers. Local topography may demand special structures.

Manholes are vertical structures that provide access to sewers for cleaning. For sewers smaller than 1.2 m, manholes should be located at changes in size, slope and direction. Drop inlets are provided if the difference between an incoming and outgoing sewer is greater than 0.5 m. The building connection is a type of appurtenance that leads from an individual building to a municipal connection. Junction chambers are provided when regular manholes cannot be used due to a higher diameter of wastewater collection pipes in trunks and mains.

Ventilation in sewers is required to avoid (i) asphyxiation of sewer maintenance personnel, (ii) the buildup of odours and (iii) explosive gases.

Box 2.8 Wastewater reuse in Japan

Tokyo is one of the cities successfully implementing wastewater reuse. The city has dual water distribution systems for (i) delivering drinking-quality water and (ii) delivering treated wastewater and stream augmentation. Stream augmentation is a process in which additional water is allowed to flow through the streams using gravity. In the Shinjuku area of Tokyo, a dual distribution system is used. Treated water from the Ochiai Municipal Wastewater Treatment Plant is disinfected and used for toilet flushing in 25 high-rise business premises and for stream augmentation. The system has been successfully supplying up to 8000 m^3/day of treated wastewater since 1984 (UNESCAP, 2007).

2.5.1.3 Pumping Requirement

A typical classification of pumps is shown in Figure 2.38. Due to the presence of rags and trash, untreated wastewater is pumped by kinetic pumps fitted with nonclog impellers. Nonclog pumps smaller than 100 mm are not recommended for handling untreated sewers. Positive displacement pumps are used for raising wastewater from building sumps and to pump sludge. With new inventions, numerous types of pumps are available on the market. A choice can be made by contacting the supplier of pumps, providing a static head and suction head along with the distance to be pumped.

Pumping systems usually account for about 50–100% of a treatment plant, even though it is possible to design treatment plants without pumps, especially in sloppy terrain where water can flow downhill through gravity. Pumps have two main purposes: (i) transfer of water from one point to another point, (ii) circulation of liquid around a system. The efficiency of existing pumping systems can be improved but it is difficult to optimize before installation. Design efforts are focused on minimizing capital costs and the chances of system failure. Lifecycle cost (LCC) consists of maintenance, energy and initial costs. Energy and maintenance costs will account for more than 50–95% of pump ownership costs. Energy losses in pumping systems occurs mainly due to (i) piping friction losses including inlet, joins, valves, bends, contractions and expansions in pipe work as well as outlet losses, (ii) pump and motor inefficiencies.

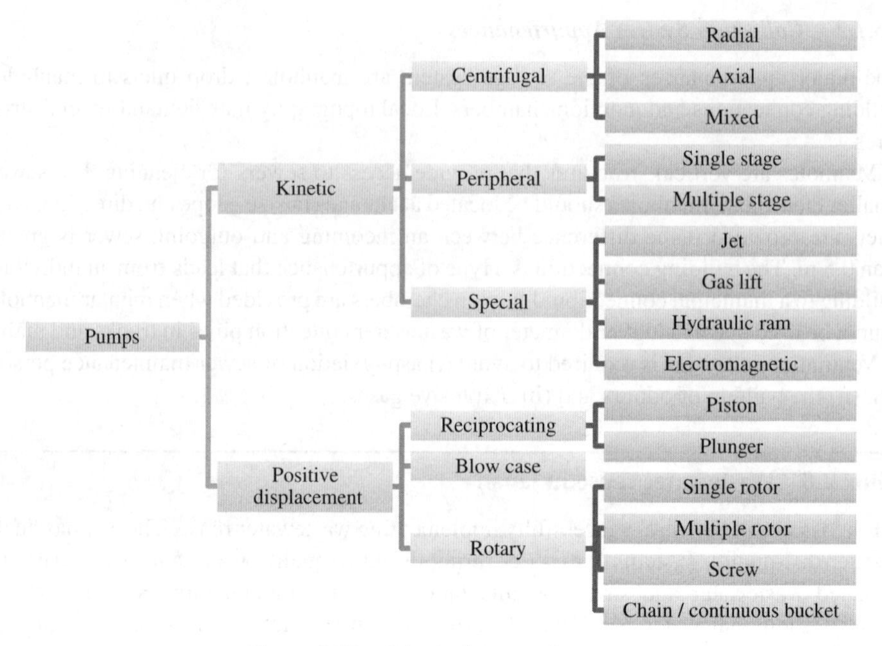

Figure 2.38 *Principal types of pumps.*

Centrifugal pumps are commonly used in industry as they provide good performance, low capital cost, low maintenance requirements and long operating lives. Centrifugal pumps are divided into radial flow, mixed flow and axial flow. Radial flow is used for a head ranging from 3 m to 1000 m for flow ranging from 0.1 to 200 m^3/s. Mixed flows are best suited for mid-range operation (flow: 3–500 m^3/sec, head: 2–20 m). Axial flow is preferred for low head and high speeds (flow: 10–1000 m^3/sec, head: 1–5 m).

In order to enhance energy efficiency, points to be considered while designing systems are: (i) ensure a whole-system approach, (ii) minimize pumping demand, (iii) reduce pumping needs, (iv) reduce leaks, (v) choose efficient pumping components. Control and operating philosophy should consider (i) variable-speed drives instead of throttling valves for flow management, (ii) installing pressure/flow sensors in the location that will help ensure process requirements without excess pumping energy and (iii) recording system trend data.

If the treatment plant has to cater for varying flows, the system should be designed with pumps of different capacity to avoid energy inefficiency during operation. Paying attention to pipe work during design will eliminate unnecessary bends and valves.

Figure 2.39 provides a view of new generation pumps in wastewater treatment. In order to enhance energy efficiency, piping configuration should be designed (i) maximizing pipe diameter, (ii) optimizing pipe layout to minimize pressure loss, (iii) minimizing pressure losses through valves and fittings, (iii) minimizing bypass flow rates. Points to be remembered when choosing motors are: (i) ensuring high motor efficiency, (ii) ensuring high pump efficiency, (iii) ensuring that the pump operates close to its best efficiency point (BEP), (iv) avoiding installing an oversized pump, (v) ensuring the right impeller size, (vi) ensuring the right pump type and (vii) ensuring compatibility with variable-speed drives.

Figure 2.39 *View of new-generation pumps in wastewater treatment plants.*

A pump is considered to be oversized when it is not operated within 20% of its BEP. To overcome this problem, impeller trimming and changing impeller are considered.

Impeller trimming is the process of reducing the diameter of an impeller by machining. The reduction in diameter decreases the energy conveyed to the water and hence the flow rate as well as system head capacity. This option is used when the impeller size needed to ensure BEP is not available from the manufacturer.

2.5.1.4 Considerations in Urban Drainage Systems

The occurrence of global climate change has resulted in previously rare environmental issues all over the world. Extreme weather events in recent years have led to destructive flash floods and are likely to occur again. Urban development and changing climate have resulted in an increase in surface runoff. Cities located in the catchment areas of the rivers will contribute to pollution and risk worse flooding further downstream. Achieving sustainable drainage is the solution to the long-term steadiness of water resources.

Fundamental requirements during design of drainage system are: (i) estimated wastewater quantity, (ii) evaluation of alternative sewers, (iii) topography and (iv) selection of sewer appurtenances. Unlike Europe many developing urban agglomerations are the outcome of haphazard development consisting of mixed land use with no definite commercial, industrial and residential area. The cities of many developing countries are also characterized by slums, narrow roads, encroachments and apartments in the places originally planned for individual dwellings.

The flow pattern in urban drainage is designed for channel flows with the most commonly used equation being the Manning equation:

$$V = \frac{1}{n} R^{2/3} S^{1/2}$$

where

V = velocity, m/s
n = friction factor
$$R = \text{hydraulic radius} = \frac{\textit{cross sectional area of flow, } m^2}{\textit{wetted perimeter, m}}$$
S = slope of energy grade line, m/m

Three types of sewers are commonly used across the world: (i) sanitary sewers, (ii) stormwater sewers (Figure 2.40) and (iii) combined sewers. Pressure and vacuum sewers are used where gravity sewers are difficult or costly. The construction of pressure and vacuum sewers also needs trained personnel.

Sewers are designed to maintain a minimum velocity of 0.6 m/sec. To prevent deposition of solids, a velocity of 0.75 m/s is adopted. To avoid damaging sewers the maximum velocity is limited to 3.0 m/s. The minimum practicable slope is limited to 0.0008 m/m. Warm weather often leads to formation of hydrogen sulfide in sewers if the flow is slower than minimum recommended values, resulting in (i) an odour problem, (ii) chemical attack on cement, (iii) precipitation of trace metals, which are necessary for micro-organisms in the wastewater treatment plants.

Built-up areas require to be drained to take out surface water. The conventional approach for surface-water drainage has been through underground pipes that convey water from the built-up region. Traditional urban drainage solutions focus on quantity and aim to remove surplus water from built-up areas quickly to avoid possible flooding incidents. They usually do not pay sufficient regard to water quality, flood control, water resources and biodiversity requirements. Such drainage systems have resulted in alteration in natural flow patterns with local effects and they cause problems in other places in the catchment area.

Figure 2.40 *Open drainage.*

2.5.2 Unsustainable Characteristics

The conventional sewerage system has evolved mainly in Europe and the United States over time and is being adopted all over the world. It has catered to the needs of advanced countries, which depend on a near-perfect management system. Adoption of such a system needs perfection in laying pipelines and attending to choking problems immediately, failing which sewage will flow onto roads, which will be picked by the tyres of moving vehicles distributing pathogens all over the city and sometimes outside the city. Such overflows will join rainwater drainage as well as flood water during floods.

Designing sewers without considering the flood/drought situation often leads to troubleshooting water flows in sewers. Infiltration of rainwater during the rainy season could end up in overflows near manholes. Exfiltration of sewage water during the lean season could lead to the settlement of solids in sewers. Poorly maintained drainage (Figure 2.41)

Figure 2.41 *Poorly maintained open drainage.*

can lead to many problems like accidents, groundwater pollution, surface-water contamination and so on.

The exfiltration of sewage often leads to groundwater contamination. The undulating terrain may often need pumping of sewage. In many cities in the developing world, sewerage and sewage treatment are considered separately and consequently financed separately too. Many cities in the developing countries, where individual houses already have dewatering systems or septic tanks, are forced to connect their sewage to a newly constructed sewerage system, which leads to nearby surface water due to the absence of an end-of-pipeline sewage treatment plant.

The invasion of roots into sewers is the most destructive single element that a wastewater collection system faces. Tree roots are the main cause of sewer stoppages, resulting in the blockage of sewers. Roots in sewers impede flow, resulting in sluggish and septic sewers enhancing grease accumulation and generating hydrogen sulfide. Roots can fracture cement sidewalks and sewer joints. They enter sewer pipes as tiny hairlike structures but soon mature, placing stress on pipes, resulting in failure of the pipes. Live roots cause stoppages, reduction in flow as well as accumulation of debris and grease, leading to hazardous atmospheres. About 25% of all sewerage systems contain roots. As roots and sewers are both underground, they are not easily detectable by the unassisted human eye. Roots suspended in a well-ventilated sewer will have a reliable source of water and nutrients year round. Mechanical methods of removing roots, like dragging, encourage regrowth, hence foaming root control is adopted in the developed countries where the material is injected into sewers as foam, which has the consistency of shaving cream. The foam is introduced to fill the air space above the wastewater and to contact all roots in pipes. The foam breaks down gradually and the dead roots degrade naturally and within a few weeks the hindrance due to roots is eliminated.

Another problem due to rapid urbanization and construction of high-rise building is increased flow in excess of the carrying capacity of sewers due to huge wastewater generation in high-rise buildings beyond the carrying capacity of sewers.

As early as the 1850s it was observed that the sewers are not capable of conveying all stormflows to a disposal point, leading to the introduction of combined sewer overflows (CSO) in England. These overflows operated when six times the flow during dry weather was reached (Myerscough and Digman, 2008). But the concept of CSO has not been adopted in most of the developing countries, resulting in sewer overflows into road.

Another difficulty in sustaining huge sewers is the freezing of sewage in places where the temperature reaches freezing point. It is often required to keep the sewage warm by proper insulation and external heating at places where sewers are not placed below the frosting level, beneath the surface of ground.

Inflow and infiltration increase the cost of wastewater treatment facilities due to an increase in pumping costs. Infiltration increases overflows due to overloading. Johnson County wastewater implemented a twin inspection program in 1985, concurrent with collection system evaluation. Under the programme more than 55 000 structures were inspected by going house to house as well as business to business. The surveys was followed by repair or replacement of 17 000 manhole structures, as well as disconnection of more than 15 600 sources of stormwater inflow that were not permitted (USEPA, 2012).

Apart from conventional wastewater sources, radioactive waste can also reduce water quality. Water contaminated with radionuclides from mining, processes in industry, nuclear power plants, medicines, defence debris and research materials has to be disposed of properly. Radioactive waste can be generated in gas/liquid/solid form and radioactivity can stay for a few hours to hundreds of years.

Between the late 1940s and the early 1970s, radioactive waste was dumped at sea or accumulated at nuclear sites and about 800 million tonnes of radioactive waste are pending for safe disposal in Central Asia (Zhunussova *et al.*, 2011).

2.5.3 Sustainable Approach

Humanity has changed the planet, altering the climate, diverting freshwater, increasing the rate of nitrogen fixation and changing land forms as well as habitat types from floodplains, forests, prairies, and deltas into cities and agricultural lands. These and other anthropogenic changes have resulted in the degradation of terrestrial and aquatic ecosystems. The man-made changes have also reduced ecosystem resilience and services. The changes have resulted in an extinction rate estimated to be a hundred to a thousand times more than prehuman rates. Since the 1960s, identification of these impacts has increased dramatically. There have been increasing numbers of efforts to protect and restore degraded/threatened habitats and ecosystems all over the world.

The process of assisting the recovery of degraded/damaged/destroyed ecology is called ecological restoration. Restoration attempts to recognize a set of historic conditions as well as their natural evolutionary development trail and return an ecosystem to that trail. Such efforts need an understanding of historic conditions as well as comparable ecosystems. Freshwater restoration projects can be as simple as removing an upstream dam in addition to recreating the stream channel to reinstate the river's former hydrograph, restore previous water temperature ranges, return sediments/nutrients to the system and allow native species to migrate. When the 35 m wide Brownsville Dam, across the Calapooia River in the Willamette Valley, United States, was removed it restored more than 60 km of habitat for many threatened species.

There are well established approaches that can be used to deal with pollution from point/nonpoint sources on a basin scale through ecohydrology and phytotechnology. These approaches are based on the interrelationships between ecological processes along with the water cycle in the catchment. They support the role of ecological processes in water-quality improvement. The application of phytotechnology such as constructed wetlands can be used for domestic sewage treatment and storm-water runoff.

References

Armstrong, E.L. (ed.) (1976) *History of Public Works in the United States: 1776–1976*, American Public Works Association, Chicago.

Bixio, D., Deheyder, B., Cikurel, H. *et al.* (2005) Municipal wastewater reclamation: where do we stand? An overview of treatment technology and management practice. *Water Supply* **5**, 77–85.

Burian, S.J., Nix, S.J., Pitt, R.E. and Durrans, S.R. (2000) Urban Wastewater Management in the United States: Past, Present, and Future. *Journal of Urban Technology* **7** (3), 33–62.

Cardiff University (2012) *Case Study: Source Separating Wastewater System in Ekoporten, Norrkoping, Sweden* www.cardiff,ac,uk/archi/programmes/cost8/case/watersewerage/sweden-ekoporten.pdf (accessed 16 December 2013).

Carr, G.M. and Neary, J.P. (2008) *Water Quality for Ecosystem and Human Health*, 2nd edn, United Nations Environment Programme Global Environment Monitoring System, Nairobi.

Chandrappa, R. and von Münch, E. (2008) Urban and Rural Sanitation Problem in India. Abstract volume of global conference, World Water Week, Stockholm 17–23 August, Stockholm International Water Institute, Stockholm.

Corcoran, E., Nellemann, C., Baker, E. *et al.* (eds) (2010) *Sick Water? The Central Role of Wastewater Management in Sustainable Development*, UN-HABITAT, Arendal.

Danish Ministry of the Environment (2013) *Water supply in Denmark*, www.geus.dk/program-areas/water/denmark/vandforsyning_artikel.pdf (accessed 16 December 2013).

Duffy, J. (1968) *A History of Public Health in New York City 1625–1866*, Russell Sage Foundation, New York.

Dworsky, L.B. and Berger, B.B. (1979) Water resources planning and public health: 1776–1976. *ASCE Journal of the Water Resources Planning and Management Division* **105**, 133–149.

Eveleigh, D.J. (2002) *Bogs, Baths and Basins: The Story of Domestic Sanitation*, Sutton, Stroud.

Eveleigh, D.J. (2008) *Privies and Waterclosets*, Shire, Botley.

Fair, G. and Geyer, J.C. (1954) *Water Supply and Waste-Water Disposal*, John Wiley & Sons, Inc., New York.

Falkus, M. (1977) The development of municipal trading in the nineteenth century. *Business History* **19** (2), 134–161

Folwell, A.P. (1916) *Sewerage: The Designing, Construction, and Maintenance of Sewerage Systems*, 7th edn, John Wiley & Sons, Inc., New York.
Hindu, The (2012) Water shortage hits healthcare at HIMS, The Hindu (June 23), p. 9.
Hoekstra, A.Y. (2011) The global dimension of water governance: Why the river basin approach is no longer sufficient and why cooperative action at global level is needed. *Water* **3** (1), 21–46, http://www.mdpi.com/2073-4441/3/1/21/ (accessed 16 December 2013).
Johnson, M. and Rix, S. (eds) (1993) *Water in Australia: Managing Economic, Environmental and Community Reform*, Pluto Press, Sydney.
Kaliampakos, D. and Benardos, A. (2013) Underground Solutions for Urban Waste Management: Status and Perspectives, The International Solid Waste Association, Vienna.
Kristiansen, K.R. (2012) *Denmark: Where Water Matters Asset Mapping of the Danish Water Sector*, Copenhagen Capacity, Copenhagen.
Moe, C.L. and Rheingans, R.D. (2008) Global challenges in water, sanitation and health. *Journal of Water and Health* **4** (1), 41–57.
Myerscough, P.E. and Digman, C.J. (2008) Combined Sewer Overflows – Do they have a Future? 11th International Conference on Urban Drainage, Edinburgh, Scotland, UK.
NRMMC EPHC and NHMRC (2008) *Australian Guidelines for Water Recycling: Augmentation of Drinking Water Supplies (Phase 2)*, Natural Resource Management Ministerial Council, Environment Protection and Heritage Council and the National Health Medical Research Council, Canberra, Australia.
Otterpohl, R. (2000) Design of highly efficient source control sanitation and practical experiences. Paper and presentation at Euro-summer School, DESAR, June 18–23, Wageningen, the Netherlands.
Rodriguez, C., Van Buynder, P., Lugg, R. *et al.* (2009) Indirect potable reuse: a sustainable water supply alternative. *International Journal of Environmental Research and Public Health* **6**, 1174–1209. doi: 10.3390/ijerph6031174
Seeger (1999) The history of German wastewater treatment. *European Water Management* **2** (5), 51–56
Sellers, D.A. (1997) Hidden Beneath Our Feet, The story of Sewerage in Leeds, Leeds City Council, Department of Highways and Transportation, www.sewerhistory.org/articles/whregion/england/The%20story%20of%20sewerage%20in%20Leeds_Sellers_1997.pdf (accessed 29 December 2013).
UNEP (2012) Status Report on The Application of Integrated Approaches to Water Resources Management, UNEP, Nairobi.
UN ESCAP (Economic and Social Commission For Asia And The Pacific) (2007) Sustainable Infrastructure In Asia. Overview and Proceedings of Seoul Initiative Policy Forum on Sustainable Infrastructure Seoul, Republic of Korea, 6–8 September 2006.
USEPA (2003) *Cross-Connection Control Manual*, USEPA, Washington.
USEPA (2012) *Water and Energy Efficiency in Water and Wastewater Facilities*, http://www.epa.gov/region9/waterinfrastructure/technology.html (accessed 17 December 2013).
van Timmeren, A., Kaptein, M., and Sidler, D. (2007) Sustainable Urban Decentralization: Case EVA Lanxmeer, Culemborg, The Netherlands. ENHR 2007 International Conference 'Sustainable Urban Areas', Rotterdam, 25–28 June.

WHO (2001) *Leakage Management and Control – A Best Training Manual*, World Health Organization, Geneva, Switzerland.

Zhunussova, T., Sneve, M., Romanenko, O. *et al.* (2011) Threat Assessment Report: Regulatory Aspects of the Remediation and Rehabilitation of Nuclear Legacy in Kazakhstan, Kyrgyzstan and Tajikistan, Norwegian Radiation Protection Authority, Østerås.

3

Water Quality Issues

Ramesha Chandrappa, Diganta Bhusan Das, Norazanita Shamsuddin and Umarat Santisukkasaem

It is not possible for humans to survive without water of adequate quality. Billions of man-days of labour are lost annually due to waterborne diseases. Many communities are approaching or have reached the limits of available water supplies. Water reclamation and reuse are necessary for survival but conventional knowledge is not sufficient to provide the desired quality of water. Due to advances in science, pollutants have spread across the globe. The pollutants released into air settle after travelling hundreds and thousands of kilometres. Currents in the ocean spread the pollutants accidentally or they are intentionally released into ocean across the globe. New chemicals are formed due to contact between reactants.

Societies across the world have yet to understand completely the issues and science related to water quality. As depicted in Figure 3.1, scientists have not fully understood *water quality science*. Hence, there is a huge amount of knowledge to be acquired apart from all the practical information, publications and patents that exist. Some knowledge is unpublished but still may exist amongst some individuals or groups of people. Efforts to understand *nature* fully are not feasible due to financial and practical aspects. It is not possible to study the acute and chronic effects of all chemicals/micro-organisms on human beings at different doses due to ethical issues. Furthermore, not all knowledge is published as some information is proprietary and patented. For financial reasons, not all the published literature is procured by people responsible for implementing environmental laws or supplying water. With limited knowledge, people form policy/legislation/action plans and these will be implemented by groups of field professionals who may not fully implement the policy/legislation/action plan.

The reasons for not using available knowledge are many and include corruption, communication gaps, lack of interest, racism, conflicts, emotions and so on. It is difficult to

---

Sustainable Water Engineering: Theory and Practice, First Edition. Ramesha Chandrappa and Diganta B. Das.

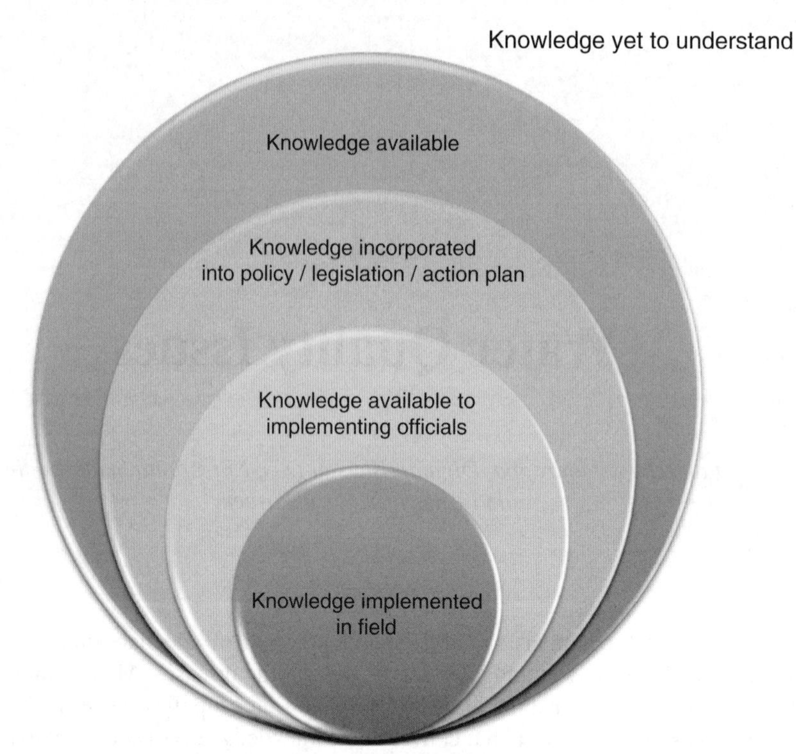

Figure 3.1 *Schematic diagram of knowledge gap in water quality.*

predict human psychology and behaviour exactly. Human behaviour varies and this may result in individuals discharging their responsibilities poorly, leading to an impact on the health of humans and on the environment.

More than 50% of the world's hospital beds are occupied by people suffering from sickness linked to contaminated water. At least 1.8 million children under five years of age die every year due to water-related disease (Corcoran *et al.*, 2010).

Physical, chemical, or biological variation in water quality has a detrimental effect on living organisms. Pollutants may be biodegradable or nonbiodegradable. Sewage, industrial chemicals, agrochemicals, industrial discharge and chemicals in soil strata are examples of materials that commonly change water quality. Water can also be polluted by deposits of pollutants from the air.

3.1 Water-Related Diseases

Bacterial contamination of water is widespread across the globe, the major pathogenic organisms being bacteria (*E. coli, Shigella, V. cholera*), parasites (*E. histolytica, Giardia,* hook worm) and viruses (hepatitis A, polio, rota virus). The entry of pathogens from haphazardly discarded infected waste from human/veterinary hospitals as well as households can transmit almost all communicable diseases. Overextraction of groundwater can often lead to an increase in fluoride/arsenic/iron/nitrate concentration. Brackishness in the groundwater of coastal areas has been on the rise due to groundwater extraction through deep tube wells leading to salinity due to seepage of seawater. Inland salinity can occur due to overextraction of groundwater.

3.1.1 Transmission Vectors

Conveyance of disease from one host to another is known as 'transmission'. An arthropod that transfers an infective agent from one host to another is referred to as a 'vector'. Vector-borne diseases are diseases transmitted by infected arthropod species like mosquitoes, ticks, sand flies, triatomine bugs and black flies. Arthropod vectors are cold blooded and sensitive to climatic factors but climate is not the only factor influencing vector distribution. Vector distribution depends on other influences like habitat destruction, pesticide application, land use and host density.

3.1.2 Field Testing and Monitoring

Contamination of a sample can occur due to entry of foreign materials, whose characteristics can vary, resulting in the collection of a nonrepresentative sample. Contamination of a sample can happen at any stage of the sampling process and affects the integrity of the sample. Many results are reported in milligrams, so even an extremely small quantity of contaminants can affect results to great extent.

Changes in natural variables like temperature and pressure can affect the integrity of a sample. A change in temperature can change chemical properties of many parameters:

Volatilization can affect quantity of volatile organic compounds (VOC) such as chlorinated hydrocarbons.

Sorption can make dissolved substances adsorb/absorb to sampling equipment/bottles. Changes in suspended solids can alter concentrations of ions, heavy metals and organic compounds.

Degassing due to an increase in temperature or drop in pressure can affect pH due to loss of carbon dioxide. pH-sensitive parameters like heavy metals, alkalinity and ammonium will vary due to variation in pH. Changes in pH can lead to precipitation of certain metals.

Precipitation caused by changes in temperature, pH, or the presence of seed particles can alter the chemical characteristics of water.

Oxidation due to introduction of oxygen into the sample will alter values of dissolved oxygen, pH and redox, thereby changing the concentrations of heavy metals, hydrogen sulfide, ammonia, COD, BOD and TOC.

Biological activity in a sample will change physical and chemical characteristics due to denitrification, respiration and photosynthesis.

Field testing and monitoring have been the challenge, especially in developing countries that lack administrative and scientific capabilities. The task of collecting samples is often left to illiterate or poorly literate personnel who do not appreciate sampling procedures. Sample collection to monitor water quality across vast jurisdictions demands extensive travelling. Understaffed monitoring organizations will usually find it hard to collect all the samples within the time frame and with the resources provided to them. Sometimes the same sample may be reported as being drawn from two points. Sometimes the sampler will not know the depth of the sample to be drawn and the preservative technique to be followed. The samples collected in unsterilized sampling jars are used for determining bacteria and no oxygen fixation is done at source, thereby allowing depletion/enhancement of dissolved oxygen. Depletion of dissolved oxygen occurs due to the presence of microbes, which use the oxygen in the sample. Enhancement occurs when groundwater comes in contact with air for a prolonged period when the samples are kept without closing the lid of the sample bottle.

Table 3.1 gives sample storage requirements. The water-quality parameters are temperature-sensitive and therefore should be kept cool (ideally at 4 °C) during transportation. Hence, it is essential that each sample is packed with ice during transit to laboratory. Immediately upon reaching the laboratory, the sample should be stored in the freezer for future use. It is quite common in many developing countries to see samples not being preserved at a lower temperature to avoid bacterial activity by placing ice around sample or being placed immediately in a refrigerator. It is quite common that many laboratory and sampling vehicles will not have refrigeration or cold-storage arrangements.

It is common for official vehicles to be used for private purposes by higher officials and their immediate family members, thereby compelling field samplers to collect samples less frequently and record incorrect dates. Bore wells are often sampled at easy-to-sample points, with pumps and approach roads. Representative bore wells distributed around the jurisdiction and near pollution points, such as industry, may not be carried out due to corruption and convenience.

Most of the monitoring and sampling organizations in developing countries usually will not have boat to collect samples of water at different points and depths. Hence sample will be collected at shore which will not be representative due to the microbial activity of soil; degrading vegetarian; presence of weeds and so on.

While collecting samples from taps, water should be allowed to pour out for few minutes before sampling. This is usually not done.

Some of the analyses are carried out at the parts per billion or trillion level. Even a small amount of dust can therefore contaminate a sample but, in many instances, sampling containers are not cleaned prior to sampling thereby contaminating samples with residues of pollutants (of samples collected earlier).

It is common to submit samples the day after they were collected, or after few days, due to holidays or staff being on leave. In many instances the laboratory will be at a distant location, which needs extensive travelling. Due to poor transportation facilities and poor staffing, the samples may become damaged and then they may not be replaced by fresh samples.

Generally in the developing countries to pile up samples for a few weeks or more before taking them for analysis, without taking steps to preserve them, by which time the characteristics of the sample would have changed completely. The purchase of poor-quality equipment and chemicals often gives misleading results. Nonavailability of spare parts and technicians to repair sophisticated analysing equipment will often compel analysers to use less sophisticated methods. Delays in decision making with respect to repairing equipment and vehicles are also major problems in many countries.

Parameter	Container			Storage	Hold time	
	Material	Number	Size (ml)		Extraction	Analysis
Alkalinity, total	Plastic	1	500	4 °C		14 d
Asbestos	Plastic	1	1000	4 °C		48 h
Chloride	Plastic	1	500	4 °C		28 d
Sulphate	Plastic	1	500	4 °C		28 d
Bromide	Plastic	1	500	4 °C		28 d
Nitrate	Plastic	1	250	4 °C		48 h
Nitrite	Plastic	1	250	4 °C		48 h
Cyanide	Plastic	1	500	4 °C, ascorbic acid (if chlorinated), NaOH, pH > 12		14 d
Fluoride	Plastic	1	250	None		28 d
Lead	Plastic	1	1000	None, preserved at laboratory with HNO_3, pH < 2		6 months
Copper	Plastic	1	1000	None, preserved at laboratory with HNO_3, pH < 2		6 months
Corrosivity	Plastic	1	250	4 °C		14 d
Total coliform	Plastic (sterile)	1	100	4 °C, $Na_2S_2O_3$		30 h
THMs	VOA vials	2	40	4 °C, ascorbic acid, HCl infield		14 d
EDB/DBCP	VOA vials	2	40	4 °C, $Na_2S_2O_3$		14 d
Pesticides	Amber glass	1	1000	4 °C, $Na_2S_2O_3$	7 d	14 d
Herbicides	Amber glass	1	1000	4 °C, $Na_2S_2O_3$	14 d	28 d
Nitrate + nitrate	Plastic	1	500	4 °C, H_2SO_4, pH < 2		28 d
Volatiles	VOA vials	2	40	4 °C, ascorbic acid, HCl in field		14 d
Semi-volatiles	Amber glass	2	1000	4 °C, Sodium sulfite, HCl in field	14 d	30 d
Carbamates	VOA vial	1	40	4 °C, $Na_2S_2O_3$, MCA in field		28 d
Glyphosate	Amber glass	1	125	4 °C, $Na_2S_2O_3$		14 d
Endothall	Amber glass	2	250	4 °C, $Na_2S_2O_3$	7 d	14 d
Diquat/paraquat	Amber plastic	1	1000	4 °C, $Na_2S_2O_3$	7 d	21 d
PAHs(PNAs)	Amber glass	2	1000	4 °C, $Na_2S_2O_3$, HCl		7 d
D/DBP	Glass vials	2	40	4 °C, ammonium chloride, pH 4.5–5		14 d
Haloacetic acids	Amber glass	1	125	4 °C, ammonium chloride		28 d

(continued)

Table 3.1 (Continued)

Parameter	Container			Storage	Hold time	
	Material	Number	Size (ml)		Extraction	Analysis
Ammonia	Plastic	1	500	4 °C, H_2SO_4, pH < 2		28 d
TKN	Plastic	1	500	4 °C, H_2SO_4, pH < 2		28 d
Total phosphorus	Plastic	1	500	4 °C, H_2SO_4, pH < 2		28 d
BOD	Plastic	1	1000	4 °C		24 d
COD	Plastic	1	500	4 °C, H_2SO_4, pH<2		28 d
Coliform, faecal	Plastic – sterile	1	125	4 °C, H_2SO_4, pH < 2		6 h
Colour	Plastic	1	500 ml	4 °C		48 h
Conductivity	Plastic	1	500 ml	4 °C		28 d
Cyanide, amenable	Plastic	1	500 ml	4 °C, NaOH pH > 12,		14 d
Cyanide, total	Plastic	1	500 ml	4 °C, NaOH pH > 12		14 d
Flashpoint	Glass	1	1000	4 °C		7 d
MBAS (surfactants)	Plastic	1	1000	4 °C		48 h
Odour	Glass	1	500	4 °C		24 h
Oil and grease	Glass	1	1000	4 °C, H_2SO_4, pH < 2		28 d
pH	Plastic	1	250	None		Immediately
Phenol	Glass	1	500	4 °C, H_2SO_4, pH < 2		28 d
Phosphorus, ortho	Plastic	1	500	Filter onsite, 4 °C		48 h
Solids, dissolved	Plastic	1	1000	4 °C		7 d
Solids, settleable	Plastic	1	1000	4 °C		7 d
Solids, suspended	Plastic	1	1000	4°C		7 d
Solids, total	Plastic	1	1000	4°C		7 d
Solids, volatile	Plastic	1	1000	4 °C		7 d
Sulfide	Plastic	1	1000	4 °C NaOH, pH > 9, Zinc acetate		7 d
TOC	Amber glass	1	250	4 °C, H_2SO_4, pH < 2		28 d
Turbidity	Plastic	1	500	4 °C		48 h
All metals except Cr^{6+} and Hg	Plastic	1	200/metal analysed	HNO_3, pH < 2		6 months
Mercury	Plastic	1	200	HNO_3, pH < 2		28 d
Chromium hexavalent	Plastic	1	1000	4 °C		24 h

Corruption in education, recruitment, transfer, promotion and deputation often leads to poor-quality technicians who may be further discouraged by slow growth opportunities and work burdens. This may lead to the preparation of analysis reports without analysis, which leads to incorrect decisions during policy formulation, poor treatment plant design, seeking external aid and correlating health with water quality incorrectly and so on.

3.1.3 Village-Level Monitoring

Villages are sparsely populated and most of the time are dominated by illiterate and less literate inhabitants who are also less influential. The location of villages could be adjacent to rivers, lakes or oceans, or villages may not have surface water bodies at all. Some of the villages in hilly or mountainous areas may not have proper access or communications systems. Many villages in drought-prone areas depend on water sources in neighbouring villages. Often the villages are victims of power shortages with only a few hours of power at night time. Bore wells, to monitor groundwater depth and quality, will not be drilled in all villages. Often the groundwater recedes beyond the depth of the bore well. In such cases the depth of water table will be reported as the depth of the bore well as '–' or 'dry'. Such data will not be useful.

3.2 Selection Options for Water Supply Source

There are many options for water supplies. Either one can capture water from the sky or siphon it from the earth but, in either case, it is essential that sustainability is maintained.

Water is a limited resource, is not elastic in nature and cannot be taken for granted considering the predictions that many countries will have more than 50% of their population in urban areas. Countries are gearing up to provide infrastructure for the projected urban population without making any effort to retain people in rural areas. PLANYAC 2007 (Box 3.1) is one key initiative that can guide cities around the world.

Dams are structures built against flowing water to store or divert water. Usual the purposes of dams are: (i) flood control, (ii) to provide drinking water, (iii) irrigation, (iv) recreation, (v) electricity generation, (vi) fishing (Chandrappa and Ravi, 2009).

Even though dams are useful for some, dams have proven to be a threat to the environment and a cause for conflict in many places. Their impact occurs prior to construction, during construction, after construction and after/during dismantling. Dams have been shown to be unsustainable over the years as they impact upstream, downstream and in the location where they are situated, the command area, the rehabilitation site and along and adjacent to transmission lines. Major impacts due to dams are given in Table 3.2. Dams are intended to change the natural distribution as well as timing of stream flows. They alter natural ecosystems by changing the pattern of the downstream flow. They change sediment/nutrient regimes and vary water temperature and chemistry. Reservoirs flood terrestrial ecosystems, displacing animals and killing terrestrial plants (Berkamp *et al.*, 2000). As many species favour valley bottoms, large-scale impoundment may eliminate unique wildlife habitats.

Table 3.2 *Impact of dams on the environment.*

Sl. no.	Location	Impact
1	Impact at dam/reservoir site	Air pollution, loss of flora/fauna, increase in vector population, decrease in water quality, reservoir induced seismicity, change in microclimate, loss of biodiversity, disturbance of habitat for flora/fauna, accumulation of sediment.
2	Impact on canal/command area	Water logging, salinity, loss of biodiversity, disturbance of habitat for flora/fauna.
3	Impact adjacent to and along power transmission line	Air pollution, radiation, loss of biodiversity, disturbance of habitat for flora/fauna, occasional fire hazard due to sparking
4	Impact at down stream	Impact on terrestrial and aquatic ecosystem due to change in water flow
5	Impact at rehabilitation location	Air pollution, loss of flora/fauna, change in microclimate, loss of biodiversity, disturbance of habitat for flora/fauna.

Box 3.1 PLANYAC 2007

PLANYC is a plan of 127 initiatives announced by mayor of New York city in 2007, addressing water, transportation, air, land, energy and climate change with the aim of helping the city become sustainable. New York city receives 1200 million gallons of water every day. The water to New York from 19 upstream river, streams, lakes and reservoirs serves the 8.2 million residents of the city. To guarantee quality and reliability of drinking water, watersheds surrounding the city are kept clean by: (i) watershed protection; (ii) construction of an ultraviolet disinfection plant, (iii) construction of a water-filtration plant, (iv) reducing citywide consumption by 60 mgd, (v) adding 245 mgd to the present water supply by increasing efficiency, (vi) construction and repair of tunnels, (vii) replacing the existing pipeline, (viii) completing the control plans of 14 New York city watersheds, (ix) reducing by more than 185 mgd the combined outflow of sewage during rainstorms, (x) increasing high-level storm sewers, (xi) increasing the extent of green, permeable surfaces to reduce stormwater runoff, (xii) avoiding localized flooding as well as septic-tank failure, (xiii) the creation of a task force to design and implement best management practices (BMP) for water quality, (xiv) providing incentives for the private development of BMPs and (xv) storing combined flows (sewer as well as stormwater) (Rodríguez, 2011).

Terrestrial ecosystems in the reservoir are replaced by lacustrine, littoral/sublittoral habitats. Mass water circulations replace riverine flows, providing new opportunities and habitats and creating a new ecological setup.

The major impact due to dams (Figure 3.2) is loss of forest. India lost around 12.6% of forest between 1980 and 2000 just because of dams; 98.54 Ha of forest was scarified for

Figure 3.2 *View of dam across river.*

transmission lines in the Uri project, India. Flash floods destroyed Coffer dam, Bakra dam and Rihad dam in India causing loss of human life and damage to property/ecology. Trees died around the Tawa reservoir, India, due to a stagnant water table. An increase in water hyacinths and other weeds reduced the storage capacity of the Loktak reservoir. Raichur district of Karnataka state in India became highly endemic for malaria after the construction of the Thugabhadra dam (Rangachari *et al.*, 2000).

3.2.1 Spring Capping

Spring capping (Figures 3.3–3.5) is a sustainable system where water is infiltrated and comes out of the ground. The flow within the soil provides natural filtration. To supply a settlement located along the banks of a river in a rolling terrain with a good groundwater table, the most important thing to look for is a spring above the settlement so that water can

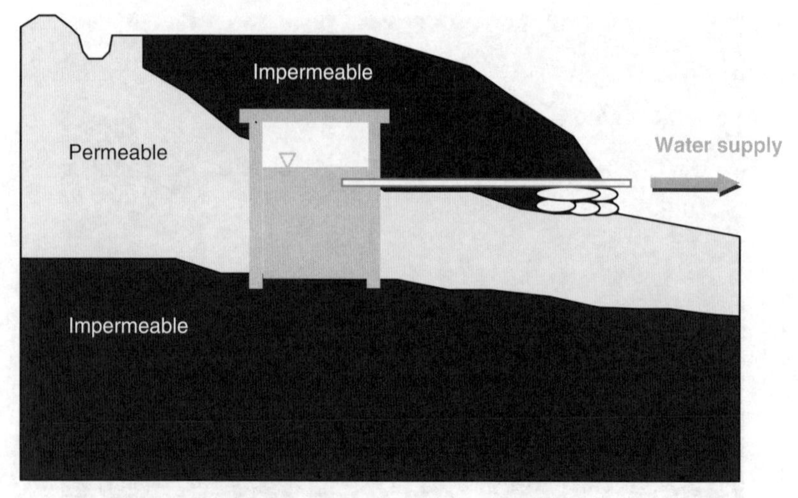

Figure 3.3 Spring capping – option 1.

be supplied by gravity. If no springs are available within a reasonable distance above the point of consumption, a well-point system or gallery at points would be best suited.

Where springs are not formed river bank filtration (RBF) has been adopted due to the advantages of the infiltration of river water through the riverbed into a well. After mixing and reducing, the infiltrated water is at its cleanest as most of contaminants present in river are removed.

For more than 100 years, RBF has been in use in Europe to supply drinking water to settlements along the Rhine, Elbe, Danube and Seine rivers. Riverbank filtration contributes around 70% of total drinking water requirements in Berlin. Water abstracted using RBF for

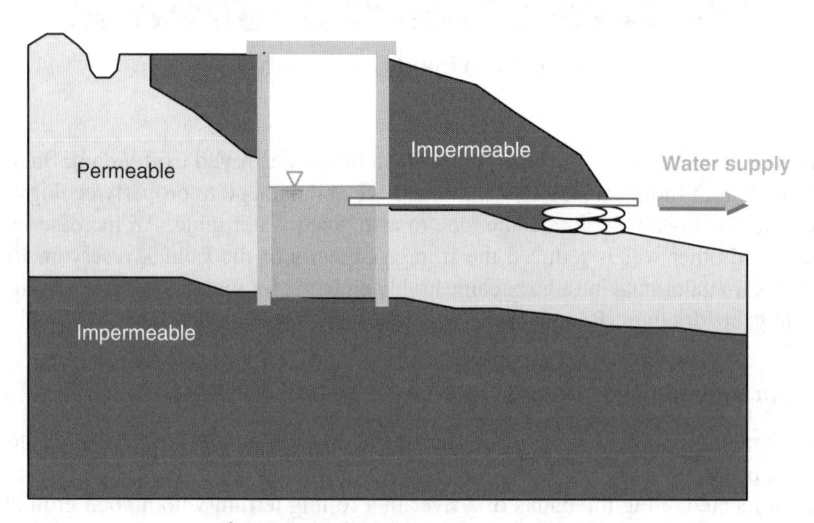

Figure 3.4 Spring capping – option 2.

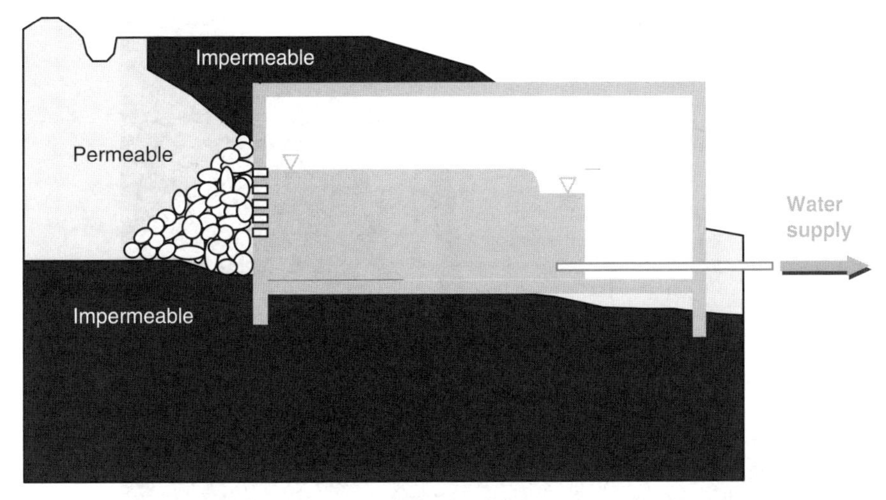

Figure 3.5 *Spring capping (spring protection chamber) – option 3.*

potable water is around 50% in the Slovak republic, 16% in Germany, 45% in Hungary and 5% in the Netherlands (Ray and Jain, 2011). In the United States, RBF systems have been supplying drinking water to many communities for nearly 50 years. Since World War II, when the rivers were tremendously polluted, RBF has been the most efficient method of generating high-quality drinking water. Today many countries like India, South Korea, China, Egypt and Jordan are using RBF.

3.2.2 Simple Tube Wells

Simple tube wells can be created by drilling (Figure 3.6) vertically to the ground. In some cases, concrete might be required to cover the external part of the pipe. These wells are easy and inexpensive to build. The length of the wells depends on the depth of the water table. In the wells, the top part is equipped with a pump to pull water up. This can be manually operated or power operated. In the downward direction, the pipe is filled with screening layers made of gravel and clay to prevent particles such as sand or silt from flowing through the pump or wearing the pipe. The lowest part is the porous bottom cap.

Karnataka, one of the southern states in India, contains much igneous and metamorphic rock, which is generally unsuitable for storage and transmission of groundwater. Availability of groundwater is estimated at 485 TMC. Groundwater resources have not been exploited uniformly throughout the state. Exploitation of groundwater in the dry region of the north and south interior Karnataka is higher compared to coastal and irrigation command areas. There is a deficiency of water for drinking, agricultural and industrial use in the dry regions of the north and south interior of Karnataka. Where adequate surface water is available, there is less use of groundwater resources. In about 43 taluks (subdistricts) there is overexploitation of groundwater resources. Further, groundwater exploitation has exceeded 50% of the available groundwater resources in 29 taluks of the state. These 72 taluks are critical taluks from the point of view of groundwater exploitation. In the 72 critical taluks, about 0.4 million wells irrigate an area of 0.75 million ha. Due to overexploitation of

Figure 3.6 *Bore well drilling under progress.*

groundwater resources, more than 0.3 million dug wells have dried. Shallow bore wells have failed and yields in deep bore wells are declining. The area irrigated by groundwater extraction structures is decreasing. Consequently, more than Rs. 200 million of investment made by individual farmers on the construction of wells, pumping equipment, pipelines, development and so on, has become nonfunctional (KSPCB, 2011). This is a small case study; similar situations can be observed in almost all the developing world. The bore well that did not exist four decades earlier became a passion for many farmers, assuming that what lies below the ground and above the ground belongs to the person who owns the ground. But when water is siphoned from the ground, water trapped in groundwater table below neighbouring land is also being pumped. This was not considered. Bangalore, which is the capital of Karnataka, had another problem as land owners drilled their own bore wells to avoid depending on the public water supply and, as a result, the water table went below 300 m in the city. People are still sinking bore wells, legally and illegally, making the situation worse. As a result many apartments and residents are depending on water supplied by tankers mounted on trucks while some of the region is receiving water from the public supply. In order to improve the water-table level, policy makers compelled property owners to collect rainwater.

3.2.3 Hand Pumps

There are many kinds of hand pumps (Figure 3.7), which depend on the range of lift or the depth of the borehole or wells as follows: suction pumps (0–7 m), low-lift and direct-action pumps (0–15 m), intermediate-lift pumps (0–25 m) and high-lift pumps (0–45 m or more). The basic principle of the pump is that when the pump handle is pulled up and pushed down, this causes the pump rods that are connected to the piston to move vertically. As the piston moves upwards, the valve is closed and thus a vacuum is created. The water is then forced into the delivery outlet.

Hand pumps are a boon to villages and locations where electricity has not reached or where the supply is not reliable. The bore well is also a reliable source of water in case of disasters when surface water is contaminated first and there is disruption of electricity; it makes it easy for the people to pump out the water. But again competition has made many bore wells with hand pumps dry, thus wasting the investment. Sustainability was also a problem in many villages where a hand pump was provided but not maintained by the public authorities and communities did not have the authority or knowledge to repair it.

3.2.4 Rainwater Harvesting

Collection of rainfall from rooftops dates back over 3000 years. The use of domestic rainwater tanks in Australia is relatively common practice, predominantly in rural and remote areas. Between 1994 and 2001, 16% of Australian houses used rainwater tanks and 13% of houses used tanks as their chief source of drinking water (Department of Health and Ageing, Government of Australia, 2004). In South Australia, 51% of households have a rainwater tank with 36% using them as the chief source of drinking water (Department of Health and Ageing, Government of Australia, 2004). In India, rainwater harvesting is not just restricted to rooftops; as much water as possible was collected in the forts (Figure 3.8) that were built above hills to safeguard against enemies as it is nearly impossible to dig a well on huge rocks where there is little soil to hold water. Boxes 3.2 and 3.3 describe sophisticated rainwater practices in Singapore and Kansas City, in the United States.

Rainwater harvesting is an anthropogenic rainwater system for collecting the rainwater for recharging aquifers or reusing the water. The main reason to harvest rainwater is to (i) avoid evaporation when it spreads over a large area, (ii) avoid contamination before it reaches surface water, (iii) provide energy and time to bring back water that has moved and stagnated in faraway places, which may be outside the political boundaries of the country or state.

Rainwater can be harvested by following means:

1. Roof catchments:
   - simple roofwater collection systems for households;
   - larger systems for educational institutions, stadiums, airports and other facilities;
   - roofwater collection systems for high-rise buildings in urbanized areas.
2. Ground catchments – ground is modified specially by creating treated earth/cemented catchment to capture rainwater.
3. Rock catchments – rainwater captured by tanks on natural, impervious outcrops.
4. Collection of stormwater in urbanized catchments.

Figure 3.7 *Different types of handpumps.*

Figure 3.8 Ancient rainwater harvesting at Chittorghar, India.

Rainwater is normally collected along the roof but collection of rainwater falling on pavements and infiltration is common. The use of feeder canals to capture water entering a neighbouring state jurisdiction is also common, which sometimes leads to conflicts.

Box 3.2 Rainwater harvesting in Singapore

More than a century ago, Singapore started collecting rainwater in catchment reservoirs, which now deliver about half of the city's needs. The majority of the rest comes from Malaysia. Political tensions affect the price and certainty of the supply. Singapore has limited land resources and about 86% of the country's population lives in tall buildings. Roofing that allows light through but not rain, dust, insects and so on, is positioned on the buildings to collect water. Separate containers are kept on the roofs for potable and nonpotable uses. As a result the city saves in terms of water, energy and capital. Larger scale rainwater harvesting and a utilization system exist at Changi Airport. Rainwater from the runways as well as the surrounding green areas is collected in two impounding reservoirs. The water is used mainly for nonpotable purposes like firefighting drills and toilet flushing.

Entry to rainwater tanks by small animals and birds can result in faecal contamination. Sometimes animals may get trapped in rainwater collection system and the tanks, leading to sewer contamination. Rainwater tanks can also provide habitats for mosquito breeding. Where water is collected from rooftops, optional filtration systems may be adopted. The filtration systems depend on whether potential pollutants exist in the area. The design of storage tank capacity depends on the rainfall volume and pattern as well as the duration of the rainy/dry season. Rainwater overflowing on contaminated areas like the waste collection centres of industries, chemical storage areas, pesticides over a sprayed area may have significant contaminants and may need further sophisticated treatment.

Wildfires generate smoke, ash and debris that settle on catchment areas and, hence, rainwater harvesting in such areas needs attention to avoid contamination from smoke/ash/debris. Some plants/trees may produce toxins and, hence, knowledge about such plants and trees is essential before rainwater harvesting. The first flush of water when rain falls washes the roof catchment and may contain accumulated dust, leaves, bird debris and animal droppings.

In rainwater harvesting practised at household/industries/commercial establishments where cleaning necessitates entering the tank, adequate ventilation is essential and all precautions required for working in confined spaces should be followed.

Box 3.3 Beacon Hill Redevelopment Project, United States

As part of the Beacon Hill Redevelopment Project, Kansas City, in the United States, a grant of US$1 474 500 was provided by US EPA to replace/relocate drinking-water mains, sanitary sewers and stormwater sewers. The project included green stormwater infrastructure like an underground detention basin, bio-retention cells and rain gardens. Rain gardens and bio-retention cells are small landscaped areas built with a special soil mix to absorb and filter runoff. Water-tolerant and low-maintenance plants are used in these rain gardens to aid the reduction of stormwater runoff, removing pollutants and refilling the aquifer.

Tokyo and Seoul have implemented large-scale rainwater harvesting systems that include public reservoirs and the latest technologies. In Tokyo, rainwater harvesting is promoted to mitigate and control floods and water shortages and secure water for emergencies. The Ryogoku Kokugikan Sumo-wrestling Arena and City Hall of Sumida use rainwater for toilet flushing and air conditioning. A unique rainwater utilization facility, 'Rojison', in the Mukojima district of Tokyo, uses rainwater collected from the rooftops of houses for gardening, firefighting and drinking in emergencies. Nearly 750 private and public buildings have introduced rainwater collection in Tokyo (UNEP, 2012).

Rainwater utilization systems in Berlin save water, control urban flooding and create a better microclimate. The Belss-Luedecke-Strasse building estate collects rainwater from all roof areas and discharges it into a public rainwater sewer where it is transferred into a cistern together with the runoff from parking spaces, streets and pathways. It is treated and used for toilet flushing and gardening. Rainwater falling on rooftops is collected in jars of capacities from 100 to 3000 l in Thailand. Rainwater collection is becoming widespread across Africa with projects in Botswana, Togo, Kenya, Namibia, Zimbabwe, Mozambique, Mali, Malawi, South Africa, Sierra Leone and Tanzania (UNEP, 2012).

3.2.5 Fog and Dew Harvesting

Fogs are condensed water droplets of water vapour at or above the Earth's surface. Dew is water droplets on exposed objects formed due to condensation of mist or fog.

Water can be harvested from fogs in favourable climatic conditions. The water droplets in the fog precipitate when they contact other objects. The common fogs that occur in the arid coasts of Peru and Chile known as *camanchacas* have the potential to provide freshwater as fog collectors. Technology can also potentially supply water in mountainous areas where water is present in stratocumulus clouds.

Fog collectors are usually of three types: (i) flat, rectangular nets of nylon arranged perpendicular to the prevailing wind direction; (ii) collection panels joined together; (iii) condensers on roofs.

The small droplets of fog join to form larger drops on fog collectors and fall into a trough or gutter at the bottom of the panel; the water is conveyed to a storage tank from where it is carried through a network of pipes. Fog harvesting has been implemented effectively in the mountainous coastal regions of Chile, Ecuador, Mexico and Peru.

Plastic sheets on slanted rooftops provide a greater yield compared to metal sheets whereas cement and tiles absorb the dew droplets thereby making it difficult to collect dues.

3.2.6 Snow Harvesting

Snow can also be harvested for an alternative supply of freshwater. Snow harvesting needs the construction of a pit of about 6 to 8 m diameter and around 10 m in depth. The collected snow of 2 to 3 m is compacted and covered with earth. A tube is placed about 50 cm above the bottom of the pit. As the snow melts, water trickles along the tube into a pot kept beneath the outlet (UNEP, 1982).

3.3 On-Site Sanitation

Sustainable sanitation techniques can be divided into: (i) separate collection of urine, (ii) separate collection of black water (urine as well as faeces), possibly together with degradable kitchen waste, (iii) local treatment as well as reuse of grey water, (iv) a combination of the above options, (v) sustainable centralized treatment (e.g. at an energy factory), (v) removal of hormones or medicine residues.

Box 3.4 Sustainable sanitation in eThekwini Municipality, in South Africa

Double-vault urine diversion dehydrating toilets (UDDTs) were considered a viable option due to their lower operating costs. Over 90 000 UDDTs were installed in eThekwini's periurban areas. Initially the dried faeces and the urine were buried and infiltrated into a soak away on site respectively. Now the eThekwini municipality is considering the potential of turning urine into a commodity (Luthi and Ingle, 2012).

3.3.1 Latrines

The word latrine refers to a toilet or a facility used as a toilet. It can be a trench in the earth, a pit, or place with a sophisticated design. The term is evolved from the Latin word *lavatrina* meaning bath. But unsophisticated toilets, like a dedicated place in a village of camps, would not be a sustainable solution as pigs and stray dogs loiter to eat human excreta. Such practices may also lead to the transfer of tapeworms from humans to pigs; they then come back to humans who eat pork. Other practices observed in the Kibera slums (Box 3.4) also raise concerns about spreading disease. Table 3.3 details different types of

Table 3.3 *Sanitation options.*

Onsite sanitation	Description	Advantages	Disadvantages
Open defecation	Defecation in open area. Sometimes a place is earmarked for this purpose.	Free	Encourages flies, hookworm larvae, which may lead to diseases. Encourages stray dogs and pigs.
Shallow pit	A small pit is made each time they defecate, which is covered with soil.	Free	Spread of hookworm larvae
Simple pit latrine	Consists of a slab over a pit.	Low cost, needs no water for operation.	Leads to fly and smell nuisance
Borehole latrine	Borehole and used to defecate.	Suitable for short-term use.	Encourages flies, may lead to groundwater pollution.
Ventilated pit latrine	Pit is ventilated by a pipe extending above the latrine roof.	Low cost, needs no water for operation, absence of smell in latrines.	Leads to groundwater contamination.
Pour-flush latrine	Latrine are fitted with a trap providing for water seal.	Control of flies, absence of smell.	A reliable water supply must be available.
Double pit	Second pit is provided when the one in use is full.	Will last for long time.	Pit contents can be used as a soil conditioner after a year or two without treatment.
Composting latrine	Excreta fall into a tank to which ash/sawdust/soil is added to avoid fly nuisance.	Can be used as manure after few months.	Arrangement should be made to divert urine.
Septic tank	Raw sewage is delivered to tank, which is kept in anaerobic condition.	May generate foul smell if not airtight and, hence, ventilation is provided in some designs.	Suitable for low-density housing.

Aqua-privy	Watertight tank is placed immediately under the latrine floor where excreta drop directly into the tank through a pipe.	May generate foul smell if not airtight and, hence, ventilation is provided in some designs.	Suitable for low-density housing.
Overhung latrine	Latrine is built over water body into which excreta drop directly.	Cheap.	Serious health risks.
Bucket latrine	Faeces are collected in a bucket/container, which are periodically removed for disposal.	Low initial cost.	Creates fly/odour nuisance.
Latrine with vaults and cesspits	Watertight tanks (called vaults) are constructed under/near latrines to store excreta prior to cleaning manually or by vacuum tanker.	Demands reliable and safe collection service.	Irregular collection leads to overflow of tank.
Latrine connected to sewerage	Discharge from latrine is directed to sewerage.	Faecal matter is transported far way continuously.	Needs huge investment. Needs treatment plant. Leakages will lead to groundwater pollution. Overflow from manholes will leads to nuisance. Settling of solids will lead to reduction in carriage capacity.

Figure 3.9 *Pay and use latrine in Sweden.*

latrine used across the globe with their advantages and disadvantages. The toilets can be as sophisticated as in the photos shown in Figure 3.9. Coin-operated self-locking public conveniences were built in Stockholm. Figure 3.10 shows a toilet where reading materials are kept.

Box 3.5 Flying Toilet

There are about 600 000 residents in the Kibera slums of Nairobi where the term 'flying toilet' originated. People use polythene bags to dispose of faeces. These are thrown onto roofs and into drains, posing a serious health hazard (Corcoran *et al.*, 2010).

Figure 3.10 *Toilet with reading material.*

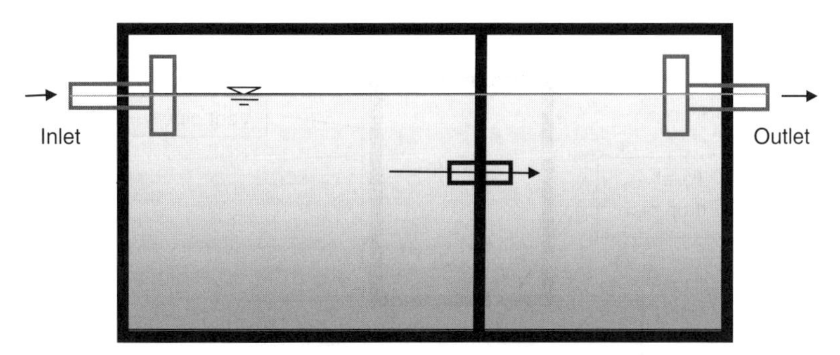

Figure 3.11 Septic tank.

3.3.2 Septic Tanks

Septic tanks (Figure 3.11) generally comprise one or two watertight tanks, which are usually rectangular cross sections. Some designers may prefer cylindrical tanks due to easily available construction material. The tanks are usually preceded by a soak pit or soak trench that distributes treated wastewater. When wastewater passes through the septic tank, heavier solids sink to the bottom and degrade anaerobically. Materials that are less dense than wastewater, such as grease and oil, float to the surface of septic tank. The effluent flows from the tanks and is let into a soak pit or trench where it may undergo natural treatment processes. Septic tanks and soak pits/trenches must be installed to ensure clear distance from: (i) the highest groundwater level, (ii) a source of water supplies, (iii) subsoil and open drainage channels, (iv) locations where vehicles will likely to drive over them as the weight of a vehicle will damage the septic tank and the vehicle may sink into the septic tank.

In poor draining soils, like clay, bigger drainage receptacles are required to increase the area of soil into which treated effluent can be absorbed. The septic tank should be cleaned once in a year (EHD, 2011).

3.3.3 Aqua Privies

Aqua privy (Figure 3.12) is a small septic tank located directly below a squatting plate. To prevent odour and insect nuisance a water seal is maintained by submerging a pipe inside water. Excreta are deposited in the tank where they are decomposed anaerobically. Housing or a shed and a vent pipe with a fly screen are installed to maintain hygiene.

Aqua privy is more useful where a water carriage system is not present. The submerged outlet pipe serves as a water seal. The settled and digested sludge is usually removed after 5–6 years. Aqua privies are filled with water before use.

3.3.4 Oxidation Pond Treatment Systems

An oxidation pond (or lagoon) treatment system is a secondary wastewater treatment process for municipal wastewater with a high BOD loading. There are many types of pond, namely (i) aerobic ponds, (ii) facultative ponds, (iii) anaerobic ponds, (iv) tertiary ponds and (v) aerated lagoons.

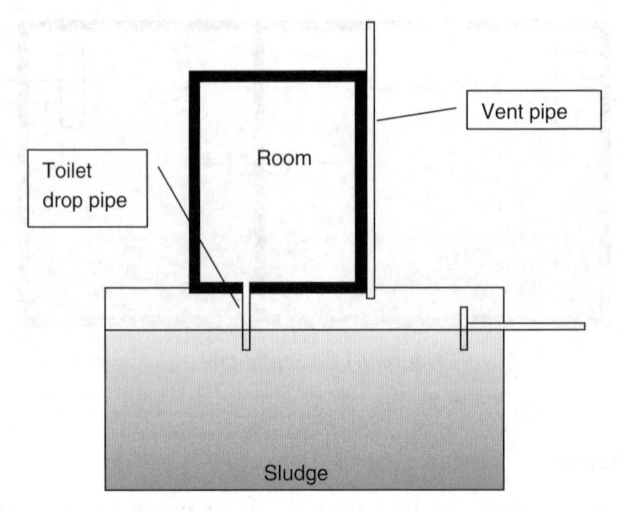

Figure 3.12 *Aqua privy.*

The aerobic pond is shallow (less than 1 m) so light can penetrate through to facilitate the growth of aerobic bacteria. The facultative pond is about 1–2.5 m in depth, composed of three different zones: the upper aerobic, middle facultative and lower anaerobic zone. The anaerobic pond is very deep, to maintain anaerobic condition for treating high organic-loading waste. The maturation or tertiary pond is used for the final treatment of biological process effluents by using dissolved oxygen through photosynthesis and surface aeration processes. The aerated lagoon is an open lagoon with surface aeration.

Oxidation ponds need less supervision but require a large amount of space. The space required depends on the amount of solar energy received by the pond. The treatment occurs due to a symbiotic relation between algae and bacteria in the oxidation pond. The oxygen released by algae is used by bacteria and the carbon dioxide released by bacteria is used by algae.

With the addition of aeration, the need for sunlight and algae can be reduced, thereby reducing the area required for the oxidation pond. The aerator can be replaced by releasing ducks (Figure 3.13) into oxidation pond (as practiced in some parts of India). The constant swimming of the ducks would transfer oxygen from the air to the water.

Box 3.6 Swimming Birds in STP at Mysore, India

Commissioned in 2002, the sewage was used for sewage farming after secondary treatment. Following the success story about treating wastewater by aerating ducks, about 750 swimming birds were released into 30 MLD STP located at Old Kesare, Mysore, in 2010 to cut down electricity bills in the range of Rs 0.175 million to Rs 2.30 million.

Figure 3.13 Swimming birds can replace surface aerators when used in large numbers.

3.3.5 Storm Drainage

A storm drainage system is built to drain the excess rainwater or groundwater from community areas. It is designed for areas that have experienced floods, where there is a risk of flood (Figure 3.14), where there is a high amount of rainfall, or for areas that are close to the coast. Normally, the drained water is discharged to rivers or streams directly without any pretreatment.

The city of Chicago, in the United States, believes that the built infrastructure alone cannot meet all of the needs for managing wastewater and stormwater. Protecting the quality of

Figure 3.14 A flooded street.

Figure 3.15 *Storm water inlet (located in the front of a car).*

water resources and managing stormwater will require upgrading the built infrastructure as well as creating a green infrastructure. By adopting innovative stormwater management techniques, the city plans to reduce the incidences of sewer overflow to the Chicago River. Rooftop gardens have helped the city to manage stormwater sustainably. The City Hall rooftop garden, with 20 000 plants, is capable of reducing stormwater drainage by 50%. The city is also encouraging large new developments to include green infrastructure into their design. The city is utilizing unique open spaces to retain water that would normally drain into the sewer system. A new kind of alley in North Side is constructed of a rigid grid system as well as gravel, allowing rainwater to percolate into the ground and reducing backyard flooding and flow in the sewers. The rain gardens in the parkways move water into the ground. Thousands of acres of undeveloped land and wet lands within Chicago help control stormwater as well as prevent flooding (Rodríguez, 2011).

The water cycle can be divided into a biological cycle (in which water as well as its useful constituents like nitrogen, carbon, and phosphorus are transported) and a technical cycle (in which harmful substances like heavy metals as well as persistent organic substances are transported) (Malmqvist, 2000). Most of the urban stormwater is unavailable for immediate reuse, particularly when it is transported to the wastewater treatment plant. Sweden has a sophisticated system for urban stormwater drainage (Figures 3.15 and 3.16) in Stockholm. Sustainable stormwater management in Sweden includes stormwater ponds, infiltration and percolation basins, wetlands, open ditches and treatment devices (Malmqvist, 2000).

Ecological stormwater management is characterized by (i) contact with sunlight, soil, air, plants and micro-organisms, (ii) utilization of natural water processes, (iii) slow runoff, (iv) source control of quantity and quality, (v) local infiltration and detention, (vi) central detention in open water ponds/courses (Malmqvist, 2000).

One of the major risks with combined sewer overflows in urban areas is bacterial contamination but stormwater in separate systems can be contaminated from droppings from birds/cat/dogs and other animals in the urban area. Many animals carry pathogenic microbes that affect human beings. In addition to animal droppings and open defecation by humans (due to absence of sanitation), toilets in trains that discharge urine and faeces on tracks (Figure 3.17) are also a source of contamination in developing countries. Crackers

Figure 3.16 *Storm water drains in narrow passage way of Stockholm.*

(Figure 3.18), household hazardous waste and biomedical waste also contaminate stormwater to a great extent.

Sustainable practice has not been followed in many countries. Urbanization and associated pollution caused a series of fish kills in Nerohalli Lake, Ulsoor Lake, Sanky Lake, Puttenahalli Lake, Lalbag Lake and Sarakki-Jaraganahalli Lake between 2005 and 2007

Figure 3.17 *Dirty railway station with faeces discharged on the railway track from trains.*

(Chandrappa and Raju, 2008). As a result of the creation of residential developments and illegal encroachments by real estate organizations, the number of lakes has reduced in Bangalore. Shoolay Lake was converted into a football stadium, Akkithimmanhalli Lake was converted into a hockey stadium, Sampangi Lake was converted into a sports stadium, Dharmambudhi Lake was converted into bus stands, Challaghatta Lake was converted into a golf course, Koramangala Lake was converted into a residential development, Nagashetthalli Lake was occupied by the Department of Space, Kadugodanahalli Lake was occupied by Ambedkar Medical College, Domlur Lake was occupied by Bangalore Development Authority, Millers Lake became a multi-facility auditorium, Kurubarahalli Lake, Kodihalli

Figure 3.18 *Burnt crackers on streets.*

Figure 3.19 *Green roofs. (For a colour version of this figure, see the colour plate section.)*

Lake, Sinivagilu Lake and Marenahalli Lake were converted into residential developments and Shivanahalli Lake was converted into a playground and bus stand (D'Souza, 2006; Chandrappa and Raju, 2008).

The sustainable urban drainage system (SUDS) approach to drainage can be divided into following categories: (i) control of rainwater at source, (ii) infiltration trenches along with filter drains, (iii) swales and basins, and (iv) ponds and wetlands.

Source control includes: (i) green roofs (Figure 3.19), which include vegetation cover on rooftops to intercept and retain precipitation, (ii) rainwater harvesting for storing rainwater for further use, (iii) permeable pavements (Figure 3.20), which provide a durable surface that allows infiltration of surface water into the soil beneath, (iv) infiltration and filtering through infiltration trenches (Figure 3.21), filtration trenches or drains.

Figure 3.20 *Permeable pavements.*

Figure 3.21 *Infiltration trenches.*

A grassed area of depression is called swale. This guides surface runoff from the source area to the storage/discharge system. Basins are designed to hold water for a few hours to ensure a settlement of solids. Basins are only temporarily in use and outside of storm periods they remain dry. They provide short-term water storage and the settlement of solid ensures water filtration, reducing contamination. Ponds and wetlands in their original form are valuable for drainage systems and will contribute to visual amenity and biodiversity.

Stormwater ponds are in use in Sweden for treating stormwater and are growing increasingly popular, especially along highways. Flemingsbergsviken's wetland was constructed in 1995 in the south of Stockholm. In addition to treatment of stormwater proper, street sweeping is essential to maintain the quality of stormwater. Gully pots trap solids and associated pollutants in Stockholm (Malmqvist, 2000).

3.4 Water Quality Characteristics of Potable Drinking Water and Wastewater Effluents

The wastewater from urban settlement, rural agriculture and livestock production constitutes a considerable challenge and is characterized by numerous contaminants originating from fertilizers, chemical runoff, human waste, industries, livestock manure and nutrients.

3.4.1 Physical Parameters

Physical parameters are those parameters that are useful in enabling us to know the characteristics of water without further sophisticated analysis. The physical parameters are easy to determine.

3.4.1.1 Suspended Solids

Any substance suspended in water qualifies as a suspended solid. It is determined by filtration followed by weighing of filter paper after drying and subtracting the weight of the filter paper. It is represented as mg/l. The suspended solids are characteristics of surface water bodies and wastewater from domestic as well as industrial/trade effluents.

3.4.1.2 Turbidity

Turbidity is the measure of the clarity of a liquid. It is contributed by both suspended solids and colloidal material in water. Most of the turbidity in surface water is due to the erosion of colloidal substances like clay, silt, rock fragments, microbes and so on. The units of measurements are the Jackson Turbidity Unit (JTU), the Formazine Turbidity Unit (FTU) and the Nephelometry Turbidity Unit (NTU). The procedures/instruments for measuring each of these units vary considerably.

3.4.1.3 Colour

Colour is commonly present in surface water due to the absorption of colour released by humus but absent in groundwater. Apart from humus, the presence of algae, soil particles and reflection of the environmental setting contributed to water bodies. With a rapidly changing environment, colour can be observed in groundwater due to entry of effluents from industries. Colour is a direct indicator of imbalance in the environment and an indication of contamination.

Colour in natural settings (Figure 3.22a–d) is not considered for recording and comparison of water quality. Colour contributed by dissolved solids after filtering out suspended matter is called *true colour*. True colour is measured by comparison with standardized coloured material. Colour comparison with a series of standards is used for direct comparison of filtered water to remove *apparent colour*. Results are expressed as *true colour units* (TCUs), where one unit is equivalent to the colour produced by 1 mg/l of platinum in the form of chlorplatinate ions. In fieldwork, instruments employing glass disks that are calibrated with colour standards are used. True colour units are used for only yellowish brown hues, which is the colour formed during degradation of humus in nature. For colours other than yellowish brown hues arising from industry, special photometric techniques are used. The analysis report is expressed at pH 7.6 as well as at the original pH in terms of *dominant wavelength*, *hue* (e.g. blue, blue-green, etc.), *luminance* (as a percentage) and *purity* (as a percentage).

3.4.1.4 Taste and Odour

Taste and odour, even if they are unacceptable, may not directly affect health by affecting biological systems – but an impact on health due to psychological factors cannot be ruled out as both body and mind are interrelated. All odour-generating substances generate taste but the reverse is not true. Taste and odour-generating substances include minerals, metals and salts and constituents of wastewater. Odour is associated with organic substances that may be carcinogenic in nature.

Odours are caused by gases produced by the decomposition of organic matter or by substances added by anthropogenic activity. The most characteristic odour of septic wastewater is that of hydrogen sulphide, produced by anaerobic micro-organisms. Industrial wastewater may contain odorous compounds depending on the activity.

Figure 3.22 (a) Colour of the water due to visibility of objects below the water; (b) Colour of the water due to reflection of sky; (c) Colour of the water due to colloidal soil particles; (d) Colour of the water in large forest is yellowish brown hues which is the colour formed during degradation of humus substance in nature. The water bodies in would look dark especially in forest with large pine trees due to colour released from pine needles on the ground. (For a colour version of this figure, see the colour plate section.)

Offensive odours can cause poor appetite for food, lowered water consumption, impaired respiration, nausea and vomiting and mental perturbation. In extreme situations, offensive odours can lead to the deterioration of personal and community pride, interfere with human relations, discourage capital investment and lower socio-economic status.

3.4.1.5 Temperature

The temperature of both ground and surface water is less than ambient temperature in natural settings, except in the case of hot water springs. Warmness in water sources indicates thermal pollution due to discharge of industrial effluents.

Temperature has an effect on reactions and on the solubility of chemicals, including gas in water. Cooler water will have wider biological diversity whereas an increase in temperature leads to growth and reproduction of only few species. Temperature also affects physical parameters like viscosity and density. Water temperature varies biological functions like spawning, migration, and metabolic rates of aquatic organisms. Changing water temperature cycles harms reproduction as well as growth patterns, resulting in a decline of the population of aquatic organisms.

3.4.1.6 Density

The density of water is the mass per unit volume of water expressed as kg/m^3. Density is an important physical characteristic of wastewater because of the potential for the formation of density currents in sedimentation tanks and in other treatment units. The density is temperature dependent and will vary with the concentration of total solids in the wastewater.

3.4.2 Chemical Parameters

Regulated chemical pollutants represent a small fraction of the chemicals that occur in nature. These targeted chemicals might be minute compared to known and yet-to-be identified chemicals; an implied assumption is that these regulated chemicals are responsible for most of the risk to the environment and to human health. As of June 2012, more than 67 million organic and inorganic substances and 63 million sequences have been indexed by the Chemical Abstracts Service (CAS) Registry of the American Chemical Society's Chemical Abstracts Service (CAS, 2012). Since the 1970s, chemical pollution has focused on lists of conventional 'priority pollutants', referred to as 'persistent, bioaccumulative toxicants' (PBTs) or 'persistent organic pollutants' (POPs) (Daughton, 2004).

Different chemicals will have different effects on different species. The effect depends on the sensitivity of the species, age, sex, body weight and health status. Similarly the impact on humans also depends on sex, age, health of individuals, sensitivity and so on. The impact of any chemical on health is studied in a laboratory, based on chronic and acute dosage via different methods of exposure (ingestion, dermal, etc.). Studies are also conducted on long-term and short-term effects. Furthermore, as humans are not exposed to toxic substances, due to ethical and legal issues, inferences regarding the toxicity of chemicals are first drawn based on experiments on animals. Even though some studies with minute dosages are done on human volunteers, this may not be sufficient to draw complete inferences. Other epidemiological studies are based on accidental exposure of humans to toxins and are not sufficient to draw conclusions about chemicals. In view of these lacunas in research, no information is available on the health effects of many chemicals on humans as well as other organisms.

The old-fashioned way of addressing emerging pollutants that 'resurface' with new concerns is at best a reactive approach. As per Daughton (2004), more resources should be devoted to studying as well as refining the message of risk to the common public. Agricultural chemicals like pesticides, herbicides, fungicides and so on are toxic and therefore are considered as significant contaminants of surface waters. These chemicals enter water from surface runoffs from the manufacture, storage, transportation and usage of these chemicals. These chemicals can result in fish kills and an accumulation of toxicity in food.

Subsequent paragraphs discuss both emerging and regulated chemicals.

3.4.2.1 Acrylamide

Acrylamide is used as a flocculant for the clarification of drinking water and wastewater and as a grouting agent (IPCS, 1985). Studies have shown convincingly that acrylamide is a cumulative neurotoxin (WHO, 2011a). Acrylamide concentrations in the water supply are controlled by limiting acrylamide content or treatment with potassium permanganate (Ma *et al.*, 1994) or ozonation (Mallevialle *et al.*, 1984). Acrylamide is readily absorbed by ingestion, by inhalation and through the skin (IPCS, 1985) and is then widely distributed in body fluids.

3.4.2.2 Alachlor

Alachlor is used to control weeds in maize, oilseed rape, peanuts, radish, cotton, brassicas, soy beans and sugar cane (Worthing 1991). Alachlor is reported to be mutagenic and carcinogenic (WHO, 2003a).

3.4.2.3 Aldicarb

Aldicarb is an insecticide used for controlling nematodes in soil as well as insects and mites on crops (WHO 2003b). It is very persistent in groundwater and remains in the environment from a few weeks to several years.

3.4.2.4 Aldrin and dieldrin

Aldrin and dieldrin are insecticides for soil-dwelling pests as well as termites and wood borers. Both aldrin and dieldrin are highly toxic to humans, the target organs being the central nervous system and the liver.

3.4.2.5 Ammonia

In the context of water pollution, the term 'ammonia' includes nonionized form (NH_3) as well as the ammonium cation (NH_4^+) unless otherwise stated. Its threshold odour concentration in water is about 1.5 mg/l. In water, ammonia forms the ammonium cation and hydroxyl ions. The extent of ionization depends on the pH, the temperature and the concentration of dissolved salts in the water. Ammonia is toxic if the intake is more than the capacity to detoxify (WHO, 2003c).

3.4.2.6 Arsenic

Arsenic is widely distributed throughout Earth's crust. It exists in oxidation states of −3, 0, 3. It occurs usually in metal arsenates or as arsenic sulfide and arsenides. In oxygenated surface waters the predominant form is arsenic (V) (Irgolic, 1982; Cui and Liu, 1988); under reducing conditions, as in deep lake sediments or groundwater, arsenic (III) is the most common form. Arsenic is introduced into water from rocks, minerals, ores and industrial effluents and by atmospheric deposition (IPCS, 1981).

The main adverse effects due to long-term ingestion of inorganic arsenic are cancer, skin lesions, cardiovascular disease, neurotoxicity and diabetes (FAO/WHO, 2011a, 2011b). Increased risks of lung and bladder cancer and of arsenic-associated skin lesions can occur with ingestion of water with concentrations $\leq$50 μg/l (IPCS, 2001).

3.4.2.7 Asbestos

Asbestos is a fibrous silicate mineral comprising magnesium, calcium or sodium. It is classified into two main groups – serpentine and amphibole. Amphibole is more resistant to acids. Asbestos is used in asbestos-cement sheets and pipes; thermal and electrical insulation; brake linings and clutch pads. Although asbestos is a carcinogen when inhaled, there is little evidence of cancer risk associated with the ingestion (WHO, 2003d).

3.4.2.8 Astrazine

Atrazine is a herbicide used to control annual weeds (Worthing, 1991). It can be degraded by photolysis and micro-organisms in surface water. Exposure to atrazine can lead to modulation of the immune system and can affect neuroendocrine function and disrupt the oestrous cycle or have developmental effects (WHO, 2011b).

3.4.2.9 Barium

Igneous and sedimentary rocks contain barium as a trace element. Barium compounds are used in the rubber, electronics, plastics, textile, ceramic, glass, brick, paper, pharmaceuticals, cosmetics, steel, oil and gas industries. Barium in water normally comes from natural sources. The solubility of barium compounds will increase as the pH level diminishes (USEPA, 1985). Barium is not vital for human nutrition (Schroeder *et al.*, 1972). It causes vasoconstriction at higher concentrations (Stokinger, 1981a). Death may occur in a few hours/days depending on the dose/solubility of the barium salt (WHO, 2004a).

3.4.2.10 Bentazone

Bentazone is a contact herbicide and can be detected in ground and surface water where it is used and is toxic for humans and animals. However, bentazone usually occurs at concentrations below those that result in toxic effects (WHO, 2004b).

3.4.2.11 Benzene

Benzene is used for the production of cumene/phenol, styrene/ethylbenzene and cyclohexane. It is added to petrol to enhance the octane number. It is carcinogenic. Acute exposure to high concentrations of benzenes by humans affects the central nervous system and exposure to 65 g/m^3 may cause death (WHO, 2003e).

3.4.2.12 Beryllium

Beryllium is a brittle metal found in silicate minerals. It is poorly soluble in water and the solubility of beryllium will increase to a small extent at higher and lower pH values. Beryllium is used in the aerospace, electronics, weapons and ceramic industries. It is carcinogenic to humans. Beryllium compounds are released into water mainly from coal burning, industries using beryllium and weathering of rocks/minerals/soils containing beryllium (WHO, 2009b). Most beryllium is adsorbed to suspended/settled solids rather than dissolved.

3.4.2.13 Boron

Boron is never present in an elemental form in nature. It enters groundwater mainly due to leaching from rocks and soils with borates and borosilicates. Boron compounds are used in glass, soaps, detergents and the flame-retardant industry. Boron is a dynamic trace element required for metabolism. The lowest reported lethal doses of boric acid are 640 mg/kg body weight (oral), 29 mg/kg body weight (intravenous injection) and 8600 mg/kg body weight (dermal). Lethal doses vary between 5 and 20 g of boric acid for adults and <5 g for infants (Stokinger, 1981b).

3.4.2.14 Bromate Ion

The bromate ion (BrO^{3-}) exists in a number of salts. It may be formed by the ozonation of water with bromide ions. Sodium and potassium bromate are used for dyeing textiles. Toxic effects of bromate salts would result in nausea, diarrhoea, anuria, central nervous system depression, vomiting, abdominal pain, haemolytic anaemia and pulmonary oedema (WHO, 2005a).

3.4.2.15 Bromide

Bromide (Br–) is the anion of bromine. It commonly exists as salts of sodium, potassium and other cations. Disinfection of water with bromide ions in water with organic matter will form brominated acetic acids. Large doses of bromide cause vomiting, nausea, abdominal pain, coma and paralysis (WHO, 2009d).

3.4.2.16 Carbaryl

Carbaryl is an insecticide used to control insect pests in plants. Its primary mode of toxicity is through inhibition of brain cholinesterase (WHO, 2008a). It is degraded by photodecomposition, hydrolysis and microbial activity with the rate of hydrolysis varying with temperature and pH. It does not appreciably bioconcentrate in fish and adsorbs to soils (WHO, 2008a).

3.4.2.17 Carbofuran

Carbofuran is used as insecticide, acaricide and nematicide. It is quickly taken up by plants through the roots and water and is translocated into the leaves. It is degraded in soil by hydrolysis, photodecomposition and microbial action. It is highly toxic through oral administration (WHO, 2004c).

3.4.2.18 Carbon Tetrachloride

Carbon tetrachloride is mainly used in the manufacture of chlorofluorocarbons, plastics, semiconductors, paints, ink and petrol additives. It is used as grain fumigant, a pesticide, in metal-cleaning solvents, in fire extinguishers and in flame retardants. It is released into the atmosphere and industrial wastewater. It is lost by volatilization from surface water within days or weeks but in anaerobic groundwater it will remain for months or years. It will be adsorbed onto organic matter in soils. It is well absorbed from the dermal, respiratory and gastrointestinal tracts in animals and humans. It is carcinogenic and distributes throughout the body, with highest concentrations in the brain, kidney, muscle liver, fat and blood (WHO, 2004d).

3.4.2.19 Chloral Hydrate

Chloral hydrate is used as a sedative and in the manufacture of DDT as well as dichloroacetic acid (DCA). The lethal blood level for chloral hydrate was estimated at 25 mg/100 ml and the toxic blood level was estimated at 10 mg/100 ml (Ellenhorn *et al.*, 1997).

3.4.2.20 Chlordane

Chlordane is a mixture of *cis*- and *trans*-chlordane. It is used as insecticide. It is very resistant to chemical and biological degradation and migrates very poorly. It is carcinogenic and human exposure can result in neurological symptoms, including dizziness, vision problems, headache, incoordination, excitability, weakness, irritability, muscle twitching and convulsions (WHO, 2004e).

3.4.2.21 Chloride

Sodium chloride is used in the production of caustic soda, chlorine, sodium chlorite and sodium hypochlorite. Chlorides are usually distributed in environment as salts of sodium (NaCl), potassium (KCl), and calcium ($CaCl_2$). Calcium chloride, sodium chloride and magnesium chloride are used in ice and snow control. Potassium chloride is used for manufacturing fertilizers. Chlorides are leached into groundwater from rocks and soil. Chloride in excess of around 250 mg/l can give a detectable taste to water. Humans may be affected by sodium chloride, leading to congestive heart failure. Increase in chloride content increases electrical conductivity and corrosivity.

3.4.2.22 Chlorine

Chlorine is used as a disinfectant and bleach in households, industries and swimming pools. It reacts with water to form hypochlorous acid and hypochlorites. At a pH above 4.0, little molecular chlorine exists in dilute water. It reacts with ammonia and amines to form chloramines. A residual chloride of 0.2–1 mg/l is maintained to act on microbes that could enter water during distribution and storage. An intake of bleach results in a burning sensation in the mouth and throat as well as irritation of the oesophagus.

3.4.2.23 Chlorite and chlorate

Chlorite and chlorate are byproducts from the use of chlorine dioxide. Chlorine dioxide is used in the disinfection of water, taste/odour control of water cleaning/detanning leather and as a bleaching agent. Sodium chlorite is used in industries as bleaching agent, for the on-site production of chlorine dioxide; for the manufacture of shellacs, waxes, varnishes, dyes, matches and explosives and in herbicides and defoliants. Chlorine dioxide decomposes into chlorates, chlorite and chloride ions in water.

3.4.2.24 *Chloroacetone*

Chloroacetone is used in perfumes, drugs, insecticides and vinyl compounds. It is also formed by an oxidation reaction between chlorine and large organic molecules. It is used in the manufacture of organic chemicals, methyl violet and as a fumigant for stored grain. It is degraded to carbon dioxide, nitrate ions and chloride ions in the presence of light.

3.4.2.25 *Chlorotoluron*

Chlorotoluron is a herbicide that will persist in water due to slow degradation. Chlorotoluron is rapidly and readily absorbed when given orally but evidence of accumulation in organs or tissue has not been reported. Further, no human cases of poisoning have been reported due to exposure to chlorotoluron (WHO, 2003h).

3.4.2.26 *Chlorpyrifos*

Chlorpyrifos is an insecticide that is poisonous to humans at an estimated dose of 300–400 mg/kg of body weight (WHO, 2004f). It is mainly used in control of flies, mosquitoes, various crop pests, household pests and aquatic larvae.

3.4.2.27 *Chromium*

Chromium is widely present in the earth's crust. It exists in oxidation states of +2 to +6. Chromium and its salts are widely used in catalysts, leather tanning, pigments, paints, fungicides, ceramics, glass, photography, chrome alloy and electroplating, and corrosion control. Ingestion of 1–5 g of 'chromate' results in severe acute effects including gastrointestinal disorders, convulsions, haemorrhagic diathesis and death (WHO, 2003g). Many parts of the developing world, where chrome plating has been extensively used without treatment, have heavily contaminated groundwater (Figure 3.23).

3.4.2.28 *Cyanazine*

Cyanazine is an herbicide capable of degrading in water and soil by micro-organisms and hydrolysis.

Many industries use the cyanide anion CN^-. Cyanogen chloride can be formed as a byproduct from the chlorination of drinking water. Cyanides are used in electroplating and hardening of metals; base metal flotation; the extraction of gold and silver and in coal gasification and fumigation. Cyanides are highly toxic and are relatively stable in nature unless they are oxidized. Cyanides in water will be affected by trace metal levels, pH, dissolved oxygen and temperature.

Acute cyanide poisoning leads to tachypnoea, headache, weak pulse, cardiac arrhythmias, vertigo, lack of motor coordination, vomiting, stupor, convulsions and coma (Ballantyne, 1983; Way, 1984; Johnson and Mellors, 1988).

3.4.2.29 *Cyanogen Chloride*

Cyanogen chloride is used in fumigant gases and tear gas. It may be formed during chloramination or chlorination of water. It is rapidly converted to cyanide in the body. It

(a)

(b)

Figure 3.23 *Chromium-contaminated groundwater. (For a colour version of this figure, see the colour plate section.)*

has been observed in finished water supplies, normally at concentrations below 10 μg/l (WHO, 2009e).

3.4.2.30 Cylindrospermopsin

Cylindrospermopsin is a toxin generated by cyanobacteria/blue-green algae. It has severe effects on the liver as well as other organs

3.4.2.31 Dichlorodiphenyltrichloroethane

Dichlorodiphenyltrichloroethane (DDT) is an insecticide banned in several countries because of ecological considerations but used in many countries for the control of vectors that spread yellow fever, sleeping sickness, malaria, typhus and other insect-transmitted diseases.

Together with its metabolites, DDT is persistent in nature and resistant to biodegradation. Dichlorodiphenyltrichloroethane and its metabolites are adsorbed onto sediments and soils, which act as sinks and sources of exposure. Dichlorodiphenyltrichloroethane and its metabolites are taken up readily by organisms in food and in the environment. Most of the uptake by aquatic organisms is from water whereas food is the major route for terrestrial fauna. It is highly lipid soluble but water solubility is low, which leads to the retention of DDT and its metabolites in fatty tissue.

3.4.2.32 Dichloroacetic Acid

Dichloroacetic acid is a type of chlorinated acetic acid formed from organic material during chlorination of water (IPCS, 2000). It is used in the synthesis of organic materials, pharmaceuticals and as a fungicide (Hawley, 1981; Toxnet, 2013). Dichloroacetate treatment significantly reduces fasting blood glucose levels, elevates plasma ketone bodies, decreases triglyceride levels, decreases plasma cholesterol levels, decreases in plasma lactate and alanine, and elevates serum uric acid levels (Stacpoole *et al.*, 1978).

3.4.2.33 Dichlorobenzines

The dichlorobenzine (DCBs) are used in industry and households as odour-masking agents, pesticides and dyes. In soils, they are slowly biodegraded under aerobic conditions. They are expected to be adsorbed onto soils of high organic content and do not leach appreciably into groundwater. In water bodies they are likely to be adsorbed onto sediments and bioaccumulate in aquatic organisms. Dichlorobenzines may biodegrade in aerobic conditions after microbial adaptation but they are not expected to biodegrade in lake sediments or groundwater (Howard, 1990).

Acute effects of DCBs include respiratory irritation, glomerulonephritis, acute haemolytic anaemia and allergic responses of the skin. Prolonged exposure to 1,4-DCB has resulted in disturbances of the reticuloendothelial system, central nervous system effects, granulomatosis, anaemia and liver damage (WHO, 2003i).

3.4.2.34 Dicofol

Dicofol is used to control spider mites as well as soft-bodied mites in cucumbers, tomatoes, apples, vines, lettuce, pears and ornamentals. It normally binds to organic matter in soil and is not likely to enter groundwater but residues in soil decrease rapidly. Dicofol is expected to adsorb to sediment in surface water and can hydrolyse to dichlorobenzophenone. It accumulates in body fat and is extensively absorbed from the gastrointestinal tract (WHO, 2007a).

3.4.2.35 Dimethoate

Dimethoate is an insecticide, highly soluble in chloroform, toluene, ethylene chloride, benzene, esters, ketones and alcohols; slightly soluble in xylene, aliphatic hydrocarbons and carbon tetrachloride and fairly soluble in water and acid solution; it is unstable in alkaline solution. Hydrolytic degradation and photochemical degradation are major causes for inactivating dimethoate in nature (WHO, 2004g).

3.4.2.36 Endosulfan

Endosulfan is a poison used to control pests and as wood preservative. It undergoes photolysis and is subject to biodegradation in soil. The lowest reported dose that resulted in death was 35 mg/kg of body weight (WHO, 2004i).

3.4.2.37 Endrin

Endrin is a foliar insecticide. It undergoes photodecomposition and bacterial degradation. Exposure to a toxic dose of endrin by humans will result in excitability and convulsions, leading to death within 2–12 h if appropriate treatment is not administered (WHO, 2004j).

3.4.2.38 Epichlorohydrin

Epichlorohydrin is used in the manufacture of glycerol, unmodified epoxy resins, elastomers, water-treatment resins, plasticizers, dyestuffs, surfactants, pharmaceutical products, oil emulsifiers, ion exchange resins, lubricants and adhesives (IPCS, 1984). It is released to the environment during manufacture/use/storage/transport/disposal. Acute toxic responses through dermal exposure lead to initial redness and an itching/burning sensation followed by tissue swelling and blisters (WHO, 2004h). Long-term effects include damage to the liver and kidneys (Schultz, 1964).

3.4.2.39 Fluoride

Fluorine is the lightest amongst the halogen group. It is one of the most reactive chemical elements and therefore is not found as fluorine in nature. It is the most electronegative amongst all the elements (Hem, 1989). Fluorine in nature is found as fluorides and represents about 0.06–0.09% of the earth's crust in the form of mineral complexes in a wide variety of minerals, including cryolite, apatite, mica, fluorspar, rock phosphate, hornblende and others (Murray, 1986).

At low concentrations in drinking-water, fluoride has beneficial effects on teeth. Excessive exposure to fluoride in drinking water and other sources can result in adverse effects such as dental fluorosis or skeletal fluorosis. The magnitude of the effect depends on exposure.

3.4.2.40 Hexachlorobutadiene

Hexachlorobutadiene is used in the production of chlorine gas, rubber compounds, as lubricant, as gyroscopic fluid, as pesticide and as a fumigant in vineyards. People exposed intermittently for four years exhibited hypotension, nervous disorder, liver function disorders, myocardial dystrophy and respiratory tract lesions. Hexachlorobutadiene may not volatilize rapidly due to its low vapour pressure (WHO, 2004l).

3.4.2.41 Heptachlor

Heptachlor is used for soil treatment and seed treatment to control insect pests in soils. It is banned in many countries or applied only by subsurface injection. It is moderately persistent in soil. Acute exposure leads to salivation, laboured respiration, irritability, muscle tremors and convulsions (WHO, 2004k).

3.4.2.42 Hydrogen sulfide

Hydrogen sulfide has a rotten-egg smell below 8 μg/m^3 (WHO, 1987), a sweet smell at concentrations of 50–150 mg/m^3, and no smell above this range (Patwardhan and Abhyankar, 1988). Hydrogen sulfide is used in the manufacture of sulfuric acid, inorganic sulfides, thioaldehydes, thioketones, thiophenes, cosmetics and dyes. It is also used in tanning, the production of wood pulp and chemical processing. It is formed due to the hydrolysis of soluble sulfides in water. It is oxidized to sulfates in aerated water and biologically oxidized to elemental sulfur. In anaerobic conditions reduction of sulfate to sulfide can occur due to microbial action. It causes nausea, vomiting, irritates mucous membranes and epigastric pain following ingestion. It is highly toxic to humans when inhaled (Gosselin *et al.*, 1984).

3.4.2.43 Iodine

Iodine is used as a disinfecting agent in hospitals/laboratories, as an antiseptic for skin wounds and for emergency disinfection of drinking water. It is also used in pharmaceuticals as well as in photographic developing materials. In water, it occurs naturally in the form of iodide (I–). It is oxidized to iodine during treatment. Chronic iodide exposure can result in iodism. Oral doses of around 30–40 mg/kg of body weight are likely to be lethal even though survival has been observed after ingestion of 10 000 mg (WHO, 2003j).

3.4.2.44 Isoproturon

Isoproturon is an herbicide with half-life of 30 days in water (DoE, 1989). Isoproturon is rapidly absorbed when ingested orally (WHO, 2003k).

3.4.2.45 Lindane

99% pure γ-hexachlorocyclohexane (γ-HCH) is called lindane. It is used as an insecticide and in the treatment of scabies in animals and humans (ATSDR, 1989; IPCS, 1991b). Many countries have limited the use of lindane. Under aerobic conditions lindane is degraded with a half-life ranging between 88 and to 1146 days. It enters water from agricultural fields where it is used and manufacturing plants.

3.4.2.46 Malathion

Malathion is used to control insects on plants, pets and humans (head and body lice). It degrades rapidly by biodegradation, hydrolysis and photolysis. In air, malathion is degraded by reaction with hydroxyl radicals. It is quickly absorbed, biotransformed and excreted in the urine and faeces (WHO, 2004n).

3.4.2.47 Methoxychlor

Methoxychlor is an insecticide used against a variety of pests. Its residues may endure in top soil for up to 14 months. Although it is poorly soluble in water, Methoxychlor has been found in surface/ground/drinking water (WHO, 2004o).

3.4.2.48 Methyl parathion

Methyl parathion is used to control insects by contact as well as stomach action. Like any other chemical distribution, the distribution of methyl parathion in the environment is influenced by several factors, which include chemical/physical and biological parameters. Methyl parathion is acutely toxic and WHO has classified it as 'extremely hazardous' (WHO, 2004p).

3.4.2.49 Methyl tertiary-butyl ether

Methyl *tertiary*-butyl ether (MTBE) is added to petrol as an oxygenate to improve combustion and lower exhaust emissions, mainly carbon monoxide. It is important with regard to water as it is resistant to chemical/microbial decomposition and has an objectionable taste/odour (WHO, 2005c).

3.4.2.50 Metolachlor

Metolachlor is a selective herbicide that photodegrades slowly when exposed to sunlight in aqueous solution. Its hydrolysis half-life is more than 200 days at 20 °C (Worthing, 1991). Signs of intoxication due to metolachlor include ataxia, dark urine, abdominal cramps, anaemia, methaemoglobinaemia, convulsions, diarrhoea, jaundice, weakness, cyanosis, hypothermia, collapse, nausea, shock, sweating, dizziness, vomiting, central nervous system depression, dyspnoea, cardiovascular failure, nephritis, liver damage, dermatitis, sensitization, corneal opacity, eye/mucous membrane irritation and reproductive effects.

3.4.2.51 Molybdenum

Molybdenum is used in special steels, nonferrous alloys, electrical contacts, X-ray tubes, filaments, spark plugs, screens and grids for radio valves, glass-to-metal seals, pigments, lubricant additive, the treatment of seeds and the formulation of fertilizers (Stokinger, 1983; Weast, 1986). Molybdenum normally occurs at low concentrations in drinking water and it is therefore not considered essential to set a formal guideline value (WHO, 2011g).

3.4.2.52 Monochlorobenzene

Monochlorobenzene (MCB) is used as a degreasing agent, as a solvent in pesticide formulations and in the synthesis of halogenated organic compounds. It will volatilize into the atmosphere and biodegrades in water. It is toxic to humans and causes headaches, dizziness, sleepiness and central nervous system disturbances (WHO, 2004q).

3.4.2.53 MX

'MX' is the common name given to 3-chloro-4-dichloromethyl-5-hydroxy-2(5H)-furanone. It is formed due to a reaction between chlorine and complex organic matter. It is present in the chlorinated waste water of pulp mills. Its presence was observed in chlorinated humic acid solutions as well as drinking-water in the United Kingdom, Finland and the United States (Hemming *et al.*, 1986; Meier *et al.*, 1989). It is a potent mutagen and some tumorigenic responses were observed during studies on rats (Komulainen *et al.*, 1997).

3.4.2.54 NDMA

N-Nitrosodimethylamine (NDMA) is produced during industrial processes that use nitrates/nitrites/amines. It may be present in effluents of industries like pesticide manufacturing, food processing, rubber manufacturing, leather tanning, dye manufacturing and foundries. It may form in sewage or in soils rich in nitrate or nitrite. Studies on rodents have shown evidence of carcinogenicity (WHO, 2008b).

3.4.2.55 Nitrite and Nitrate

Nitrate and nitrite occur naturally and are part of the nitrogen cycle.

The nitrate ion (NO_3^-) is stable and chemically unreactive but it can be reduced due to microbial action. Chemical and biological processes can oxidize nitrite to nitrate or reduce it to various compounds.

Nitrate is used in inorganic fertilizers, explosives, glass making and food preservative (especially in cured meats). It can reach surface water/groundwater due to agricultural activity, wastewater and solid waste. Nitrite can be formed in distribution pipes due to the presence of nitrosomonas.

The major effect of nitrite in humans is the oxidation of haemoglobin (Hb) to methomoglobin (metHb). MetHb is incapable of transporting oxygen to body tissues. When metHb concentrations reach 10% of more of usual Hb concentrations, the condition is called methaemoglobinaemia, which causes cyanosis and asphyxia. Nitrite reacts with nitrosatable compounds in environment and the human stomach leading to formation of *N*-nitroso compounds. Many of the *N*-nitroso compounds are carcinogenic (WHO, 2007b).

3.4.2.56 Parathoin

Parathion is an insecticide that is used to control insects by contact as well as stomach action. Parathion will adsorb strongly to the top soil and does not leach significantly. Parathion is degraded rapidly in the environment.

3.4.2.57 Pendimethalin

Pendimethalin is a selective herbicide that is usually stable in alkaline and acidic conditions. It is moderately persistent and can give rise to metabolites, mainly by photodegradation (WHO, 2003l).

3.4.2.58 *Pentachlorophenol*

Pentachlorophenol (PCP) and other chlorophenols are used for protecting wood from fungus. It has been discontinued completely in many developed countries but it is still an important pesticide in many developing countries due to its low cost and broad spectrum. It even enters countries where PCP has been discontinued, because many materials that are imported are treated with it. Pentachlorophenol and other higher chlorinated phenols inhibit ATPase and other enzymes (Jorens and Schepens, 1993). Symptoms of acute poisoning due to PCP exposure include dyspnoea, central nervous system disorders and hyperpyrexia leading to cardiac arrest.

3.4.2.59 *Permethrin*

Permethrin is a contact insecticide and it is not classifiable with regard to its carcinogenicity in humans (IARC, 1991). Permethrin is photodegraded in environment. It is readily degraded by hydrolysis and microbial action under aerobic conditions (WHO, 2011h).

3.4.2.60 *Potassium*

Potassium is a vital element in humans. It is rarely found in water at levels that could be a concern for humans. It occurs in all natural water. It can occur in drinking water as a result of the use of potassium permanganate in water treatment and the use of potassium chloride in ion exchange. Potassium is present in animal as well as plant tissues. Adverse health effects due to consumption of potassium from drinking-water are not likely to occur for healthy people (WHO, 2009f).

3.4.2.61 *Propanil*

Propanil is a contact herbicide. In water it is rapidly degraded by sunlight to phenolic compounds. The possible oral lethal dose of Proponil is 0.5–5 g/kg of body weight. Exposure to this chemical causes local irritation as well as central nervous system depression. Ingestion causes local irritation, burning sensation of the mouth, oesophagus and stomach, gagging, coughing, nausea and vomiting, headache, dizziness, drowsiness as well as confusion (WHO, 2004r).

3.4.2.62 *Polynuclear Aromatic Hydrocarbons*

Polynuclear aromatic hydrocarbons (PAHs) are a group of organic compounds having more than one fused aromatic ring of carbon as well as hydrogen atoms. They are formed from the burning of fossil fuels. They are normally not found in water in significant concentrations. Polynuclear aromatic hydrocarbons are stable to hydrolysis and are slowly biodegradable under aerobic conditions (WHO, 2003m).

3.4.2.63 *pH*

pH is a measure of the quantity of free hydrogen ions in water. It is defined as the negative logarithm of the molar concentration of hydrogen ions.

$$pH = -\log 10[H^+]$$

Table 3.4 *Relation between ion concentration and pH value.*

pH	Hydrogen ion concentration (moles/l) $[H^+]$	Hydroxide ion concentration (moles/L) $[OH^-]$
0	1	1×10^{-14}
1	0.1	1×10^{-13}
2	0.01	1×10^{-12}
3	0.001	1×10^{-11}
4	0.0001	1×10^{-10}
5	0.00001	1×10^{-9}
6	1×10^{-6}	1×10^{-8}
7	1×10^{-7}	1×10^{-7}
8	1×10^{-8}	1×10^{-6}
9	1×10^{-9}	0.00 001
10	1×10^{-10}	0.0001
11	1×10^{-11}	0.001
12	1×10^{-12}	0.01
13	1×10^{-13}	0.1
14	1×10^{-14}	1

Table 3.4 gives the relation between ion concentration and pH value. Since pH is measured on a logarithmic scale, a rise in one unit indicates an augment of ten times the quantity of hydrogen ions. A pH less than 7 is considered as acidic. A pH of 7 is considered to be neutral and a pH of more than 7 is considered as alkaline. Measurement of pH is done by pH metre (Figure 3.24) and pH paper. pH paper provides indicative values whereas the pH metre provides accurate pH values.

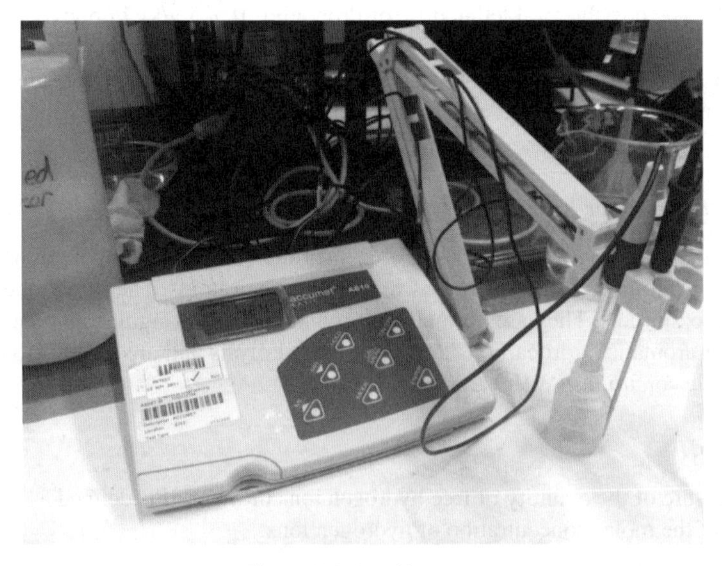

Figure 3.24 *pH metre.*

The pH in natural waters is controlled by the carbon dioxide–bicarbonate–carbonate equilibrium system. An increase in CO_2 concentration will lower pH and a decrease in CO_2 will raise the pH. Temperature will affect the carbon dioxide–bicarbonate–carbonate equilibrium, thereby altering pH. In pure water, pH will decrease by about 0.45 with increase in 25 °C. The pH of drinking water lies within the range 6.5–8.5. The pH also affects corrosivity of the water: the lower the pH higher the potential level of corrosion.

3.4.2.64 Volatile Organic Compounds (VOCs)

Organic compounds with a boiling point less than 100 °C and/or vapour pressure greater than 1 mm Hg at 25 °C are considered as volatile organic compounds (VOCs). They are of great concern because (i) they are likely to enter the atmosphere as vapours; (ii) some of these compounds in the atmosphere pose a significant health risk; (iii) they contribute to an increase in reactive hydrocarbons in the atmosphere, resulting in the formation of photochemical oxidants.

Sophistication in instrumentation has enabled scientist to analyse water for contaminants. The atomic absorption spectrometer (Figure 3.25), the gas chromatograph (Figure 3.26), and high performance liquid chromatography (Figure 3.27) have been very useful for the analysis of many chemicals.

3.4.3 Solids in Water

The total solids content of water is the matter that remains as residue after evaporation at 103–105 °C. Matter with significant vapour pressure at this temperature is lost during evaporation and does not qualify as a solid.

The size of the solids in water varies widely depending on source. Wastes from ships include food packaging, oily sludge and food waste. Ports and harbours can generate a variety of wastes, which vary from dust to damaged finished products. Parts of shipments that lie in the ports for long time may partially have entered water streams. Plastic pellets used for manufacturing enter the sea due to accidental spillages. Each year container ships lose more than 10 000 containers in marine environments, usually during storms, adding to the waste entered from activities on land (Podsada, 2001).

Suspended solids enter fresh water bodies through sewage, industrial waste water, runoff or flocs of microbes formed during the degradation of organic matter. Figure 3.28 shows electron microscope pictures of China clay, which gives an idea of the shape of suspended particles in water.

Settleable solids are solids that will settle due to gravity. They are measured with a cone-shaped container called an Imhoff cone. A sample is allowed to settle for 60 min and settled solids are measured in ml/l. This gives an approximate measure of the quantity of the sludge that will be removed at the primary sedimentation stage of water/waste-water treatment.

Total solids upon evaporation can be classified further as nonfilterable (suspended) or filterable (dissolved) by passing a known quantity of water through a filter. A glass-fibre filter with a nominal pore-size of about 1.2 μm is most commonly used for this separation step.

Filterable solid fractions consists of colloidal and dissolved solids. The colloidal fraction consists of the particulate matter with an approximate size range of from 0.001 to 1 μm. The colloidal fraction cannot be removed by settling.

Figure 3.25 Atomic absorption spectrometer used to measure concentration of contaminants.

The dissolved solids consist of both organic and inorganic fractions that are present in true solutions in water. These categories of solids can identified on the basis of their volatility at 550 ± 50 °C. The organic fraction will be oxidized and driven off as gas at this temperature, and the inorganic fraction remains as ash. Thus, the terms 'volatile suspended solids' and 'fixed suspended solids' are quite often used to refer to the organic and inorganic content of the suspended solids. At 550 ± 50 °C, the decomposition salts are restricted to magnesium carbonate, which decomposes into magnesium oxide and carbon dioxide at 350 °C.

3.4.3.1 Acidity

Acidity increases as pH values decrease and alkalinity increases as pH values increase. Most natural waters are buffered by a carbon dioxide-bicarbonate system because the

Figure 3.26 *Gas chromatograph used to analyse volatile organic compounds.*

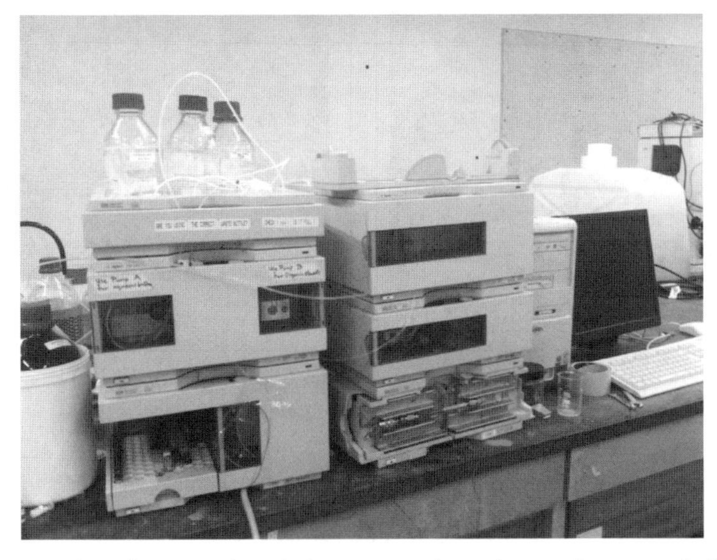

Figure 3.27 *High-performance liquid chromatograph used to analyse nonvolatile chemical and biological compounds.*

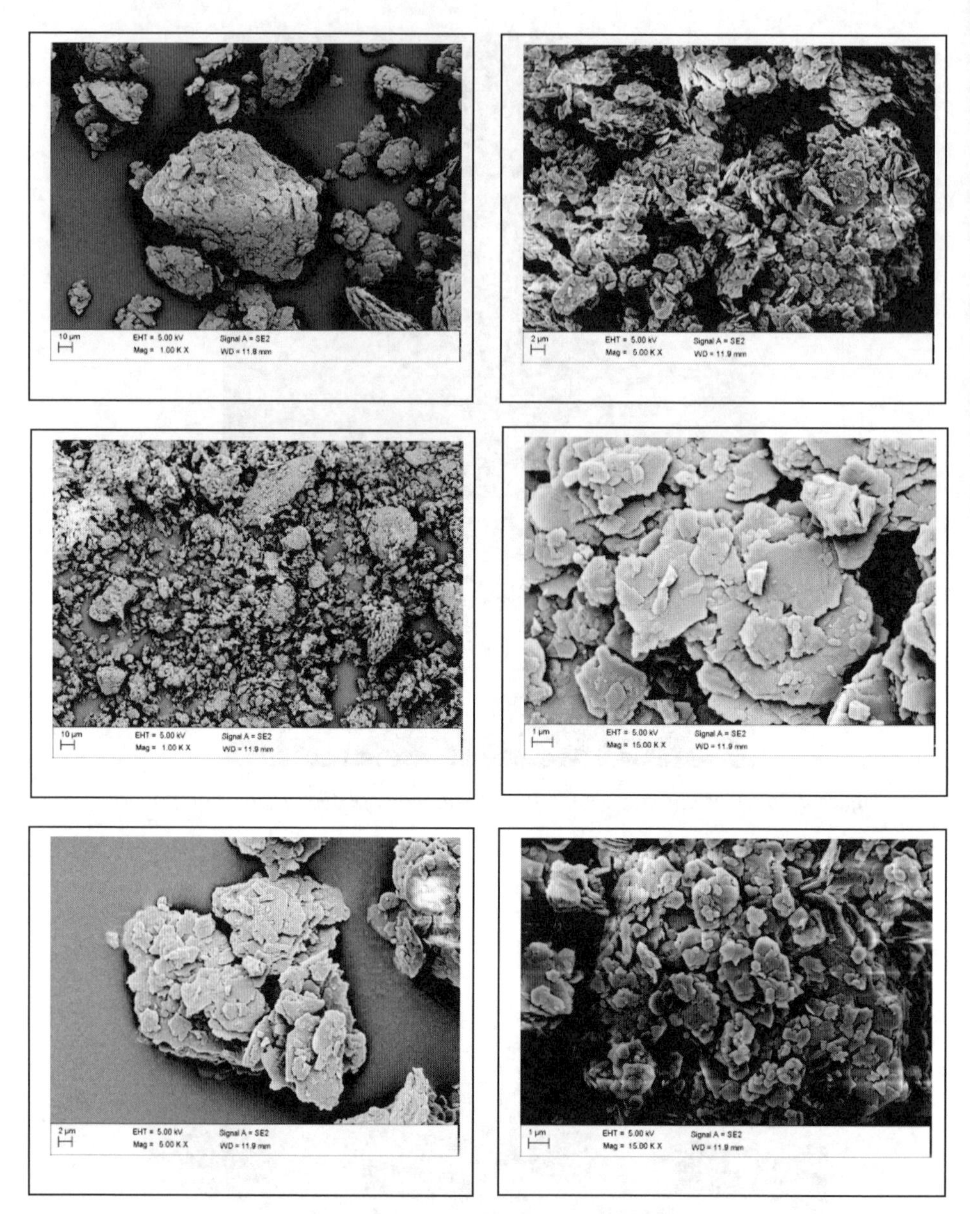

Figure 3.28 *Electron microscope pictures of China clay.*

carbon dioxide in the atmosphere serves as a source of carbonic acid.

$$H_2CO_2 \rightarrow HCO_3 + H + pK \sim 7.5$$

This reaction tends to keep pH of most waters around 7–7.5, unless large amounts of acid or base are added to the water. Most streams draining coniferous woodlands tend to be slightly acidic (6.8 to 6.5) due to organic acids produced by the decaying of organic matter. Natural waters in the Piedmont of Georgia also receive acidity from the soils. In waters with high algal concentrations, pH varies diurnally, reaching values as high as 10 during the day when algae are using carbon dioxide in photosynthesis. pH drops during the night when the algae respire and produce carbon dioxide. The pH of water affects the solubility of many toxic and nutritive chemicals; the availability of these substances to aquatic organisms is therefore affected. As acidity increases, most metals become more water soluble and more toxic. Toxicity of cyanides and sulfides also increases with a decrease in pH (increase in acidity). Ammonia, however, becomes more toxic with only a slight increase in pH.

3.4.3.2 Alkalinity

An alkali is a basic, ionic salt of an alkaline earth or alkali metal element. Alkalinity measures the ability of a solution to neutralize acids. Alkalinity is sometimes incorrectly used to mean basicity. A base is a substance that ionizes to give a hydroxide ion. Carbonate, borate, hydroxide, phosphate, silicate, nitrate, dissolved ammonia, sulfide and conjugate bases of some organic acids contribute to the alkalinity of water. Alkalinity is measured in mEq/L (milliequivalent per litre) or ppm.

As rain falls on the earth, water droplets become saturated with carbon dioxide and the pH is lowered. The carbon dioxide concentration will increase due to bacterial processes in the soils as they react with rock formations containing carbonates, resulting in increased alkalinity, pH and hardness.

3.4.3.3 Dissolved Oxygen

Aquatic plants and animals depend on dissolved oxygen (DO) to live; the quantity of DO in surface water depends on the water temperature, sediment in the stream, decaying organisms, photosynthesizing plants and oxygen transfer from the surface of water. Dissolved oxygen is measured in milligrams per litre (mg/l). Most warm water fish need DO in excess of 2 mg/l but some species need more than 5 mg/l.

3.4.3.4 Biochemical Oxygen Demand (BOD)

Over the years, various tests have been developed to quantify the organic content of wastewater. Trace concentrations in the range of 10^{-6} to 10^{-3} g/l are determined using instrumental methods including gas chromatography and mass spectroscopy. Organic matter greater than 1.0 mg/l in wastewater is quantified by: (i) biochemical oxygen demand (BOD); (ii) chemical oxygen demand (COD); (iii) total organic carbon (TOC).

Biochemical oxygen demand (BOD) is a measure of the quantity of oxygen that microbes will consume while decomposing biodegradable matter under aerobic conditions. The BOD is determined by incubating a sample of water for five days followed by measuring the loss of oxygen at the end of the test. Samples are usually diluted prior to incubation as bacteria will deplete all the oxygen in the bottle even prior to the completion of the test. The main aim of wastewater treatment plants is to decrease the BOD in the effluent and this is measured in mg/l. The BOD is usually not measured in groundwater and drinking surface water sources.

The most widely used parameter of organic pollution applied to both wastewater and surface water is the 5-day BOD (BOD_5).This involves the measurement of the dissolved oxygen used by micro-organisms in the biochemical oxidation of organic matter. Despite the widespread use of the BOD test, it has a number of limitations (which are discussed later in this section). It is hoped that, through the continued efforts of workers in the field, one of the other measures of organic content, or perhaps a new measure, will ultimately be used in its place. The BOD test's results are used:

1. to determine the approximate quantity of oxygen that will be required to stabilize the organic matter present biologically;
2. to determine the size of waste-treatment facilities;
3. to measure the efficiency of treatment processes;
4. to determine compliance with the terms of wastewater discharge permits.

It is likely that the BOD tests will continue to be used for some time so it is important to know as much as possible about the test and its limitations. To ensure that meaningful results are obtained, the sample must be suitably diluted with specially prepared dilution water so that adequate nutrients and oxygen will be available during the incubation period. Normally, several dilutions are prepared to cover the complete range of possible of values. The sample of wastewater is diluted with distilled water in order to measure BOD of samples with high BOD.

When the sample contains a large population of micro-organisms (untreated wastewater, for example), seeding is not necessary. If it is required, the dilution water is 'seeded' with a bacterial culture that has been acclimated to the organic matter or other materials that may be present in the wastewater. The seed culture that is used to prepare the dilution water for the BOD test is a mixed culture. Such cultures contain large numbers of saprophytic bacteria and other organisms that oxidize the organic matter. They also contain certain autographic bacteria that oxidize noncarbonaceous matter. A variety of commercial seed preparations are also available.

The incubation period is usually five days at 20 °C, but other lengths of time and other temperatures can be used. Longer time periods (typically seven days), which correspond to work schedules, are often used, especially in small plants where the laboratory staff is not available on the weekends. The temperature, however, should be constant throughout the test. The dissolved oxygen of the sample is measured before and after incubation, and the BOD is calculated using the following equation:

When dilution water is not seeded:

$$BOD = \frac{D1 - D2}{P}$$

When dilution water is seeded:

$$BOD = \frac{(D1 - D2) - (B1 - B2)f}{P}$$

where

D1 = dissolved oxygen of diluted sample immediately after preparation, mg/l;
D2 = dissolved oxygen of diluted sample after 5 day incubation at 200 °C, mg/l;
P = decimal volumetric fraction of sample used;
B1 = dissolved oxygen of seed control before incubation, mg/l;
B2 = dissolved oxygen of seed control after incubation, mg/l;
f = ratio of seed in sample to seed in control
= (% seed in D1)/(% seed in B1).

Biochemical oxidation is a slow process and theoretically takes an infinite time to go to completion. Within a 20-day period, the oxidation of the carbonaceous organic matter is about 95 to 99% complete. In the 5-day period used for the BOD test, oxidation is from 60 to 70% complete. The 20 °C temperature used is an average value for slow-moving streams in temperate climates and is easily duplicated in an incubator. Different results would be obtained at different temperatures because biochemical reaction rates are temperature dependent.

3.4.3.5 Chemical Oxygen Demand (COD)

Chemical oxygen demand (COD) is a measure of the total quantity of oxygen required to oxidize all organic matter into carbon dioxide and water. It is analysed by adding a fixed volume with a known excess amount of the oxidant to the sample of the solution that needs to be analysed. After a refluxing digestion step, the initial concentration of organic content in the sample is calculated by a titrimetric/spectrophotometric determination of the oxidant remaining in the sample. The unit of COD is mg/l.

3.4.3.6 Hardness

Hardness is the concentration of multivalent metallic cations in water. Hardness cations will react with anions under supersaturated conditions. Hardness can be classified as carbonate hardness and noncarbonated hardness. Multivalent metallic ions that are abundant in water are calcium, magnesium, iron, manganese, strontium and aluminium. As other multivalent metallic ions are present in small quantities, for all practical purpose hardness is defined as the sum of calcium and magnesium ions.

Calcium and magnesium are necessary for the human body. Inadequate intake of calcium will lead to risks of osteoporosis, kidney stones, coronary artery disease, colorectal cancer, hypertension, stroke, insulin resistance and obesity. Hardness contributes to the taste of water to varying degrees, increased soap consumption as well as scale deposition in heated water applications, the water distribution system and boilers. Excessively hardness of water can lead to corrosion (WHO, 2011e).

3.4.3.7 Metals

Heavy metals can enter the water when they seep into groundwater or dissolve into surface water. Heavy metal pollution can be due to solubility of metals in soil, mine waste, industrial processes, solid waste disposal or scrubbing of metal in air. Heavy metals pose significant risks to health and environment at high concentrations.

3.4.3.7.1 Aluminium

Aluminium constitutes nearly 8% of the Earth's crust and occurs naturally as silicates, oxides and hydroxides. A positive relationship between Alzheimer disease and aluminium in drinking water has been demonstrated in several studies (WHO, 1997, 2010).

3.4.3.7.2 Antimony

Antimony is an inflexible metal. It occurs mostly in the form of antimony trioxide (ATO), from coal burning (Nriagu and Pacyna, 1988). Soluble forms of antimony are mobile in water. Less soluble species are adsorbed onto soil particles and sediments (Crecelius *et al.*, 1975). Antimony (III) is more toxic than antimony (V) and inorganic antimony compounds are more toxic than organic compounds (Stemmer, 1976).

3.4.3.7.3 Cadmium

Cadmium is a metal electroplated onto steel mainly due to its anticorrosive nature. It has an oxidation state of +2. Fertilizers manufactured from phosphate ores are the major source of cadmium pollution (WHO, 2011c). Cadmium is used for producing pigments for plastics, electronic components, electric batteries and nuclear reactors (Friberg *et al.*, 1986; Ros and Slooff, 1987). Contamination of drinking water may occur due to the presence of cadmium in the zinc of galvanized pipes, water heaters, cadmium-containing solders, water coolers and taps. A dose of 3 mg of cadmium does not show an acute effect in adults, but 350–3500 mg is considered to be a lethal oral dose for humans (Krajnc *et al.*, 1987).

3.4.3.7.4 Calcium

Calcium is essential to human health and more than 99% of total body calcium is found in bones and teeth. An inadequate intake of calcium would increase the risk of osteoporosis, kidney stones, coronary artery disease, colorectal cancer, hypertension, stroke, insulin resistance and obesity.

Desalinated seawater/brackish water is stabilized prior to distribution by adding minerals and alkalinity like limestone or by blending seawater or brackish groundwater to avoid corrosion of distribution pipes. This stabilization modifies the water composition by reintroducing sodium chloride and other salts. Stabilization practices should ensure that the total intake of nutrients like calcium, magnesium and fluoride will be in the required quantity. Hence, normally 1% seawater is used for post-treatment blending, thereby introducing magnesium of around 12–17 mg/l and calcium of about 4–5 mg/l into the stabilized water (WHO, 2009c).

3.4.3.7.5 Iron

Iron is the second most common metal in the Earth's crust. Fe^{2+} and Fe^{3+} readily combine with oxygen/sulfur compounds to form oxides, hydroxides, sulfides and carbonates. It is commonly found in the form of oxides (Knepper, 1981; Elinder, 1986). As iron (II) salts are unstable, they are precipitated as insoluble iron (III) hydroxide when directly pumped from a well. Iron also promotes growth of 'iron bacteria' resulting in a slimy coating on the piping (DNHW, 1990). Precipitation may also be developed in piped systems at levels above 0.05–0.1/mg/litre. Staining of laundry and plumbing may happen at concentrations above 0.3 mg/litre (DNHW, 1990).

Iron dextran complex repeatedly injected intramuscularly or subcutaneously was carcinogenic to animals (IARC, 1987). Iron being an essential element for human nutrition, humans have a minimum daily requirement of about 10 to 50 mg/day. Even though the mean lethal dose of iron is 200–250 mg/kg of body weight, death has occurred after the ingestion of doses of 40 mg/kg of body weight (NRC, 1979).

3.4.3.7.6 Manganese

Manganese is one of the more abundant metals in the Earth's crust, normally occurring with iron. Even though it is one of the components of over 100 minerals, it is not found in its pure elemental form in nature (ATSDR, 2000a, 2002b). It is essential for proper functioning of humans and animals (IPCS, 2002). Manganese is used in alloys, batteries, glass and fireworks, varnish, livestock feeding supplement, fertilizers, varnish and fungicide. In surface waters, manganese occurs in both dissolved and suspended forms. Anaerobic groundwater often contains elevated levels of dissolved manganese. It is an essential element for living organisms, including humans. Manganese in water will be objectionable if it is deposited in the water mains, resulting in water discoloration.

Manganese concentrations in seawater range from 0.4 to 10 μg/l (ATSDR, 2000a, 2002b). Adults consume manganese in the order of 0.7 to 10.9 mg/day. The presence of manganese in water will be objectionable if it is deposited in pipes resulting in water discoloration. Concentrations less than 0.05 mg/l are acceptable, although this may show a discrepancy with local circumstances (WHO, 2011d).

3.4.3.7.7 Mercury

Mercury is a metallic element distributed by natural processes like volcanic activity. Mercury is used in the production of chlorine, caustic soda, lamps, mercury cells, arc rectifiers, switches, thermometers, barometers, antiseptics, fungicides, preservatives, pharmaceuticals, electrodes, reagents and dental amalgams. Elemental mercury vapour is insoluble in water, whereas mercury (II) chloride is easily soluble and mercury (I) chloride is less soluble. Mercury sulfide has very low solubility. Some bacteria that are capable of methane synthetase are capable of mercury methylation. The released methyl mercury enters the food chain.

Mercury will disrupt tissue with which it comes into contact in sufficient concentrations. Acute exposure to toxic doses of any form of mercury through ingestion will result in cardiovascular collapse, shock, acute renal failure and severe gastrointestinal damage. Acute oral poisoning ultimately damages the kidney. Long-term exposure to mercury results in mental disturbances and gingivitis (WHO, 2005b).

3.4.3.7.8 Nickel

Nickel is a lustrous hard, white, ferromagnetic metal that can enter water by leaching from pipes, fittings and nickel ore-bearing rocks in ground. Effects on kidney function have been reported (IPCS, 1991a). Nickle can cause allergic contact dermatitis. Occupational exposure to sulfidic or oxidic nickel at high concentrations can cause lung and nasal cancers (WHO, 2005d).

3.4.3.7.9 Tin

Tin is a silvery white metal used in plating and pigments. Its concentration in surface water is usually less than 5 μg/litre. The use of organo-tin biocides results in higher concentrations.

The low toxicity of tin is mainly due to its low tissue accumulation, low absorption and rapid excretion. Due to low toxicity, tin in drinking-water does not pose a hazard to human health (WHO, 2004m).

3.4.3.7.10 Lead

Lead is used in lead acid batteries, cable sheathing, solder, pigments, ammunition, rust inhibitors, glazes, plastic stabilizers and plumbing fittings. It may enter tap water due to dissolution from natural sources and may be present in plumbing work. Polyvinyl chloride (PVC) pipes also have lead compounds that can contribute to concentrations in drinking water. The quantity of lead dissolved from the plumbing depends on pH, water softness, the presence of chloride and dissolved oxygen, temperature and the standing time of the water. Lead is a cumulative poison. Foetuses and pregnant women are most susceptible to adverse health effects. It affects the central nervous system. Signs of acute intoxication include dullness, restlessness, headaches, muscle tremor, irritability, poor attention span, abdominal cramps, hallucinations, kidney damage, loss of memory as well as encephalopathy occur at blood lead levels of 80–100 μg/dl in children and 100–120 μg/dl in adults. Signs of chronic lead toxicity include sleeplessness, irritability, tiredness, headaches, joint pain and gastrointestinal symptoms; these may occur in adults with lead levels of 50–80 μg/dl in blood. Exposure to lead can lead to neurological and behavioural effects, impaired renal function, mortality due to cardiovascular diseases, impaired fertility, hypertension, delayed sexual maturation, adverse pregnancy outcomes and impaired dental health (WHO, 2011f).

3.4.3.7.11 Silver

Silver is a white metal used in photographic materials, alkaline batteries, electrical equipment, alloys, mirrors, catalysts, coins, utensils, jewellery, antiseptic agents, bacteriostatic agents and as disinfectants. Silver may be absorbed via the lungs, mucous membranes, the gastrointestinal tract and skin lesions (USEPA, 1980). The important silver compounds with regard to drinking water are silver nitrate and silver chloride. Much of the silver in blood is attached to globulins (USEPA, 1980). All silver consumed will not be absorbed by human body. Retention rates vary between 0 and 10% (USEPA, 1980). Oral LD50 values have been observed for various silver salts in mice between 50 and 100 mg per kg of body weight (Goldberg *et al.*, 1950).

Mice that had received silver in the dose of 4.5 mg/kg of body for 125 days showed hypoactive behaviour (Rungby and Danscher, 1984). Olcott (1950) observed that after 218 days of exposure, rats receiving silver in quantities of about 60 mg/kg of body weight/day via drinking-water showed a slight greyish pigmentation in the eyes, which further intensified. The estimated acute lethal dose for humans with respect to silver nitrate is about 10 g (Hill and Pillsbury, 1939). Chronic silver intoxication could cause argyria.

3.4.3.7.12 Sodium

Metallic sodium is used in electric power cables, the manufacture of tetraethyl lead and sodium hydride, as a catalyst for synthetic rubber, in titanium production, as a coolant in nuclear reactors, as a laboratory reagent, in nonglare lighting for roads and in solar-powered electric generators (Sax and Lewis, 1987). Sodium salts are used in water softening, disinfection, cooking, corrosion control, glass, soap, pH adjustment, road deicing and the paper, pharmaceutical, chemical and food industries. Sodium salts are highly soluble in water and, hence, they leach into groundwater/surface water. As sodium salts are nonvolatile, they will be found in the atmosphere only with particulate matter.

Sodium is naturally present in all foods and may be added during food processing. Fresh vegetables and fruit contain sodium in the range <10–1000 mg/kg; cereals as well cheese may contain 10–20 g/kg and cow and human milk contain 770 and 180 mg/l, respectively (Diem and Lentner, 1970; WHO, 2003n).

A daily intake of about 120–400 mg is required for infants and children, as compared to about 500 mg for adults (NRC, 1989).

Even though people in Western Europe and North America consume a high sodium salt diet, they do not exhibit persistent hypertension till the fourth decade whereas 'nonwesternized' populations consume low sodium and the occurrence of hypertension is low and blood pressure usually does not rise with age.

The effects of sodium on infants and adults are different due to the immaturity of kidneys in infants. Infants with gastrointestinal infections can experience fluid loss leading to dehydration as well as hypernatraemia (WHO, 2003n).

Even though sodium salts are usually not acutely toxic due to the efficiency of kidneys in excreting sodium, acute effects and death may occur in some individuals. Acute effects include nausea, convulsions, vomiting, cerebral and pulmonary oedema and muscular twitching and rigidity (Elton *et al.*, 1963; DNHW, 1992).

3.4.3.8 Organics

In a wastewater of medium strength, about 75% of the suspended solids and 40% of the filterable solids are organic in nature (see, e.g. Figure 2.2). These solids are derived from both the animal and plant kingdoms and the activities of man as related to the synthesis of organic compounds. Organic compounds are normally composed of a combination of carbon, hydrogen, and oxygen, together with nitrogen in some cases. Other important elements, such as sulfur, phosphorus and iron, may also be present. The principal groups of organic substances found in wastewater are proteins (40–60%), carbohydrates (25–50%) and fats and oils (10%). Urea, the chief constituent of urine, is another important organic compound contributing to wastewater. Because it decomposes so rapidly, undecomposed urea is seldom found other than in very fresh wastewater.

Along with the proteins, carbohydrates, fats and oils and urea, wastewater contains small quantities of a large number of different synthetic organic molecules ranging from simple to extremely complex in structure.

3.4.3.8.1 Protein

Proteins are complex in chemical structure and unstable, being subject to many forms of decomposition. Some are soluble in water; others are insoluble. Protein is the principal constituent of the animal organism. It occurs to a lesser extent in plants. The chemistry of the formation of proteins involves the combination or linking together of a large number of amino acids. The molecular weights of proteins range from about 20 000 to 20 million.

All proteins contain carbon, hydrogen, oxygen and nitrogen. In many cases, sulfur, phosphorus, and iron are also constituents. Urea and proteins are the chief sources of nitrogen in wastewater. When proteins are present in large quantities, extremely foul odours are apt to be produced by their decomposition.

3.4.3.8.2 *Carbohydrates*

Carbohydrates are compounds comprising carbon, hydrogen and oxygen. Examples of carbohydrates include sugars, starches, cellulose and wood fibre. Some carbohydrates, notably the sugars, are soluble in water; others, such as the starches, are insoluble. The sugars tend to decompose; the enzymes of certain bacteria and yeasts set up fermentation with the production of alcohol and carbon dioxide. The starches, on the other hand, are more stable but are converted into sugars by microbial activity and by dilute mineral acids. Cellulose is the most important carbohydrate found in wastewater. The destruction of cellulose in the soil largely occurs as a result of the activity of various fungi, particularly when acid conditions prevail.

3.4.3.8.3 *Fats, Oils and Grease*

The term 'grease', as commonly used, includes the fats, oils, waxes and other related constituents found in wastewater. Fats and oils are compounds (esters) of alcohol or glycerol (glycerine) with fatty acids. The glycerides of fatty acids that are liquid at ordinary temperatures are called oils, and those that are solids are called fats. They are quite similar, chemically, being composed of carbon, hydrogen and oxygen in varying proportions. Fats and oils are present in butter, vegetable oils, and meats. Fats are amongst the more stable organic compounds and are not easily decomposed by bacteria. Mineral acids attack them, however, resulting in the formation of glycerine and fatty acid.

Kerosene, lubricating oil and road oils are derived from petroleum and coal tar and contain essentially carbon and hydrogen. These oils sometimes reach the sewers in considerable volume from shops, garages and the streets. Fats, oil and grease float on the wastewater and a portion is carried into the sludge.

3.4.3.8.4 *Surfactants*

Surfactants, or surface-active agents, are large organic molecules that are slightly soluble in water and cause foaming in wastewater treatment plants and in the surface waters into which waste effluent is discharged. Surfactants tend to collect at the air-water interface. During aeration of wastewater, these compounds collect on the surface of the air bubbles and thus create very stable foam.

3.4.3.9 Nutrients

Nutrients, rather than carbon or energy sources, may at times be the limiting material for microbial cell synthesis and growth. The principal inorganic nutrients needed by microorganisms are N, S, P, K, Mg, Ca, Fe, Na and Cl. Important minor nutrients include Zn, Mn, Mo, Se, Co, Cu, Ni, V, and W.

In addition to the inorganic nutrients cited above, organic nutrients may also be needed by some organisms. Required organic nutrients, known as 'growth factors' are compounds needed by an organism as precursors or constituents of organic cell materials that cannot be synthesized from other carbon sources. Although growth factor requirements differ from one organism to another, the major growth factors fall into the following three classes: (i) amino acids, (ii) purines and pyrimidines and (iii) vitamins.

3.4.4 Biological Parameters

Biological pathogens come from the entry of pathogenic microbes from faeces and infected waste that originate from open defecation, healthcare establishments, slaughter houses, animal yards in the catchment area of surface water sources. These sources may contribute to contamination of groundwater sources as well. Guinea worm, cyanobacteria, *Legionella,* roundworms and flatworms can be transmitted from animals/humans through drinking water. Most microbial pathogens transmitted by water infect the gastrointestinal tract. There are also bacterial pathogens that can grow in water and soil. In addition to ingestion, these pathogens are transmitted through inhalation and by contact with infections occurring in skin, the respiratory tract or in the brain.

3.4.4.1 Micro-Organisms

Micro-organisms (derived from the Greek *mikrós* meaning 'small' and *organismós* meaning 'organism') are microscopic organisms, which could be single-cell organisms or multicellular ones. Figure 3.29 is an electron microscope picture of microbial floc and Figure 3.30 is an electron microscope picture of microbes and disintegrated cells. Micro-organisms can be prokaryotes (groups of organisms whose cells lack a cell nucleus) or eukaryotes (groups of organisms with cells having cell nucleus).

3.4.4.1.1 Bacteria

Bacteria are microscopic organisms with single cell having neither a membrane-enclosed nucleus nor membrane-enclosed organelles. They are prokaryotic and do not have mitochondria or chloroplasts. They process single chromosomes.

- ***Acinetobacter.*** *Acinetobacter* sp. species are Gram-negative, nonmotile coccobacilli. The *Acinetobacter calcoaceticus baumannii* complex includes all subgroups of this

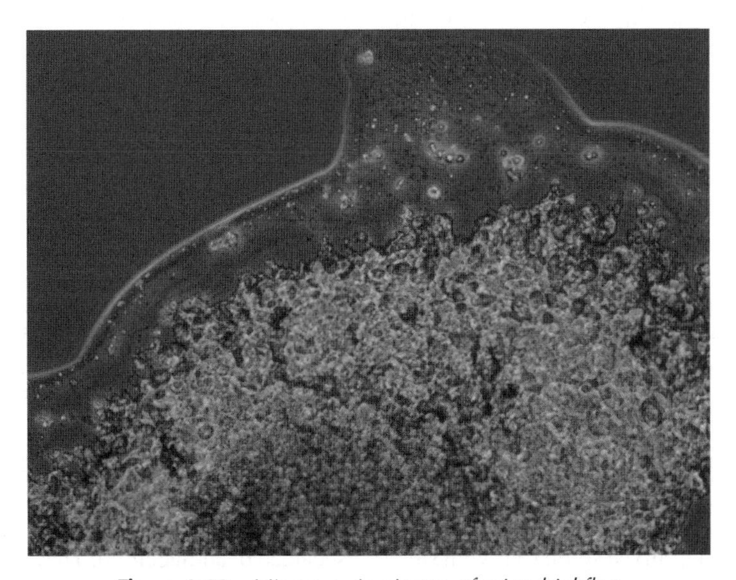

Figure 3.29 Microscopic picture of microbial floc.

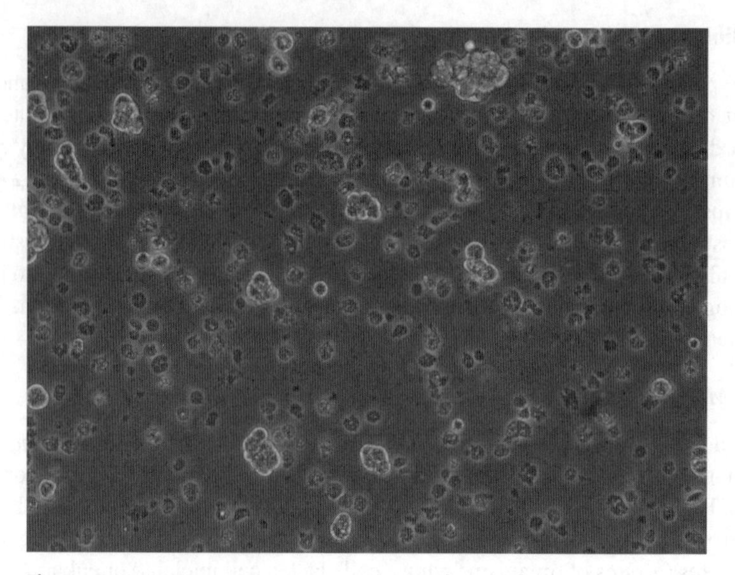

Figure 3.30 *Microscope picture of microbes and disintegrated cells.*

species. The *Acinetobacter* species cause urinary tract infections, bacteraemia, pneumonia, secondary meningitis and wound infections. The *Acinetobacter* species are universal inhabitants of soil, water and sewage. *Acinetobacter* species are present on skin as well as occasionally in the respiratory tract of healthy people. Thermotolerant coliforms or *E. coli* cannot be used as an index for *Acinetobacter* species.

- ***Aeromonas.*** *Aeromonas* species are non-spore-forming, Gram-negative, facultative anaerobic bacilli. These bacteria occur in water, soil and meat and milk. *Aeromonas* species cause infections in humans. Wound infections have been associated during swimming, diving, boating and fishing. *E. coli* or thermo tolerant coliforms cannot be used as an index for *Aeromonas* species.
- ***Bacillus.*** *Bacillus* species are 4–10 mm Gram-positive, aerobic or facultatively anaerobic encapsulated bacilli. They produce spores that are resistant to unfavourable conditions. Most *Bacillus* species are harmless but few of them are pathogenic to humans and animals.
- ***Bacillus thuringiensis.*** *Bacillus thuringiensis* (Bt) is a facultative anaerobic bacterium. The subspecies *israelensis* (Bti) is widely used for controlling mosquitoes, chironomids and blackflies. Bti is often applied directly to water. Bti is not considered to cause a danger to humans through drinking water (WHO, 2009a).
- ***Burkholderia pseudomallei.*** *Burkholderia pseudomallei* are Gram-negative bacillus found in soil and muddy water, mainly in tropical regions. The organism can survive in water for prolonged time in the absence of nutrients. *Burkholderia pseudomallei* can cause melioidosis.
- ***Campylobacter.*** *Campylobacter* species are Gram-negative, microaerophilic and capnophilic, curved spiral rods with single unsheathed polar flagellum. They are the most important causes of acute gastroenteritis worldwide. They occur in wild and domestic

Figure 3.31 *Colour of water that has turned green due to Eutrophication. (For a colour version of this figure, see the colour plate section.)*

animals. Meat and unpasteurized milk are important sources of *Campylobacter* infections. The occurrence of these organisms in surface waters depends on rainfall, water temperature as well as the presence of waterfowl.

- **Cyanobacteria.** The cyanobacteria accumulate in surface water as 'blooms' and may form 'scums'. Cyanobacteria are photosynthetic bacteria. Many cyanobacteria produce potent toxins, with specific properties and with different concerns including liver damage, neurotoxicity as well as tumour promotion. They are widespread and can be observed in a diverse range of environments, which include soils, seawater and freshwater environments. Sunlight, low turbulence, warm weather and high nutrient levels promote growth. Depending on the species water may become greenish with a high density of cyanobacteria. In some cases, they form of surface scums leading to high toxin concentrations. They have been causing animal as well as human poisoning in several parts of the world for more than 100 years. Exposure to the toxins occurs through ingestion, recreation and showering. Human fatalities have occurred due to the use of water containing cyanotoxins. Eutrophication (Figure 3.31) can support cyanobacterial blooms. Some species of cyanobacteria generate toxins, which are classified as neurotoxins, hepatotoxins, skin irritants and other toxins. Acute effects include gastrointenstinal disorders, fever, skin irritation, eyes irritation, throat irritation and respiratory tract irritation. The occurrence of a specific genus and species of cyan bacteria is influenced by chemistry and climatic conditions. *Cylindrospermopsis* blooms occur in tropical waters but not in temperate climates whereas *Anabaena* and *Microcystis* blooms occur in the temperate regions (AWWA, 1995). Turbulence and water flows are unfavourable to cyanobacteria. Heavy rainstorms can increase nutrient levels and encourage the formation of blooms. Fatalities in animals have been reported following the consumption of water with large numbers of cyanobacteria (Beasley *et al.*, 1989; Carmichael, 1992).

- ***Escherichia coli* pathogenic strains.** *Escherichia coli* are present in huge numbers in the intestines of humans and animals, where they do not normally cause any harm. In other parts of the body they can cause serious diseases like urinary-tract infections, bacteraemia and meningitis. Some enteropathogenic strains cause acute diarrhoea. *E. coli* O157:H7 as well as *E. coli* O111 cause diarrhoea. Waterborne transmissions of pathogenic *E. coli* predominantly occur in recreational waters and contaminated drinking water. Conventional testing for *E. coli* or thermo tolerant coliform bacteria provides an indication for the entry of pathogenic serotypes in water.
- ***Helicobacter pylori.*** *Helicobacter pylori* is a Gram-negative, spiral-shaped, microaerophilic, motile bacterium. It is found in the stomach and can cause complications like peptic and duodenal ulcer disease as well as gastric cancer. The infections are more predominant in the developing countries. Consumption of contaminated water is the main source of infection. *Escherichia coli* or, alternatively, thermotolerant coliforms is not a reliable index of this organism.
- ***Klebsiella.*** *Klebsiella* species are Gram-negative, nonmotile bacilli, which can cause serious infections like destructive pneumonia. They can multiply in water distribution systems. *Klebsiella* species are excreted in the faeces of healthy humans and animals.
- ***Legionella.*** Legionellae are Gram-negative, non-spore-forming, rod-shaped bacteria found in a wide range of water environments and are potentially pathogenic for humans. *E. coli* or thermo-tolerant coliforms are not suitable indices for this organism.
- ***Mycobacterium.*** 'Typical' species of *Mycobacterium* are not transmitted by water. 'Atypical' species of *Mycobacterium* are inhabitants of water environments. Atypical *Mycobacterium* spp. can cause diseases involving the skeleton, lymph nodes, skin, soft tissues, respiratory, gastrointestinal and genitourinary tracts. *E. coli* or thermotolerant coliforms are not suitable indices for this organism.
- ***Pseudomonas aeruginosa.*** *Pseudomonas aeruginosa* is aerobic, polarly flagellated, Gram-negative rod. *Pseudomonas aeruginosa* is present in water and can cause destructive lesions or septicaemia as well as meningitis.
- ***Salmonella.*** *Salmonella* species are motile, gram-negative bacilli that typically cause gastroenteritis, bacteraemia orsepticaemia, typhoid fever/enteric fever. They infect humans and animals. *Escherichia coli* or, alternatively, thermotolerant coliforms are a reliable index for *Salmonella* species.
- ***Shigella.*** *Shigella* species are Gram-negative, nonmotile, non-spore-forming, rodlike bacteria. *Shigella* spp. can cause serious intestinal diseases that include bacillary dysentery. More than 2 million infections occur every year, resulting in nearly 600 000 deaths.
- ***Staphylococcus aureus.*** *Staphylococcus aureus* is a nonmotile, catalase- and coagulase-positive, non-spore-forming, Gram-positive coccus. Multiplication of bacteria in tissues can result in boils, endocarditis, enteric infections, septicaemia, skin sepsis, postoperative wound infections, osteomyelitis and pneumonia. Gastrointestinal disease is characterized by projectile vomiting, fever, abdominal cramps, diarrhoea, electrolyte imbalance and loss of fluids.
- ***Tsukamurella.*** *Tsukamurella* species are gram-positive, nonmotile, obligate aerobic, weakly or variably acid-fast, irregular rod-shaped bacteria. Infections with *Tsukamurella* species have been linked with immune suppression, chronic lung diseases, and post-operative wound infections. *Tsukamurella* species exist in soil, water and the foam of activated sludge.

- ***Vibrio.*** *Vibrio* species are small, curved, Gram-negative bacteria having a single polar flagellum. Nontoxigenic *V. cholera* is extensively distributed in water environments compared to toxigenic strains. The presence of *V. cholerae* O1 as well as O139 serotypes in drinking water is of major public health importance as it results in cholera.
- ***Yersinia.*** *Yersinia* species are Gram-negative bacilli that are not motile at 37 °C but motile at 25 °C. The genus *Yersinia* comprises seven species, of which *Y. pestis*, *Y. pseudotuberculosis* and certain serotypes of *Y. Enterocolitica* are pathogens to humans. *Yersinia pestis* causes bubonic plague through contact with rodents as well as their fleas. *E. coli* orthermotolerant coliforms are not suitable indices for these organisms.

3.4.4.1.2 Viral Pathogens

Viruses can cause a variety of infections. Except for hepatitis E, humans are the only source of human infectious virus. Enteric viruses normally cause acute diseases with a short incubation period.

- **Denoviruses**. The family Adenoviridae is divided into the two genera, *Aviadenovirus* (avian hosts) and *Mastadenovirus* (mammal hosts). Adenoviruses are widespread in the environment, infecting birds, mammals and amphibians. Human adenoviruses (HAds) cause a wide range of infections, which include gastroenteritis, acute respiratory diseases, pneumonia, pharyngoconjunctival fevercervicitis, urethritis, haemorrhagic cystitis, epidemic keratoconjunctivitis and pharyngoconjunctival fever. Adenoviruses are excreted in human faeces and occur in sewage, raw water sources as well as treated drinking-water supplies. *E. coli* orthermotolerant coliforms is not a reliable index of HAds in drinking-water supplies.
- **Astroviruses**. Eight serotypes of human astroviruses (HAstVs) have been identified. HAstVs cause gastroenteritis. *E. coli* or thermotolerant coliforms are not a reliable index of stVs in drinking-water supplies.
- **Caliciviruses**. The family Caliciviridae comprises four genera. Human caliciviruses (HuCVs) include the genera *Norovirus* and *Sapovirus*. HuCVs are cause of acute viral gastroenteritis and symptoms include nausea, vomiting and abdominal cramps. Since infected individuals excrete HuCVs in faeces, domestic wastewater and faecally contaminated food/water would have these viruses.
- **Enteroviruses**. The genus *Enterovirus* consists of 69 species that infect humans. Enteroviruses causes myocarditis, herpangina, hand-foot-and-mouth disease, meningoencephalitis, poliomyelitis, and neonatal multiorgan failure. *E. coli* or, alternatively, thermotolerant coliforms is not a reliable index of enteroviruses in drinking-water supplies.
- **Hepatitis A virus.** Hepatitis A virus (HAV) is highly infectious and causes the disease hepatitis A. *E. coli* or, alternatively, thermotolerant coliforms are not a reliable index of HAV.
- **Hepatitis E virus.** Hepatitis E virus (HEV) causes hepatitis similar to that caused by HAV. E. coli or thermotolerant coliforms are not a reliable index of HEV in water.
- **Rotaviruses and orthoreoviruses.** Nearly 50–60% of acute gastroenteritis amongst hospitalized children throughout the world are caused by Human rotaviruses (HRVs) and they are the important cause of infant death. *Orthoreovirus* are typical ‘orphan viruses’

that are not associated with meaningful disease. *E. coli,* or thermotolerant coliforms, are not a reliable index of HRVs in water.

3.4.4.1.3 Protozoan Pathogens

Protozoa and helminthes are amongst the common causes of infection and disease in humans and animals. Controlling transmission is a real challenge as most of the pathogens generate oocysts, cysts, or eggs that are resistant to disinfectants. Some of the protozoa are responsible for 'emerging diseases'. A notable examples of such a disease is cryptosporidiosis.

- ***Acanthamoeba.*** *Acanthamoeba* species are free-living amoebae commonly found in water environments and soil, amongst which *A. castellanii*, *A. polyphaga* and *A. culbertsoni* are human pathogens.
- ***Balantidium coli.*** *Balantidium coli* are protozoan parasites and are known to infect humans. *E. coli* thermotolerant coliforms are not a reliable index for *B. coli in water.*
- ***Cryptosporidium.*** *Cryptosporidium* is a parasite that is responsible for most human infections. It generally causes self-limiting diarrhoea. Oocysts occur in recreational waters. *E. coli* or thermotolerant coliforms cannot be used as an index for the *Cryptosporidium* oocysts in water.
- ***Cyclospora cayetanensis.*** *Cyclospora cayetanensis* is a parasite. Oocysts, when ingested, penetrate epithelial cells in the small intestine and cause cyclosporiasis. *E. coli* or, alternatively, thermotolerant coliforms cannot be used as an index for *Cyclospora* in water.
- ***Entamoeba histolytica.*** *Entamoeba histolytica* is an intestinal protozoan pathogen that causes acute intestinal Amoebiasis. *Entamoeba histolytic* are present in sewage and contaminated water and cysts remain in suitable environments for several months. The possibility of waterborne transmission is more in the tropics compared to more temperate regions. *E. coli* or thermotolerant coliforms cannot be used as an index of the *E. histolytica.*
- ***Giardia intestinalis.*** *Giardia* spp. parasitize the gastrointestinal tracts of certain animals and humans and are responsible for giardiasis. *Giardia* can multiply in these animals and humans. These organisms are capable of forming cysts that are present in recreational water as well as contaminated food. *E. coli* or thermotolerant coliforms cannot be used as an index of *Giardia* in water.
- ***Isospora belli.*** Out of many species of *Isospora,* only *I. belli* is known to infect humans. It causes illness similar to that caused by *Giardia* and *Cryptosporidium.* Oocysts are resistant to disinfectants. Hence, *E. coli* or thermotolerant coliforms cannot be used as an index of *I. belli* in water supplies.
- ***Microsporidia.*** The term 'microsporidia' is a designation commonly used for a group of protozoa that belongs to the phylum Microspora. A number of genera are responsible for human infections, which include *Enterocytozoon, Nosema, Pleistophora, Encephalitozoon* (including *Septata*), *Vittaforma* and *Trachipleistophora.* Microsporidia are emerging human pathogens. Spores are excreted in faeces, urine and respiratory secretions.
- ***Naegleria fowleri.*** *Naegleria* are free-living amoeboflagellates. *Naegleria fowleri* causes amoebic meningoencephalitis (PAM). Patients usually die within 5–10 days. Treatment is difficult. *E. coli* or, alternatively, thermotolerant coliforms cannot be used as an index for the *N. fowleri* in water.

- ***Toxoplasma gondii.*** Of many species of *Toxoplasma* and *Toxoplasma*-like microbes only *T. gondii* is known to cause human infection. Usually humans contract toxoplasmosis by consumption of undercooked or raw meat containing *T. gondii* cysts. Contaminated drinking water could lead to toxoplasmosis outbreaks. The reliability of *E. coli* or, alternatively, thermotolerant coliforms as indicator organisms in water supplies is unknown.

3.4.4.1.4 Indicator and Index Organisms

Due to issues pertaining to cost, complexity and timeliness of obtaining results, testing for definite pathogens is usually limited to discovering whether a treatment is effective in eliminating target organisms. However, microbial testing is usually limited to indicator organisms as an index of faecal contamination or to measure the efficiency of control measures.

The use of indicator organisms as a hint of faecal pollution is established practice. The criteria determined for such organisms are (i) they should not be pathogens themselves, (ii) they should be universally present in human/animal faeces, (iii) they should not multiply in natural waters, (iv) they should be preserved in water in a similar way to faecal pathogens, (v) they should be present in larger numbers than faecal pathogens, (vi) they should respond to treatment processes in a similar way to faecal pathogens and (vii) they should be easily detected by simple and inexpensive methods.

It has become evident that no one indicator fulfils all of the above criteria. The shortcomings of traditional indicators like *E. coli* has led to examination of other organisms.

- ***Total coliform bacteria.*** These bacteria include a wide variety of aerobic and facultatively Gram-negative, anaerobic, non-spore-forming bacilli. *E. coli* and thermotolerant coliforms are a subgroup of the total coliform group capable of fermenting lactose at higher temperatures. Conventionally, coliform bacteria were considered to fit into the genera *Escherichia*, *Citrobacter*, *Klebsiella* and *Enterobacter*, but the group is heterogeneous and includes a range of genera, like *Serratia* and *Hafnia*. The total coliform includes both faecal as well as environmental species. Total coliforms include microbes that can survive and grow in water. They are therefore not used as an index of faecal pathogens. However they can be used as indicator of treatment effectiveness/cleanliness/integrity of distribution systems as well as the potential presence of biofilms. Total coliform bacteria excluding *E. coli* occur in sewage as well as natural waters. Some of these total coliform bacteria are excreted in human and animal faeces but many coliforms are able to multiply in water as well as soil environments. The existence of total coliforms in stored water supplies and distribution systems indicates contamination through entrance of foreign material or biofilm formation. The presence of total coliforms after disinfection indicates inadequate treatment.
- ***Escherichia coli and thermotolerant coliform bacteria.*** *E. coli* is recognized as the most suitable index with respect to faecal contamination as it provides evidence of the latest faecal contamination. Its detection should lead to deliberation of further action, like additional sampling and investigation. Total coliform bacteria capable of fermenting lactose at 44–45 °C are called thermotolerant coliforms. The predominant genus normally present is *Escherichia*. Other thermotolerant coliform include *Citrobacter*, *Klebsiella* and *Enterobacter*. *E. coli* is present in large numbers in human and animal faeces and is rarely

found in the water bodies in the absence of faecal pollution, although it may grow in tropical soils.

- ***Heterotrophic plate counts.*** The heterotrophic plate counts (HPCs) detect a variety of heterotrophic micro-organisms including bacteria and fungi that are sensitive to disinfection. Hence, the test is useful for monitoring treatment/disinfectant efficiency. Heterotrophic plate counts can include 'opportunistic' pathogens like *Acinetobacter, Flavobacterium, Klebsiella, Moraxella, Aeromonas, Serratia, Pseudomonas* and *Xanthomonas*.
- ***Intestinal enterococci.*** Intestinal enterococci are group within larger group of microbes called faecal streptococci. Intestinal enterococci consist of the species *Enterococcus faecal is, E. durans, E. faeciumas* well as *E. hirae.* The presence of intestinal enterococci indicates recent faecal contamination.
- ***Clostridium perfringens.*** *Clostridium* species produce pores that are extremely resistant to unfavourable conditions in water. Hence, they have been proposed as an index of enteric viruses and protozoa.
- ***Coliphages.*** Bacteriophages are viruses that infect and replicate in bacteria. Coliphages use *E. coli* and other related species as hosts. Hence, there is high possibility of release from faeces of humans as well as other warm-blooded animals.
- ***Bacteroides fragilis phages.*** The bacterial genus *Bacteroides* is present in the human gastrointestinal tract in numbers greater than *E. coli. Bacteroides* are inactivated by environmental oxygen but *Bacteroides* bacteriophages are resistant to this. Hence, *Bacteroides* bacteriophages have been suggested as a possible index of faecal contamination.
- ***Enteric viruses.*** Well known viruses that infect the human gastrointestinal tract, conveyed by the faecal-oral route, include the entero viruses, orthoreo viruses, rotaviruses, astro viruses, entericadeno viruses, calici viruses and hepatitis A and E viruses. The survival of faecal bacteria during treatment differs noticeably from that of enteric viruses. Hence, water should be monitored for contamination by enteric viruses. Enteric viruses are excreted by worldwide by humans in huge numbers. However, the presence of different viruses varies due to variations in rates of infection as well as excretion.

3.4.4.1.5 Multi-Cellular Organisms

Multicellular organisms are living beings with multiple cells. A group of cells that performs a special function in a multicelluar organism is known as an organ.

- **Helminth pathogens.** The word 'helminth' refers to all types of worms. The key parasitic worms are classified mainly in the phylum Nematoda and the phylum Platyhelminthes. Helminth parasites infect people and animals worldwide.
- ***Dracunculus medinensis.*** *Dracunculus medinensis* (the guinea worm) inhabits the cutaneous and subcutaneous tissues of infected people. Its presence is restricted to a central belt of sub-Saharan Africa. The female can reach a length of up to 700 mm, whereas the male is 25 mm. The number of infected people all over the world came down from 3.3 million cases in 1986 to 60 000 cases in 2002 due to an eradication programme. The female guinea worm releases large numbers of larvae (when it is ready to do so) when the affected body part is immersed in water. The larvae can move in water for about 3 days. Symptoms of infection include vomiting, pruritus and giddiness. The only route

of acquaintance is the consumption water containing *Cyclops* species carrying infectious guinea worm larvae.

- ***Fasciola* species.** Fascioliasis is caused by *F. hepatica*, and *F. gigantica.* Eggs of these species are excreted by the host human/animal. The parasites live in the large biliary passages as well as the gall-bladder. Human cases have been on rise in 51 countries of five continents affecting around more than 2.4 to 17 million people. Humans can contract fascioliasis by eating contaminated raw plants and drinking contaminated water.

3.5 Standards and Consents

Different water use requires different standards. Water of higher quality will be required for certain industries like pharmaceuticals and electronic equipment. Water for fisheries, floor washing, agriculture, recreation, water sports requires different water quality. In reality it is very difficult to meet and maintain these standards as it requires considerable resources, which include manpower and money.

Standards are set by an international/national/local standard-setting agency or by a regulatory agency, based on guidance from experts in the field. Figure 3.32 shows the evolution of standards. The guidance is based on existing literature.

Consents are agreements between two people/organizations. Not all industries or projects can be issued with permits with the same set of standards and conditions. Hence, based on the information provided by the project or industry, regulatory authorities issue consent. Even though the word 'consent' means 'agreement', in many parts of world regulatory authorities issue consent without obtaining the countersignature of stakeholder to whom it is issued. The stakeholder may approach the issuing authority for a change in condition. If the issuing authority does not agree, the legislation would usually specify an authority to which affected party can appeal before approaching a court of law.

The discharge standards are maximum permissible limits set by legislation. But laws across the world usually allow the enforcing authority to make standards more or less stringent from case to case. For example, even if the maximum permissible limit for BOD for on land discharge is 100 mg/l as per statute, a regulator may insist that an industry adopt 30 mg/l.

3.5.1 Potable Water Standards

Safe drinking water should not pose a health risk. The greatest risk of waterborne disease occurs to infants, young children, the elderly, the sick and people living in unsanitary condition. Safe drinking water should be suitable for all domestic purposes.

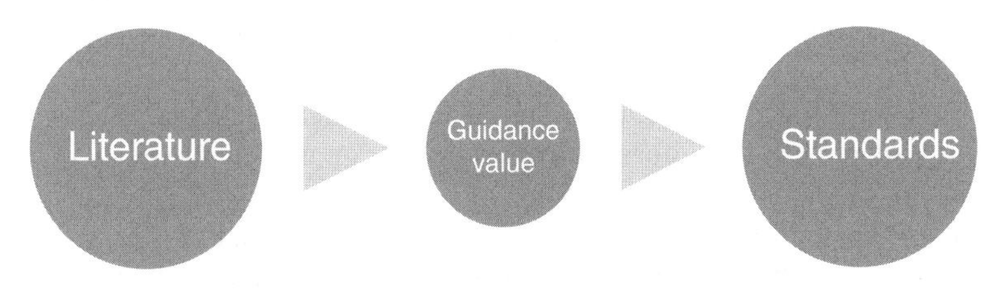

Figure 3.32 Evolution of standards.

International standards are not promoted by international organizations. In the process of developing standards and regulations, nations should ensure that scarce resources are not diverted to develop and monitor standards for substances of minor importance to public health (WHO, 2006).

Common aspects that need to be considered while fixing standards are: (i) microbial aspects, (ii) chemical aspects, (iii) radiological aspects and (iv) acceptability aspects:

1. ***Microbial aspects.*** The process of securing microbial safety with respect to drinking water supplies from catchment to consumer should consider preventing contamination of water resources, proper treatment and distribution systems. In the absence of any formal treatment/distribution infrastructure, safety from microbes is achieved by boiling water to disinfect water.
2. ***Chemical aspects.*** There are few chemicals that can cause health problems due to a single exposure. Treating and monitoring these chemicals is more expensive. Hence, sources of such chemicals should be identified and their entry should be stopped. Other chemicals that cause health problem due to chronic exposure should be brought down to safe limits. Bringing down the chemical concentration to safe limits requires treatment and monitoring, which is usually absent in the developing world.
3. ***Radiological aspects.*** Although the contamination of water due to radio nuclides from natural and anthropogenic sources is very small, it should not be neglected. Values have not been set with respect to individual radio nuclides in drinking water. The approach used is based on gross alpha and gross beta radiation activity. The absence of sophisticated labs and personnel to analyse water samples for radioactive levels across the globe has led to its significance only in literature and research rather than practice. Further stringent information confidentiality is maintained by governments across the globe with respect to radioactive substances.
4. ***Acceptability aspects.*** Water should be devoid of tastes and odours that are objectionable to consumers. Hence, many standards across the world stipulate that drinking water should not have an objectionable odour and taste rather than stipulating threshold limit values.

3.5.2 Wastewater Effluent Standards

Many countries have national standards whereas some countries have standards at state/regional/provincial level.

Technological standards specify that the treatment should include certain technologies or processes. Technological standards are fixed based on (i) best conventional pollutant control technology (BCT), (ii) best practicable control technology currently available (BPT), (iii) best available technology economically achievable (BAT) (iv) new source performance standards (NSPS).

Effluent standards specify physical, biological and chemical quality of water/wastewater. This type of standard can be uniform effluent standards or ambient or stream quality standards. Uniform standards stipulate standards of pollutants across the entire country/state/province/area. This uniform effluent standards approach may lead to excessive treatment in some cases and insufficient treatment in others. Ambient or stream quality standards can lead to stipulation of unfair standards due to corruption or political pressure.

Once the desired stream quality is known, environmental scientists can then 'work backwards' to determine the maximum concentrations of each pollutant to be allowed in each wastewater discharge; these concentrations become the standards for that discharge. Under this arrangement, different discharges in different environmental contexts must meet different effluent standards.

Ambient standards have the advantage that they ensure that resources are efficiently allocated to address local environmental conditions. The major disadvantages are that the approach requires considerable planning and environmental science. Soundly developed and enforced ambient standards offer a far more efficient and cost-effective approach to improving the environment.

Where rigorous water quality modelling is not possible or practical, the establishment of watershed or river basin management boards can still help to prioritize investment in treatment in a systematic fashion. It makes little sense to invest large amounts of money in wastewater treatment for a single city if pollution from other cities or from agricultural runoff will nullify any beneficial impact. It will usually make sense to concentrate efforts where they can make a difference.

The comparative advantage of a watershed management board is not necessarily its technical capacity but its perspective; the board is responsible for looking at the appropriate use of water across the watershed. Such boards often serve as a forum where political and economic conflicts over water quality and treatment can be analysed, debated and resolved.

3.6 Kinetics of Biochemical Oxygen Demand

The kinetics of the BOD reaction are, for practical purposes, formulated in accordance with first-order reaction kinetics and may be expressed as

$$\frac{dL_t}{dt} = -KL_t$$

where, L_t is the amount of the first-stage BOD remaining in the water at time t and k is the reaction rate constant. This equation can be integrated and the result will be

$$L_t = L(e^{-kt})$$

where L is the BOD remaining at time t = 0.

For polluted water and wastewater a typical value of k (base e, 20 °C) is 0.23 d^{-1}.

3.7 Water Management for Wildlife Conservation

Water is a critical factor in determining the distribution and abundance of wildlife. Impact varies by habitat, species and season. The loss of water resources can threaten wildlife. Wildlife was affected since the beginning of commercial grazing and agriculture. All life requires water for survival. Access to surface water affects the population dynamics of species depending on water. Species inhabiting arid environments are coupled with unpredictable availability of water. Many species of wildlife live in arid regions to improve habitat. In recent decades, degradation of water sources from anthropogenic factors has

Figure 3.33 *Wildlife and water.*

decreased surface water availability to wildlife. In South Africa, pollution on the Olifants River due to tourism has resulted in wildlife mortality (Oberholster, 2009).

Sustainability cannot happen in the absence of species diversity. Wildlife (Figure 3.33) does need quality water for survival and a healthy ecosystem. Any epidemics in the wild would not be localized – as observed in recent avian flu and swine flu epidemics. The Middle Rio Grande Basin of New Mexico is growing rapidly and holds more than 50% of the state's population. Dams and constructed river channels avoid spring floods and nearly 50% of the wetlands in the drainage vanished in just half a century. Invasion by non-native trees such as saltcedar and Russian-olive altered riparian forest composition (Jackson *et al.*, 2001).

The water ecosystems existed is wetlands, watershed, rivers and lakes. The main management methods are restoration, enhancement and construction of new water resources.

Wetland can be natural or manmade. Several activities can be conducted to protect natural wetland, restore wetland and constructed wetland. Basically, laws and regulations are set up to protect and support the conservation of the wetland. The protective measures are the creation of buffer zones – that is, vegetation areas along the wetland perimeter.

Watershed management is site specific; therefore, the management system is mainly the PDCA process (plan-do-check-act). The overall process involves defining the scope and setting goals (i.e. existing regulations, target contaminant concentration), gathering data and characterizing pollutants/pollutant sources, identifying management strategies and implementing and measuring progress (EPA, 2008).

Rivers flow through many areas so river management requires collaboration and cooperation from every stakeholder. In principle, each stakeholder has his or her own management plan, related to his or her responsibility. Common activities are to protect and improve the ecological condition of water, to comply with the regulations and promote the sustainable use of water resources (Environment Agency, 2009).

Figure 3.34 *Fishing in heart of the city in Stockholm.*

The problems of lakes are associated with their depth. The fundamental task of lake management is to consider the main factors: (i) light and temperature, (ii) nutrients and (iii) oxygen. There are only two ways that those main factors will cause problems: either there is an excess of them or they are lacking. General solutions are to minimize light and heat, reduce runoff to reduce nutrients and install aerators or fountains to increase levels of oxygen.

Studies by Worm *et al.* (2006) suggested that marine species currently fished will reduce by 100% in the year 2048 in a 'business-as-usual' scenario. Constant fishing (Figure 3.34) and unsustainable living practices may deplete fish species even before 2048 due to diminishing marine/fresh water quality, and decreasing fresh water resources. Similarly most of the species have reached endangered species status or are rapidly approaching endangered species status. Hence, sustainable water management should take conservation of wild life into account.

Most desert birds do not have mechanisms for water conservation within the body and, hence, water is critically important for resident birds (Krausman *et al.* 2006; Lynn *et al.* 2006; O'Brien *et al.*, 2006). The quality of water available for wildlife is an important management consideration (Simpson *et al.*, 2011) as poor water quality could affect the health of wildlife due to electrolyte imbalances, physiological distress, dehydration and being noxious or toxic (Broyles 1995). Wildlife professionals have used water developments since the 1940s to enhance wildlife habitats (Rosenstock *et al.* 1999, 2004; Bleich *et al.* 2005; O'Brien *et al.*, 2006). As animals are attracted to surface water, it acts as a 'predation trap' or 'predation sink' where animals are likely to be trapped by predators (Rosenstock *et al.* 1999, 2004; DeStefano *et al.* 2000).

Even the desert-adapted bat needs water periodically and loss of water source can threaten the local population. To get water, they fly down and scoop up a drink from the water surface and fly away from the pool. Obstacles in the flight path prove deadly and bats are vulnerable to drowning if trapped in water tanks. The size of the water feature and the swoop zone required varies according to the species' flight characteristics. Most common watering structures for wildlife can be divided into three types: *troughs and drinkers*; *storage tanks* and *open reservoirs* (Taylor and Tuttle, 2007).

3.8 Water-Quality Deterioration

Apart from drinking and domestic use, people come into direct contact with water for recreational use. The World Tourism Organization has predicted that 346 million tourists per annum will visit Mediterranean destinations by 2026. Water-quality issues may affect the people's health and the economy due to waterborne diseases. A decline in tourists to Lake Malawi in Southern Africa resulted from reports about schistosomiasis cases (WHO, 2003f).

Sewage discharged onto land without treatment (Figure 3.35) and into water bodies has been constantly threatening the environment and fresh water sources that act as sources of drinking water.

Industrial activities, mining as well as power production using fossil fuels, can cause acidification of freshwater systems locally. Acid rain, due to emissions from fossil fuel and atmospheric processes, can affect larger areas. Acidification affects young organisms (both flora and fauna), which are less tolerant to a low pH. Further, lower pH can mobilize metals in soils, resulting in additional stresses/fatalities amongst aquatic species. Acidification is widespread downwind of power plants, emitting huge quantities of oxides of nitrogen and sulfur. Downstream of mines, contaminated groundwater causes stress to the environment. Freshwater flora and fauna usually do not tolerate high salinity. Actions such as agricultural drainage from high-salt areas, groundwater discharge from oil/gas and other pumping operations, numerous industrial activities and some municipal water-treatment operations can change the salinity of soil (UNEP, 2010).

The increasing incidence of invasive species displacing endemic species and altering water quality and local food webs affects freshwater systems and needs to be considered a water-quality issue (Carr and Neary, 2008). Aquatic species have been introduced, in many cases deliberately, into far-away ecosystems for recreational/economic/other purposes

Figure 3.35 Sewage discharged on land without treatment.

(UNEP, 2010). In many situations, these introductions have devastated endemic aquatic organisms.

In the trash water of commercial boats, or on the exteriors of recreational water craft, for example, invasive species like zebra mussels (*Dreissena polymorpha*) and quagga mussels (*D. bugensis*) have distressed local ecosystems, changing nutrient cycles and stressing endemic species to the edge of extinction (UNEP, 2010). Mussels also threaten infrastructure, clogging pumps/intakes and choking canals, resulting in continuous costly maintenance challenges. Invasive plant species in South Africa have altered water quality, decreased water quantity and enhanced evapotranspiration rates in watersheds. Invasive alien species in South Africa are harming the country's economy and are the biggest threat to the nation's biodiversity. Since 1995, more than a million hectares of invasive alien flora have been cleared. In the United States, the invasion of certain species of mussel has affected local ecosystems and caused additional costs to the water power industry, exceeding one billion dollars per year.

References

ATSDR (1989) *Toxicological Profile for Alpha-, Beta-, Gamma- and Delta-Hexachlorocyclohexane*, US Department of Health and Human Services, Public Health Service, Agency for Toxic Substances and Disease Registry, Atlanta, GA.

ATSDR (2002a) *Toxicological Profile for Copper (Draft for Public Comment)*, US Department of Health and Human Services, Public Health Service, Agency for Toxic Substances and Disease Registry (Subcontract No. ATSDR-205-1999-00024), Atlanta, GA.

ATSDR (2000b) *Toxicological Profile for Manganese*, United States Department of Health and Human Services, Public Health Service, Agency for Toxic Substances and Disease Registry, Atlanta, GA.

AWWA (1995) *Cyanobacterial (Blue-Green Algal) Toxins: A Resource Guide*, AWWA Research Foundation and American WaterWorks Association, Denver, CO.

Ballantyne, B. (1983) Acute systemic toxicity of cyanides by topical application to the eye. *Journal of Toxicology – Cutaneous and Ocular Toxicology* **2**, 119–129.

Beasley, V.R., Cook, W.O., Dahlem, A.M. *et al.* (1989) Algae intoxication in livestock and waterfowl. *Veterinary Clinics of North America, Food Animal Practice* **5**, 345–361.

Berkamp, G., McCartney, M., Dugan, P. *et al.* (2000) *Dams, Ecosystem Functions and Environmental Restoration.* http://intranet.iucn.org/webfiles/doc/archive/2001/IUCN913.pdf (accessed 1 January 2014).

Bleich, V.C., Kie, J.G., Loft, E.R., *et al.* (2005) Managing rangelands for wildlife, in *Techniques for Wildlife Investigations and Management*, 6th edn (ed. C.E. Braun), The Wildlife Society, Bethesda, MD, pp. 873–897.

Broyles, B. (1995) Desert wildlife water developments: questioning use in the southwest. *Wildlife Society Bulletin* **23**, 663–675.

Carmichael, W.W. (1992) A review: cyanobacteria secondary metabolites – the cyanotoxins. *Journal of Applied Bacteriology* **72**, 445–459.

Carr, G.M. and Neary, J.P. (2008) *WaterQuality for Ecosystemand Human Health*, 2nd edn., United Nations Environment Programme Global Environment Monitoring System, Burlington, Ontario.

CAS (2012) CAS Registry and CAS Registry Numbers. http://www.cas.org/expertise/cascontent/registry/regsys.html (accessed 21 December 2013).

Chandrappa, R. and Raju, N.R. (2008) *Policy and Its Implementations to Abate Impact of Urbanization on Bangalore Lakes, Conservation and Management*, ICFAI University Press, Lakes.

Chandrappa, R. and Ravi, D.R. (2009) *Environmental Issues, Law and Technology – An Indian Perspective*, Research India Publications, Delhi.

Corcoran, E., Nellemann, C., Baker, E. *et al.* (eds) (2010) *Sick Water? The Central Role of Wastewater Management in Sustainable Development*, A Rapid Response Assessment. United Nations Environment Programme, UN-HABITAT, Arendal.

Crecelius, E.A., Bothner, M.H. and Carpenter, R. (1975) Geochemistries of arsenic, antimony, mercury, and related elements in sediments of Puget Sound. *Environmental Science and Technology* **9**, 325–333.

Cui, C.G. and Liu, Z.H. (1988) Chemical speciation and distribution of arsenic in water, suspended solids and sediment of Xiangjiang River, China. *The Science of the Total Environment* **77**, 69–82.

Daughton, C.G. (2004) Non-regulated water contaminants: emerging research. *Environmental Impact Assessment Review* **24**, 711–732. doi: 10.1016/j.eiar.2004.06.003

Department of Health and Ageing, Government of Australia (2004) Guidance on Use of Rainwater Tanks, Department of Health and Ageing, Canberra.

DeStefano, S., Schmidt, S.L. and deVos, J.D. Jr. (2000) Observations of predator activity at wildlife water developments in southern Arizona. *Journal of Range Management* **53**, 255–258.

Diem, K. and Lentner, C. (eds) (1970) *Documenta Geigy. Scientific Tables*, 7th edn., Ciba-Geigy, Basel, p. 688.

DNHW (Department of National Health and Welfare, Canada) (1990) Nutrition Recommendations. The Report of the Scientific Review Committee, DNHW, Ottawa.

DNHW (Department of National Health and Welfare, Canada) (1992) Guidelines for Canadian Drinking Water Quality. Supporting Documentation, DNHW, Ottawa.

DoE (Department of the Environment) (1989) Pesticides in Water Supplies, Reference: WS/45/1/1, DoE, London.

D'Souza, R. (2006) Impact of Privatisation of lakes in Bangalore, Final report for Fellowship, Centre for Education and Documentation, Bangalore.

EHD (2011) Understanding Septic Tank Systems, Environmental Health Directorate, Perth, Australia.

Elinder, C.G. (1986) Iron, in *Handbook on the Toxicology of Metals*, Vol. **II** (eds L. Friberg, G.F. Nordberg and V.B. Vouk), Elsevier, Amsterdam, pp. 276–297.

Ellenhorn, M.J. *et al.* (1997) *Ellenhorn's Medical Toxicology: Diagnosis and Treatment of Human Poisoning*, 2nd edn. Williams and Wilkins, Baltimore, MD.

Elton, N.W., Elton, W.J. and Narzareno, J.P. (1963) Pathology of acute salt poisoning in infants. *American Journal of Clinical Pathology* **39**, 252–264.

Environment Agency (2009) *River Basin Management Plan, Anglian River Basin District*, http://a0768b4a8a31e106d8b0-50dc802554eb38a24458b98ff72d550b.r19.cf3.rackcdn.com/gean0910bspm-e-e.pdf (accessed 21 December 2013).

EPA (2008) *Handbook for Developing Watershed Plans to Restore and Protect Our Waters* (EPA 841-B-08-002), http://water.epa.gov/polwaste/nps/upload/2008_04_18_NPS_watershed_handbook_handbook.pdf (accessed 21 December 2013).

FAO/WHO (2011a) Evaluation of Certain Contaminants in Food. Seventy-second Report of the Joint FAO/WHO Expert Committee on Food Additives, Technical Report Series, No. 959), WHO, Geneva.

FAO/WHO (2011b) Safety Evaluation of Certain Contaminants in Food, World Health Organization, Geneva.

Friberg, L., Nordberg, G.F. and Vouk, V.B. (eds) (1986) *Handbook of the Toxicology of Metals*, Vol. **II**, Elsevier, Amsterdam, pp. 130–184.

Goldberg, A.A., Shapiro, M. and Wilder, E. (1950) Antibacterial colloidal electrolytes: the potentiation of the activities of mercuric-, phenylmercuric- and silver ions by a colloidal sulphonic anion. *Journal of Pharmacy and Pharmacology* **2**, 20–26.

Gosselin, R.E., Smith, R.P. and Hodge, H.C. (1984) *Clinical Toxicology of Commercial Products*, 5th edn., Williams & Wilkins, Baltimore, MD.

Hawley, G.G. (1981) *The Condensed Chemical Dictionary*, 10th edn., Van Nostrand Reinhold, New York, NY, p. 241.

Hem, J.D. (1989) *Study and Interpretation of the Chemical Characteristics of Natural Water.* Water Supply Paper 2254, 3rd edn. US Geological Survey, Washington, D.C., p. 263.

Hemming, J., Holmbom, B., Reunanen, M. and Kronberg, L. (1986) Determination of the strong mutagen 3-chloro-4-(dichloromethyl)-5-hydroxy-2(5H)-furanone in chlorinated drinking and humic waters. *Chemosphere* **15**, 549–556.

Hill, W.R. and Pillsbury, D.M. (1939) *Argyria, the Pharmacology of Silver*, Williams & Wilkins, Baltimore, MD.

Howard, P.H. (1990) *Handbook of Environmental Fate and Exposure Data for Organic Chemicals*, vol. 1, Lewis Publishers Inc., Chelsea, MI.

IARC (International Agency for Research on Cancer) (1987) Overall Evaluations of Carcinogenicity: An Updating of IARC Monographs Volumes 1–42, IARC, Lyon.

IARC (1991) Occupational Exposure in Insecticide Application, and Some Pesticides, International Agency for Research on Cancer, IARC, Lyon, pp. 329–349

IPCS (1981) Arsenic, World Health Organization, International Programme on Chemical Safety, Geneva.

IPCS (1984) Epichlorohydrin, World Health Organization, International Programme on Chemical Safety, Geneva.

IPCS (1985) Acrylamide, World Health Organization, International Programme on Chemical Safety, Geneva.

IPCS (1991a) Nickel, World Health Organization, International Programme on Chemical Safety, Geneva.

IPCS (1991b) Lindane, World Health Organization, International Programme on Chemical Safety, Geneva.

IPCS (2000) Disinfectants and Disinfectant By-Products, World Health Organization, International Programme on Chemical Safety, Geneva.

IPCS (2001) Arsenic and Arsenic Compounds, World Health Organization, International Programme on Chemical Safety, Geneva.

IPCS (2002) Principles and Methods for the Assessment of Risk from Essential Trace Elements, World Health Organization, International Programme on Chemical Safety, Geneva.

Irgolic, K.J. (1982) *Speciation of Arsenic Compounds in Water Supplies*, United States Environmental Protection Agency, Research Triangle Park, NC.

Jackson, R.B., Carpenter, S.R., Dahm, C.N. *et al.* (2001) Water in a changing world. *Issues in Ecology* **11**(4), 1027–1045.

Johnson, R.P. and Mellors, J.W. (1988) Arteriolization of venous blood gases: a clue to the diagnosis of cyanide poisoning. *Journal of Emergency Medicine* **6**, 401–404.

Jorens, P.G. and Schepens, P.J. (1993) Human pentachlorophenol poisoning. *Human and Experimental Toxicology* **12**, 479–495.

Knepper, W.A. (1981) Iron, in *Kirk-Othmer Encyclopedia of Chemical Technology*, vol. **13**, Wiley Interscience, New York, NY, pp. 735–753.

Komulainen, H., Vaittinen, S-L., Vartiainen, T. and Tuomisto, J. (1997) Carcinogenicity of the drinking water mutagen 3-chloro-4-(dichloromethyl)-5-hydroxy-2(5H)-furanone (MX). *Journal of the National Cancer Institute* **89**, 848–856.

KSPCB (2011) *Karnataka State Environmental Atlas 2011*, Karnataka State Pollution Control Board, Bangalore.

Krajnc, E.I. *et al.* (1987) *Integrated Criteria Document. Cadmium — Effects. Appendix.* National Institute of Public Health and Environmental Protection (Report No. 758476004), Bilthoven.

Krausman, P.R., Rosenstock, S.S. and Cain, J.W. (2006) Developed waters for wildlife: science, perception, values, and controversy. *Wildlife Society Bulletin* **34**, 563–569.

Luthi, C. and Ingle, R. (2012) *Sustainable Cities*, Green Media Ltd, London.

Lynn, J.C., Chambers, C.L. and Rosenstock, S.S. (2006) Use of wildlife water developments by birds in southwestern Arizona during migration. *Wildlife Society Bulletin* **34**, 592–601.

Ma, J., Li, G. and Graham, N.J.D. (1994) Efficiency and mechanism of acrylamide removal by potassium permanganate. *Aqua*, **43**(6), 287–295.

Mallevialle, J., Bruchet, A. and Fiessinger, F. (1984) How safe are organic polymers in water treatment. *Journal of the American WaterWorks Association* **76**(8), 87–93.

Malmqvist, P.-A. (2000) Sustainable storm water management -some Swedish experiences. *Journal of Environmental Science and Health, Part A* **35**(8), 1251–1266.

Meier, J.R., DeAngelo, A.B., Daniel, F.B. *et al.* (1989) Genotoxic and carcinogenic properties of chlorinated furanones – important byproducts of water chlorination, in *Genetic Toxicology of Complex Mixtures: Short Term Bioassays in the Analysis of Complex Environmental Mixtures* (ed. M.D. Waters, D.F. Bernard, J. Lewtas *et al.*), Plenum Press, New York, NY, pp. 185–195.

Murray, J.J. (ed.) (1986) *Appropriate Use of Fluorides for Human Health*, World Health Organization, Geneva.

NRC (National Research Council) (1979) *Iron*. University Park Press, Baltimore, MD.

NRC (National Research Council) (1989) *Recommended Dietary Allowances*, 10th edn., National Academy Press, Washington, DC.

Nriagu, J.O. and Pacyna, J.M. (1988) Quantitative assessment of worldwide contamination of air, water and soils by trace metals. *Nature* **333**, 134–139.

Oberholster, P.J. (2009) Impact on ecotourism by waterpollution in the Olifants River catchment, South Olifants River catchment. *SIL News*, (55), 8–9. http://researchspace.csir.co.za/dspace/bitstream/10204/3841/3/oberholster4_2009.pdf (accessed 21 December 2013).

O'Brien, C.S., Waddell, R.B., Rosenstock, S.S. and Rabe, M.J. (2006) Wildlife use of water catchments in southwestern Arizona. *Wildlife Society Bulletin* **34**, 582–591.

Olcott, C.T. (1950) Experimental argyrosis. V, Hypertrophy of the left ventricle of the heart. *Archives of Pathology* **49**, 138–149.

Patwardhan, S.A. and Abhyankar, S.M. (1988) Toxic and hazardous gases. IV. *Colourage* **35**(12), 15–18.

Podsada, J. (2001) Lost Sea Cargo: beach bounty or junk? National Geographic News. http://news.nationalgeographic.com/news/2001/06/0619_seacargo.html (accessed 21 December 2013).

Rangachari, R., Sengupta, N., Iyer, R.R. *et al.* (2000) Large Dams: India's Experience, Final report prepared for the World Commission on Dams, Cape Town.

Ray, C. and Jain, R. (2011) Drinking water treatment technology – comparative analysis, in *Drinking Water Treatment, Strategies for Sustainability* (eds C. Ray and R. Jain), Springer, Heidelberg. doi:10.1007/978-94-007-1104-4 2

Rodríguez, R.S. (2011) Eco-Efficient and Sustainable Urban Infrastructure Development in Asia and Latin America. Eco-efficiency and Sustainable Infrastructure in the United States and Canada. Project document, Economic and Social Commission for Asia and the Pacific (ESCAP) and United Nations Economic Commission for Latin America and the Caribbean (ECLAC) in association with the United Nations Human Settlements Programme (UN-HABITAT), Santiago, Chile. www.eclac.cl/dmaah/noticias/paginas/3/44973/Ecoefficient_and_sustainable_urban_infrastructura__USA_and_Canada.pdf (accessed 21 December 2013).

Ros, J.P.M. and Slooff, W. (eds) (1987) *Integrated Criteria Document. Cadmium*, National Institute of Public Health and Environmental Protection, Bilthoven (Report No. 758476004).

Rosenstock, S.S., Ballard, W.B. and deVos, J. C. Jr. (1999) Viewpoint: benefits and impacts of wildlife water development. *Journal of Range Management* **52**, 302–311.

Rosenstock, S.S., O'Brien, C.S., Waddell, R.B. and Rabe, M.J. (2004) Studies of wildlife water developments in southwestern Arizona: wildlife use, water quality, wildlife disease, wildlife mortalities and influences on native pollinators. Technical Guidance Bulletin No. 8. Arizona Game and Fish Department, Phoenix, USA.

Rungby, J. and Danscher, G. (1984) Hypoactivity in silver exposed mice. *Acta Pharmacologica et Toxicologica* **55**, 398–401.

Sax, N.I. and Lewis, R.J. (eds) (1987) *Hawley's Condensed Chemical Dictionary*, 11th edn., Van Nostrand Reinhold, New York, NY, pp. 1050–1051.

Schroeder, H.A., Tipton, I.H. and Nason, P. (1972) Trace metals in man: strontium and barium. *Journal of Chronic Diseases* **25**, 491–517.

Schultz, V.C. (1964) Fettleber und chronisch-asthmoide Bronchitis nach Inhalation eines Farbenlosungsmittels (Epichlorohydrin). (Fatty liver and chronic asthmoid bronchitis following inhalation of a paint solvent (epichlorohydrin).) *Deutsche Mediziniche Wochenschrift* **89**, 1342–1344.

Simpson, N.O., Stewart, K.M. and Bleich, V.C. (2011) What have we learned about water developments for wildlife? Not enough! *California Fish and Game* **97**(4), 190–209.

Stacpoole, P.W., Moore, G.W. and Kronauser, D.M. (1978) Metabolic effects of dichloroacetate in patients with diabetes mellitus and hyperlipoproteinemia. *New England Journal of Medicine* **298**, 526–530.

Stemmer, K.L. (1976) Pharmacology and toxicology of heavy metals: antimony. *Pharmacology and Therapeutics Part A* **1**, 157–160.

Stokinger, H.E. (1981a) The metals, in *Patty's Industrial Hygiene and Toxicology*, vol. **2A**, 3rd edn. (eds G.D. Clayton and F.E. Clayton) John Wiley & Sons, Inc., New York, NY, pp. 1493–2060.

Stokinger, H.E. (1981b) Boron, in *Patty's Industrial Hygiene and Toxicology. Vol. 2B. Toxicology*, 3rd edn. (eds G.D. Clayton and F.E. Clayton), John Wiley & Sons, Inc., New York, NY, pp. 2978–3005.

Stokinger, H.E. (1983) *Encyclopedia of Occupational Health and Safety*, 3rd rev. edn., vol. **2**, International Labour Organisation, Geneva, pp. 1403–1404.

Taylor, D.A.R. and Tuttle, M.D. (2007) *Water for Wild Life, A Hand Book For Ranchers and Range Managers*, Bat Conservation International, Austin, TX.

Toxnet (2013) *Dichloroacetic Acid*, http://toxnet.nlm.nih.gov/cgi-bin/sis/search/f?./temp/~c4RDxU:1 (accessed 28 December 2013)

UNEP (1982) *Rain and Storm Water Harvesting in Rural Areas*, Tycooly International Publishing Ltd, Dublin.

UNEP (2010) Clearing the Waters A focus on water quality solutions.

UNEP (2012) *Rainwater Harvesting and Utilisation, An Environmentally Sound Approach for Sustainable Urban Water Management: An Introductory Guide for Decision-Makers*. http://www.unep.or.jp/ietc/publications/urban/urbanenv-2/9.asp (accessed 21 December 2013.

USEPA (1980) *Ambient Water Quality Criteria for Silver* (EPA 440/5-80-071), US Environmental Protection Agency, Washington, DC.

USEPA (1985) *Health Advisory – Barium*, US Environmental Protection Agency, Office of Drinking Water, Washington, DC.

Way, J.L. (1984) Cyanide intoxication and its mechanism of antagonism. *Annual Review of Pharmacology and Toxicology* **24**, 451–481.

Weast, R.C. (ed.) (1986) *Handbook of Chemistry and Physics*, 67th edn., CRC Press, Cleveland, OH.

WHO (1987) *Air Quality Guidelines for Europe*, WHO Regional Office for Europe, Copenhagen.

WHO (1997) *Aluminium*, International Programme on Chemical Safety (Environmental Health Criteria 194), World Health Organization, Geneva.

WHO (2003a) Alachlor in Drinking Water. Background document for development of WHO Guidelines for Drinking Water Quality, WHO, Geneva.

WHO (2003b) Aldicarb in Drinking Water, Background document for development of WHO Guidelines for Drinking Water Quality, WHO, Geneva.

WHO (2003c) Ammonia in Drinking Water, Background document for development of WHO Guidelines for Drinking Water Quality, WHO, Geneva.

WHO (2003d) Asbestos in Drinking Water, Background document for development of WHO Guidelines for Drinking Water Quality, WHO, Geneva.

WHO (2003e) Benzene in Drinking Water, Background document for development of WHO Guidelines for Drinking Water Quality, WHO, Geneva.

WHO (2003f) *Guidelines for Safe Recreational Water Environments*, vol. 1, *Coastal and Freshwaters*, WHO, Geneva, Switzerland.

WHO (2003g) Chromium in Drinking Water, Background document for development of WHO Guidelines for Drinking Water Quality, WHO, Geneva.

WHO (2003h) Chlorotoluron in Drinking Water, Background document for development of WHO Guidelines for Drinking Water Quality, WHO, Geneva.
WHO (2003i) Dichlorobenzenes in Drinking Water, Background document for development of WHO Guidelines for Drinking Water Quality, WHO, Geneva.
WHO (2003j) Iodine in Drinking Water, Background document for development of WHO Guidelines for Drinking Water Quality, WHO, Geneva.
WHO (2003k) Isoproturon in Drinking Water, Background document for development of WHO Guidelines for Drinking Water Quality, WHO, Geneva.
WHO (2003l) Pendimethalin in Drinking Water, Background document for development of WHO Guidelines for Drinking Water Quality, WHO, Geneva.
WHO (2003m) Polynuclear Aromatic Hydrocarbons in Drinking Water, Background document for development of WHO Guidelines for Drinking Water Quality, WHO, Geneva.
WHO (2003n) Sodium in Drinking Water, Background document for development of WHO Guidelines for Drinking Water Quality, WHO, Geneva.
WHO (2004a) Barium in Drinking Water, Background document for development of WHO Guidelines for Drinking Water Quality, WHO, Geneva.
WHO (2004b) Bentazone in Drinking Water, Background document for development of WHO Guidelines for Drinking Water Quality, WHO, Geneva.
WHO (2004c) Carbofuran in Drinking Water, Background document for development of WHO Guidelines for Drinking Water Quality, WHO, Geneva.
WHO (2004d) Carbon Tetrachloride in Drinking Water, Background document for development of WHO Guidelines for Drinking Water Quality, WHO, Geneva.
WHO (2004e) Chlordane in Drinking Water, Background document for development of WHO Guidelines for Drinking Water Quality, WHO, Geneva.
WHO (2004f) Chlorpyrifos in Drinking Water, Background document for development of WHO Guidelines for Drinking Water Quality, WHO, Geneva.
WHO (2004g) Dimethoate in Drinking Water, Background document for development of WHO Guidelines for Drinking Water Quality, WHO, Geneva.
WHO (2004h) Epichlorohydrin in Drinking Water, Background document for development of WHO Guidelines for Drinking Water Quality, WHO, Geneva.
WHO (2004i) Endosulfan in Drinking Water, Background document for development of WHO Guidelines for Drinking Water Quality, WHO, Geneva.
WHO (2004j) Endrin in Drinking Water, Background document for development of WHO Guidelines for Drinking Water Quality, WHO, Geneva.
WHO (2004k) Heptachlor and Heptachlor Epoxide in Drinking Water, Background document for development of WHO Guidelines for Drinking Water Quality, WHO, Geneva.
WHO (2004l) Hexachlorobutadiene in Drinking Water, Background document for development of WHO Guidelines for Drinking Water Quality, WHO, Geneva.
WHO (2004m) Inorganic Tin in Drinking Water, Background document for development of WHO Guidelines for Drinking Water Quality, WHO, Geneva.
WHO (2004n) Malathion in Drinking Water, Background document for development of WHO Guidelines for Drinking Water Quality, WHO, Geneva.
WHO (2004o) Methoxychlor in Drinking Water, Background document for development of WHO Guidelines for Drinking Water Quality, WHO, Geneva.

WHO (2004p) Methyl Parathion in Drinking Water, Background document for development of WHO Guidelines for Drinking Water Quality, WHO, Geneva.
WHO (2004q) Monochlorobenzene in Drinking Water, Background document for development of WHO Guidelines for Drinking Water Quality, WHO, Geneva.
WHO (2004r) Propanil in Drinking Water, Background document for development of WHO Guidelines for Drinking Water Quality, WHO, Geneva.
WHO (2005a) Bromate in Drinking Water, Background document for development of WHO Guidelines for Drinking Water Quality, WHO, Geneva.
WHO (2005b) Mercury in Drinking Water, Background document for development of WHO Guidelines for Drinking Water Quality, WHO, Geneva.
WHO (2005c) Methyl tertiary-Butyl Ether (MTBE) in Drinking Water, Background document for development of WHO Guidelines for Drinking Water Quality, WHO, Geneva.
WHO (2005d) Nickel in Drinking Water, Background document for development of WHO Guidelines for Drinking Water Quality, WHO, Geneva.
WHO (2006) Guidelines for Drinking Water Quality, First Addendum to Third Edition, Volume 1, Recommendations, WHO, Geneva.
WHO (2007a) Dicofol in Drinking Water, Background document for development of WHO Guidelines for Drinking Water Quality, WHO, Geneva.
WHO (2007b) Nitrate and Nitrite in Drinking Water, Background document for development of WHO Guidelines for Drinking Water Quality, WHO, Geneva.
WHO (2008a) Carbaryl in Drinking Water, Background document for development of WHO Guidelines for Drinking Water Quality, WHO, Geneva.
WHO (2008b) N-Nitrosodimethylamine in Drinking Water, Background document for development of WHO Guidelines for Drinking Water Quality, WHO, Geneva.
WHO (2009a) *Bacillus thuringiensis israelensis* (Bti) in Drinking Water, Background document for development of WHO Guidelines for Drinking Water Quality, WHO, Geneva.
WHO (2009b) Beryllium in Drinking Water, Background document for development of WHO Guidelines for Drinking Water Quality, WHO, Geneva.
WHO (2009c) Calcium and Magnesium in Drinking Water : Public Health Significance, WHO, Geneva.
WHO (2009d) Bromide in Drinking Water, Background document for development of WHO Guidelines for Drinking Water Quality, WHO, Geneva.
WHO (2009e) Cyanogen Chloride in Drinking Water, Background document for development of WHO Guidelines for Drinking Water Quality, WHO, Geneva.
WHO (2009f) Potassium in Drinking Water, Background document for development of WHO Guidelines for Drinking Water Quality, WHO, Geneva.
WHO (2010) Aluminium in Drinking Water, Background document for development of WHO Guidelines for Drinking Water Quality, WHO, Geneva.
WHO (2011a) Acrylamide in Drinking Water, Background document for development of WHO Guidelines for Drinking Water Quality, WHO, Geneva.
WHO (2011b) Atrazine and Its Metabolites in Drinking Water, Background document for development of WHO Guidelines for Drinking Water Quality, WHO, Geneva.
WHO (2011c) Cadmium in Drinking Water, Background document for development of Guidelines for Drinking Water Quality, WHO, Geneva.

WHO (2011d) Manganese in Drinking Water, Background document for development of WHO Guidelines for Drinking Water Quality, WHO, Geneva.
WHO (2011e) Hardness in Drinking Water, Background document for development of WHO Guidelines for Drinking Water Quality, WHO, Geneva.
WHO (2011f) Lead in Drinking Water, Background document for development of WHO Guidelines for Drinking Water Quality, WHO, Geneva.
WHO (2011g) Molybdenum in Drinking Water, Background document for development of WHO Guidelines for Drinking Water Quality, WHO, Geneva.
WHO (2011h) Permethrin in Drinking Water: Use for Vector Control in Drinking Water Sources and Containers, WHO, Geneva.
Worm, B., Barbier, E.B., Beaumont, N., *et al.* (2006) Impacts of biodiversityloss on ocean ecosystem services. *Science* **314**(1), 787–790. doi: 10.1126/science.1132294
Worthing, C.R. (1991) *The Pesticide Manual*, 9th edn., British Crop Protection Council, Farnham.

4

Fundamentals of Treatment and Process Design, and Sustainability

Every community produces both liquid and solid wastes. The liquid waste is essentially the water after it has been fouled due to the generation of a variety of wastes. If untreated wastewater is discharged into the environment this can lead to malodorous gases. Furthermore, untreated wastewater usually contains many pathogenic micro-organisms. Wastewater also consists of nutrients, which can enhance growth of aquatic plants. For these reasons, wastewater treatment and disposal is desirable and necessary in an industrialized society.

Wastewater engineering is that branch of engineering in which the fundamental principles of science and engineering are applied to control water pollution problems. The ultimate goal, namely, wastewater management, is to protect the environment.

A variety of technologies address the challenge of the UN target that aims to provide safe drinking water and hygienic sanitation to all people on the Earth by the year 2025 but these solutions do not provide an integral solution and lead to problems like eutrophication, water shortages, heavy metals in sludge, loss of fertility, scarcity of nutrients and biomagnification due to traces of medicine and chemicals in the water and disruption of the nitrogen/phosphorus cycle. Sustainable solutions should therefore balance the use of different resources, including their environmental, economical, and social-cultural aspects (van der Vleuten-Balkema, 2003).

Theoretically, it is possible to close the water cycle on a household, industry, institution and urban scale. When the water cycle includes a nutrient cycle (like carbon, potassium, calcium, nitrogen, phosphorus and magnesium), the nutrients end up in surface water or sludge or as gaseous emissions, in contrast to the bio-geo cycles in the natural ecosystems.

Conventional solutions like water-flush toilets, combined sewerage and centralized treatment dilute the waste stream, making it difficult to separate the toxic components from the valuable components (van der Vleuten-Balkema, 2003). This led to experiments with rainwater harvesting, sewerage systems, rainwater infiltration, rain/household water for

Sustainable Water Engineering: Theory and Practice, First Edition. Ramesha Chandrappa and Diganta B. Das.

toilet flushing, low flush/vacuum toilets, the separate treatment of black/grey/yellow water (urine), anaerobic digestion, and so on.

The end-of-pipe approach found in some centralized systems is inadequate and increasingly sophisticated treatment is required as laws become stricter, making treatment expensive. The monitoring and control of centralized treatment is easier than a large number of small treatment units. Decentralized systems can keep different wastewater streams separate, use various treatment techniques, close water/nutrient cycles locally and eliminate the need to transport wastewater over long distances.

While the expansion of the central sewerage systems has come to an end in the Netherlands (van der Vleuten-Balkema, 2003), it has gained momentum in developing countries like India where most of the urban population is not connected to sewers and hence many projects funded by international agencies are in the pipeline. The central treatment systems are growing in scale and complexity, providing business for solution providers and contributing to the GDP of the country.

Box 4.1 Christie Walk EcoCity Model

Adelaide's Christie Walk EcoCity Project in Australia is the first to have onsite sewage treatment and is able to provide treated wastewater for the irrigation of public parklands. Other features of the EcoCity are: (i) photovoltaic panels that deliver power to the grids and hot water to all dwellings, (ii) a roof garden, (iii) community garden produce and (iv) underground stormwater tanks. It was awarded a silver prize in the Ryutaro Hashimoto Awards (UN ESCAP, 2007).

4.1 History of Water and Wastewater Treatment Regulatory Issues across the World

Initially, wastewater treatment was done in irrigation fields. This was followed by the construction of sewage treatment plants with screens, grit removal, strainers and settling tanks. The first sewage treatment plant was constructed in Frankfurt/Main in Germany in 1887 (Seeger, 1999).

Sedimentation of wastewater received most attention in Europe until the end of the Second World War. Whenever supplemental biological treatment was required, wastewater was trickled through irrigation fields. Around the end of the nineteenth century, trickling filtration became widespread and it demanded less space and operational costs compared to other treatment methods through irrigation fields. The trickling filter was soon followed by the introduction of oxidation ponds for the treatment of wastewater with the first of its kind in Germany for use in Berlin in 1898 (Seeger, 1999).

Water treatment originally concentrated on improving the taste and odour of water and was recorded around 4000 BC. Sanskrit and Greek writings describe water treatment by filtering through charcoal, boiling, straining and exposure to sunlight. Egyptians used alum as early as 1500 BC to clarify water. Filtration was used regularly in Europe by the early 1800s. Disinfectants like chlorine were used to control waterborne disease outbreaks during 1900s in the United States (USEPA, 2000).

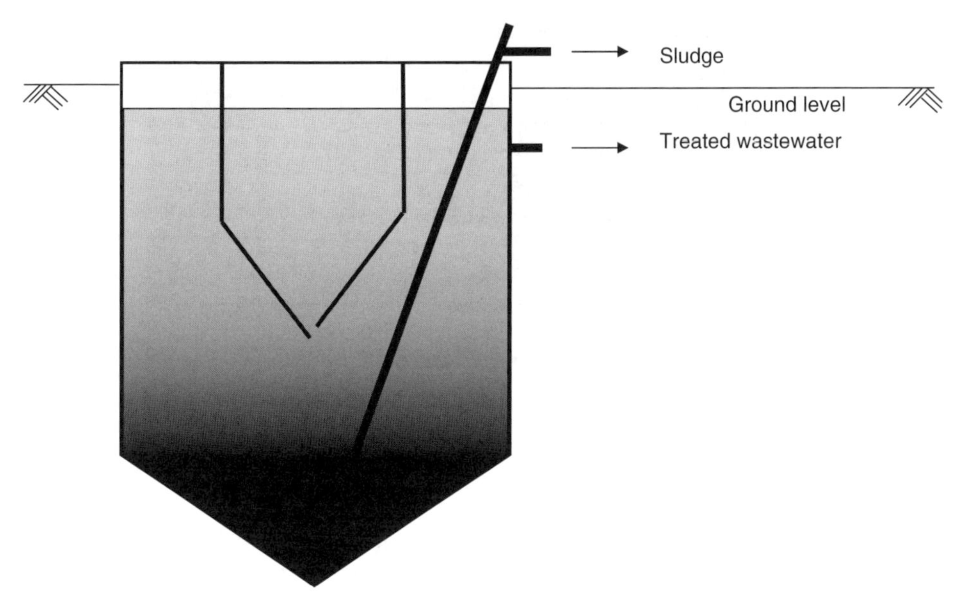

Figure 4.1 *Imhoff tank.*

The problem of sludge handling was addressed for the first time by the introduction of the Imhoff tank (Figure 4.1), which was patented in 1906. Due to the economic problems that followed the war, sewage treatment plants were given lower priority and the construction of such plants in Europe was prevented. By the mid-1920s, as the situation improved, activated sludge processing was invented and the first plant of its kind in Germany was constructed in Essen-Recklinghausen in 1925. The activated sludge process became popular due to its lack of stench and flies were no longer a problem; it was therefore possible to construct such plants within a city. Automatic sludge-scraping systems were developed in 1927 with sludge digestion being done in separate digestion tanks (Seeger, 1999). Besides the developments in sewage treatment, industrial wastewater was also treated with chemical and physical methods during this period.

Despite so much development, several parts of the world still practise farming using sewage. The limitations of the solution that Europe adopted during the Industrial Revolution and before the First World War has now, in the developing countries, ended up polluting groundwater and surface water. The practice has also placed an extra burden on these countries due to increased expenditure on health, packaged drinking water and portable water treatment units at the point of use. But such expenses have enhanced gross domestic product (GDP), employment opportunities, foreign direct investment (FDI) and a rise in stock value indices, thereby making the politicians and economists happy.

4.1.1 Low-Tech versus Hi-Tech

Water and wastewater treatment is not in its early stages. It has blended itself with other technologies, including those from such fields as construction materials, process control, construction technology, mechanical devices, combustion technology and wireless

Figure 4.2 Control room of a large sewage treatment plant without process control engineers.

technology. Figure 4.2 shows a control room without people operating it in a sewage-treatment plant in Stockholm, indicating the extent of automation possible. The process in the plant can be controlled by mobile phone, without requiring any people to stay in the plant.

The use of sophisticated technology is sustainable if the institutions running wastewater/water-treatment plants have quality manpower with good management so that they can retain their staff. The failure of technology adoption and absorption in developing countries is not because technical solutions are not available on the market. Such failure is largely due to poor institutional management or human resources. Many technology-transfer projects have failed immediately after commissioning of a project or withdrawal of support from international agencies as the leadership in the institutions has changed frequently or the management have had other priorities or did not understand the intricacies of the technology. Many departments responsible for water and wastewater treatment are headed by nontechnocrats who are political nominees or administrators.

Many small and medium-scale industries do not integrate wastewater treatment with information technology and depend mainly on humans who may not have the capability to understand fundamental processes. The small- and medium-scale industries that are operated by families just because they have inherited them or to fulfil financial obligation to return loans will often fail to adopt new technology either in their products or processes. In case of a new large-scale plant like a petroleum complex or pulp-and-paper plant, the wastewater and water treatment usually come as part of the process plant and is well integrated with information technology.

Figure 4.3 shows a biogas filling station where fuel for public transport is generated from sewage/organic solid waste. Such a solution can be adopted in any part of the world provided the institutions responsible for treating sewage have the capacity. Many institutions across the globe are burdened with people who are not motivated to learn new technologies.

Biogas has become a common fuel in the bus services of Stockholm (Figure 4.4). Bus depots have biogas filling stations, with gas lines from wastewater treatment facilities. Use of biogas instead of fossil fuels means that emissions and the impact on climate are largely reduced. By using biogas for buses, carbon dioxide emissions have been reduced

Figure 4.3 Biogas filling station for public transport in Stockholm, Sweden.

around by 3100 tons per year, carbon monoxide emission have been reduced by 384 kg per year, nitrogen oxides emission have decreased by nearly 21 tons per year and particulate emissions have come down by 311 kg per year. Noise from these buses is low and this has resulted in lower traffic noise and reduced quantities of ground-level ozone due to a drop in hydrocarbons and nitrogen oxides (Swedish EPA, 2011).

4.1.2 Low Cost versus High Cost

High cost does not necessarily give the best results. The cost of a treatment plant depends on the cost of manpower, raw material and expenditure on construction equipment. Hence, a plant designed by a consultant and a consulting firm in Europe will cost more because the manpower costs are 40 times more than in India or China. Similarly, the cost of raw materials depends on the distance of raw material from the construction site.

Energy requirements are higher in cold regions and lower in warmer regions, which will determine the expenditure on the construction and operation of a wastewater

Figure 4.4 Bus fuelled by biogas generated from sewage sludge and organic solid waste in Stockholm, Sweden.

Figure 4.5 *Rapid sand filters in sewage treatment plant at Bromma, Stockholm, Sweden.*

treatment facility. Figure 4.5 shows rapid sand filters in a sewage treatment plant at Bromma, Stockholm, Sweden, which is constructed inside a tunnel. This serves two important purposes, namely, (i) it saves land and (ii) it avoids freezing during winter.

4.2 Design Principles for Sustainable Treatment Systems

Climate and climate change have the potential to increase water consumption as people take more showers and drink more water; hence, there would be more evaporation losses. For example, although Singapore has a demand management approach, water intake per household is still relatively high at about 156 l per day (Lee, 2005) as people take several showers a day due to the climate. Hence, it is essential that cities and industries are prepared for days when there is water stress. Due to the water demand in south-east London, Thames Water, the water supply company, constructed 140 000 tonnes/day desalination plants, with renewable energy, to cater for the city in an emergency (Green, 2013).

4.2.1 Low Carbon

Sewage sludge and food waste in Sweden are increasingly treated via anaerobic digestion. This produces digestate, which is an excellent fertilizer and biogas (carbon dioxide and methane). Biogas has proved to be effective in reducing greenhouse-gas emissions (there was a total reduction in Sweden of 170 000 metric tons CO_2 equivalents during 2003–2010) mainly by being a renewable alternative to fossil fuel. This amount is equivalent to the emissions of 56 000 cars annually. The emission of carbon dioxide per energy unit is estimated to be reduced by around 90% if biogas replaces fossil fuels. When biogas replaces gasoline or diesel it also reduces air pollutants. For heavy vehicles like buses, the emissions of particles can be reduced by around 20–25%.

4.2.2 Low Energy

Providing reliable wastewater services and safe drinking water is an energy-intensive activity. Historically windmills (Figure 4.6) were used to pump water, either for land drainage

Figure 4.6 *Windmill used for pumping.*

or from groundwater but, after the invention of electricity, such sustainable practice took back stage.

Reasons for the poor efficiency for water pumping include: (i) changes in the water/wastewater system, (ii) major additions of new technology or features to older systems, (iii) changes in operating practices/schedules, (iv) tender documents do not insist on efficient pumps, (v) comparison of tenders on the lowest capital investment but not on lifecycle cost, (vi) no design-based approach, (vii) use of second-hand equipment, and (viii) inadequate metering and monitoring facilities.

Energy represents a considerable cost in wastewater treatment and distribution. Water and wastewater systems consume about 3% to 4% of total US electricity consumption, used for the movement and wastewater and treatment of water (EPRI, 2002; Galbraith, 2011). The percentage of electricity consumed in this way varies from country to country and statistics are not maintained in all countries, especially in the developing world. The cost of energy constitutes 30% of operational costs in conventional treatment. Aeration consumes most energy, with nearly 50–60% of the total energy demand for treatment. Pumping is the second most energy-consuming activity in cases of aerobic treatment involving aeration. The area required for wastewater treatment is inversely proportional to energy input in the case of biodegradable effluents. Avoiding pumping and aeration can greatly reduce the energy demands of a wastewater treatment plant. Water-energy issues are also important in the context of material costs, water shortages, higher energy and a changing climate (Daw *et al.*, 2012). Energy auditing at wastewater/water treatment facilities is one way to save money, energy and water.

The primary duties of operators and managers of water/wastewater facilities include: (i) complying with regulatory requirements, (ii) balancing repair and replacement needs, (iii) providing enhanced service at reasonable rates and (iv) optimizing operations and maintenance to decrease costs and ensure longevity of assets.

Water supply and demand vary with terrain, geological location, season and culture. Many industries and local bodies do not operate waste-treatment plants to save energy. Consideration of energy requirements in the planning stage could provide solutions to unsustainable practice.

A summary of energy efficiency strategies is shown in Table 4.1. Figure 4.7 provides a comparison of the energy and the footprints of various treatment processes located at a place that receives sufficient sunlight to sustain an oxidation pond and constructed wetlands. Aiming for zero-energy wastewater treatment at planning is the best way to achieve energy-efficient wastewater treatment. If water/wastewater treatment has to be on

Table 4.1 *Energy efficiency strategies for wastewater treatment plants.*

Sl. no.	Parameter	Strategy
1	Process energy	Focus on the biggest energy consuming unit/operation
2	Operational controls	Change operations to meet seasonal/daily changes
3	Quality versus energy	Balance water quality and energy needs
4	Repair and replacement	Consider equipment life and energy usages, replace all faulty diffusers, pipes, fixtures and equipment, etc.
5	Bio solids	Consider energy efficient disposal
6	Infiltration/inflow	Address infiltration and inflow to reduce treatment energy
7	Leaks and breaks	Address leaks and breaks to decrease pumping energy
8	Onsite renewable energy	Consider opportunities for onsite generation of energy
9	Conservation	Educate the community to consume less water
10	Flow fluctuation	Provide equalization tank, install adjustable speed drivers on pumps
11	Maintaining DO	Install DO monitoring and control to maintain DO at desired level as excess DO does not improve efficiency and it only consumes energy
12	Checks and balance	Conduct frequent auditing, install electric load-monitoring device, install capacitors to improve power factor, avoid reduce pumping operations; avoid oversized/under sized equipment/fixtures.
13	Chemical use	Use optimum dosage. Overdosage does not guarantee improve in efficiency. Efficiency of some operations like flocculation will reduce due to use of chemicals beyond optimal dosage.

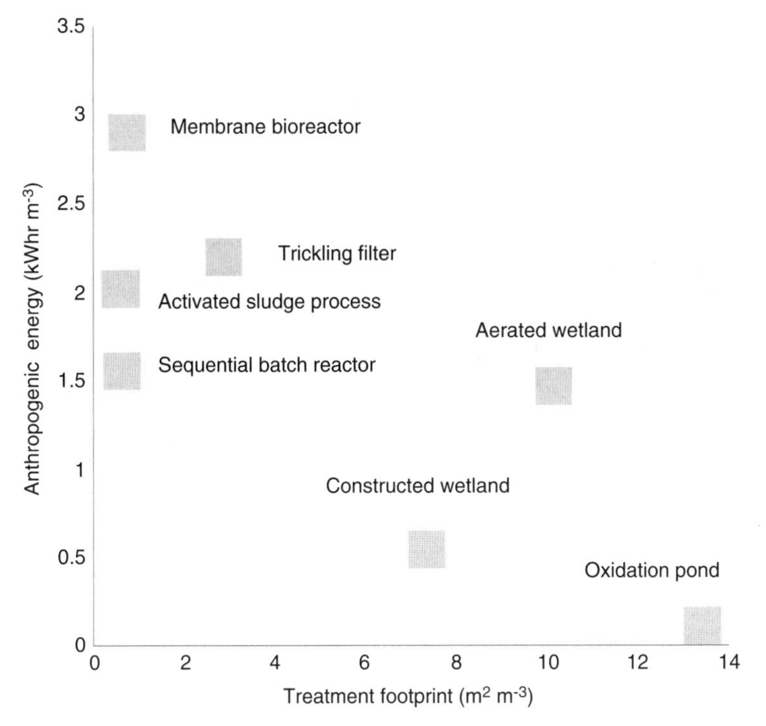

Figure 4.7 *Comparison of energy and footprints of various treatment processes located at places that receive sufficient sunlight to sustain oxidation pond and constructed wetlands.*

steep terrain, unnecessary pumping can be avoided by designing the treatment plant so that water will flow by gravity. Pumping for the recirculation of sludge can be avoided by adopting a membrane bioreactor of packed-bed reactors. Energy requirements in windy areas can be met by erecting wind turbines. If the location receives sufficient sunlight, solar panels can be fixed to trap solar energy. If land is not a constraint, it is better to adopt constructed wetland or wastewater treatment ponds.

Generating energy with bio-solids generated after treatment with sludge digestion (Figure 4.8) is one of the treatment options to attain nondependency on external energy. But the plants should be sophisticated and have knowledgeable staff to run the plant safely and efficiently. Box 4.2 explains efforts made by Delhi Jal Board an organization responsible for supply of water and treating wastewater in Delhi.

Box 4.2 Sewage treatment plant in Okhla, Delhi, India

The Delhi Jal Board, an agency responsible for water supply and sewage treatment, inaugurated a 30 MGD (136.5 m^3/d) sewage-treatment plant in December 2012, financed largely by the Japan International Cooperation Agency. It is capable of generating 1105 kWh of power, which enables the wastewater treatment plant to be self-sufficient in energy.

Figure 4.8 *Sludge digester and generator building at Okhla, Delhi, India.*

4.2.3 Low Chemical Use

Chemicals are hardly used for treating sewage, apart from disinfection with chlorine or other chemicals. The use of chlorine can be avoided by using UV light or water pasteurization. Treatment of industrial effluents with chemical precipitation/reaction should be done after process optimization.

The city of Joensuu, Finland adopted nonchemical water treatment (proprietary in nature provided by commercial establishment) resulting in reduced repair and maintenance expenses for the plumbing and heating ventilation and air-conditioning systems (Sustainable Cities, 2012).

4.2.4 Modelling of Treatment Processes to Attain Sustainability

Models describe phenomena/processes in the language of mathematics. The majority of interacting systems are far too complicated, hence, the most important parts of the system are considered for modelling.

Models that describe processes are *mechanistic models* and models based on experiments are called *empirical models*. Deterministic models ignore random variation whereas stochastic models are based on statistics. On the other hand *black-box* models do not describe changes in the process mathematically.

Creating a new model includes the following steps: (i) problem specification, (ii) reviewing prior knowledge in the field, (iii) setting up a verbal model, (iv) translation into mathematical model, (v) implementing into modelling software and solving the equations, (vi) determination of parameter sensitivity, (vii) experimental design, (viii) parameter estimation, (ix) model validation, (x) application of the model to solve the problem that was originally specified (Morgenroth *et al.*, 2002).

A growing global population, increasing scarcity of water, urbanization, and rising agrochemical applications are the driving forces for efficient methods of wastewater and water treatment. Due to the complexity of water/wastewater treatment, the use of artificial neural networks (ANNs) has been gaining popularity to predict the performance of water/wastewater treatment systems. Artificial neural networks do not need mathematical

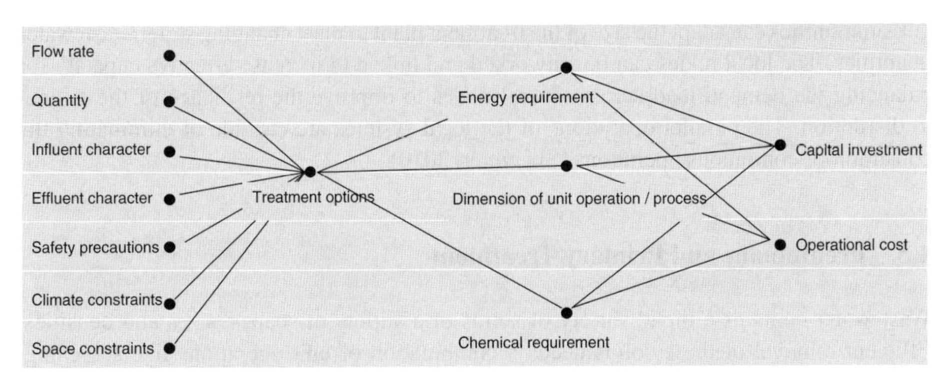

Figure 4.9 *Interrelationship between various parameters of water/wastewater treatment plant.*

equations for describing the process (Khataee and Kasiri, 2011). They are computer-based systems designed to simulate the learning process by mimicking the neurons of the human brain.

For example:

quantity of waste trapped in screen = function of velocity of water, clear space, size of waste Substance

$$Q_s = f(v, d, D)$$

quantity of grit separated = function of velocity of water, weight of particles

$$Qg = f(v, W)$$

quantity of organic material
converted to biomass aeration tank = function of temperature, MLSS, influent BOD, Mean cell residence time, dissolved oxygen in effluent

$$Q_o = f(t, X, \theta, O)$$

oxygen in aeration tank O = function of volume of air blown, size of bubble $f(V_a, V_b)$
overall efficiency of wastewater treatment = $f(Qs, Q_g, Q_o, \ldots)$

The output of one equation will become the input for another equation at the intermediate stages of calculation in a model. Mathematical modelling of wastewater treatment has become popular in recent years. Figure 4.9 shows the interrelationship between various parameters of water/wastewater treatment plants. During the design or upgrading of a wastewater treatment plant, various process alternatives and operating strategies can be evaluated by using computer software (McGhee *et al.* 1983; Spearing, 1987).

4.2.5 Operation, Management, Financial, Socio-Economic Aspect

Centralized large-scale, water-treatment systems cannot accommodate rapid changes in supply or demand as there are too many perceived risks including technical, financial, system, and organizational issues. Localized water-treatment systems can obtain better access to drinking water through reduced technical and financial risks, reducing the probability of total system failure and providing greater organizational capacity (Slaughter, 2010).

Communities can adapt the size of the treatment plant to meet changing water/wastewater quantities. The local nodes can be networked and linked to increase effective capacity for balancing the demand load across several nodes to improve the resilience of the system to disruption. The parallel operations of the local systems are capable of eliminating the constraint of continuous operations (Slaughter, 2010).

4.3 Preliminary and Primary Treatment

Wastewater contains a broad variety of solids of a variety of shapes, sizes and densities. Efficient removal of these solids needs a combination of unit operations like screening, grinding and settling. Even though no material is removed, flow measurement devices are necessary for the operation of wastewater treatment and they are usually included in the primary system. Operations to remove large objects and grits and flow measurement are usually referred to as preliminary treatment.

4.3.1 Screening

Screening is done to prevent entry of solid articles above a certain size – objects like cigarette butts, leaves, plastic cups, sachets, paper dishes, polythene bags, sanitary napkins, cloth pieces and so on – into the treatment plant. These items are removed to avoid clogging of pumps, aerators, stirrers and other electromechanical devices. Screening is usually achieved with a screen of vertical bars, 10 to 25 mm apart, placed across the effluent flow.

Two screens may be used in series with different amounts of space between bars. The use of moving screens (Figures 4.10 and 4.11) is common in large effluent-treatment plants to allow continuous simultaneous cleaning during the operation. Small effluent-treatment plants can install provisions for manual cleaning to cut down costs.

The screen chamber must have sufficient cross-sectional openings between the bars and it should be designed to allow passage of effluent at peak flow rate at a velocity of 0.8 to 1.0 m/s. The screen must extend 30 cm above the maximum level of effluent in the chamber (Kodavasal, 2011).

4.3.2 Coarse-Solid Reduction

Coarse solids that escape screens cause mechanical abrasion to pipes and moving parts in wastewater-treatment facilities. The course solids or grits also occupy aeration tanks, sedimentation tanks and other treatment units and operations. Coarse-solid reduction is essential and positioned after screens but prior to other treatment units/operations. Grit chambers (Figure 4.12) are either designed for a single purpose or for both grit separation and scum/oil/grease removal.

4.3.3 Grease Removal Chamber

As discussed in Section 4.3.2, grease and grit can be separated in the same chamber or different chambers. For example, the grease-and-grit trap (Figure 4.13) is provided at the point of discharge from canteen/kitchen to arrest grit and oil/grease. The grease-and-grit trap is also essential in industries like engineering, where oil emulsion is used for cooling during

Figure 4.10 Mechanically cleaned screen.

Figure 4.11 Fully automated self-cleaning screen with enclosure.

Figure 4.12 A grit-removal chamber with scrapper.

cutting/drilling operations and where significant grease and grit are expected. Typically, grease traps are shallow traps to allow the rapid rise of fats, oils and grease (FOG) to the surface. The length of the trap is usually two times its depth and the residence time of wastewater in the trap is 5–20 minutes at peak flow. As a rule of thumb, the surface area of the trap in m^2 is kept about 1.5 to 2 times the depth of the trap in m.

The outlet of the incoming pipe is placed below the water level to avoid disturbance of the upper floating layer of FOG. The floating film of FOG is collected frequently.

The recovered FOG can be used for producing bio-fuel. It can be used to recover energy and recycled waste oils. In addition to clogging pipes, FOG can clog pipes in the public sewers and in wastewater-treatment facilities. Fats, oils and grease collected from publicly owned treatment works (POTW) can generate an income in many ways. The Pacifica Wastewater Treatment Facility in Pacifica, California, United States, is installing a biodiesel facility to operate its Calera Creek Wastewater Plant by converting fryer oil collected from restaurants into biodiesel (USEPA, 2013).

The San Francisco Public Utility Commission located at San Francisco, California, is collecting FOG for free in the city and the biodiesel generated from FOG is used as fuel

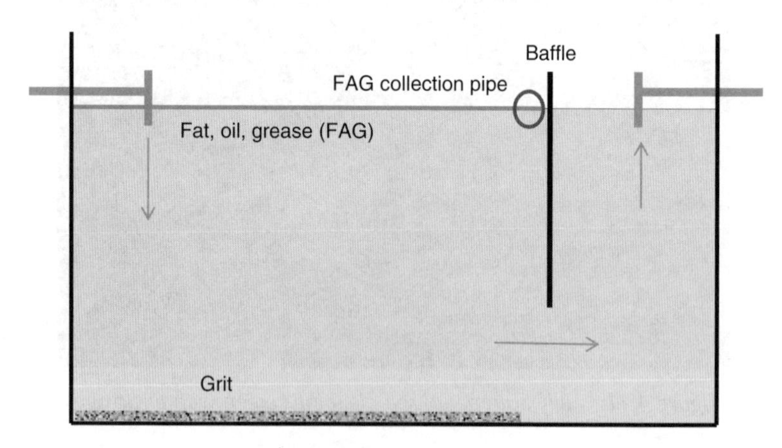

Figure 4.13 Grease trap.

for fire trucks, in airports, and in public transit (USEPA, 2013). Eastern Municipal Water District of Perris in California is exploring a project to convert grease into biodiesel.

4.3.4 Flow Equalization

The wastewater from the bar screen chamber and the FOG and grit trap is collected in the equalization tank. Its main function is to supply a uniform flow rate to wastewater treatment plants to avoid shock loads (hydraulic shock loads and organic shock loads). During the peak hours, sewage comes at a high rate. The equalization tank stores the wastewater during high-flow periods (like peak hours in towns/township/apartment and emptying/washing tanks in industries) and lets it out during lean/no-flow periods.

The equalization tank must have the volume to hold the peak-time inflow. For residential complexes, maximum flow occurs in the morning when residents are using kitchens, bathrooms and toilets. Sewage generation may be higher during the weekends in countries like India, whereas it may be the opposite in Europe where people may not stay at home during the weekend. In the case of a commercial/industry/software park, peak flows occur during lunch/dinner times from the canteen. The equalization tank should therefore have the capacity to hold the maximum variance between the outflow and the inflow (Figure 4.14).

The incoming influent line to the equalization tank is normally gravity fed and is hence likely to be at a substantial depth below the ground. The equalization tank should therefore not be too deep to avoid deep excavations and costly construction. Greater depth also makes maintenance difficult and hazardous. Compressed air is usually purged to avoid septicity, to suppress odour generation and to keep solids in suspension. As a rule of thumb, the air volume required per hour is 2.5–3.0 m^3/m^2 of floor area or 1.2–1.5 times the capacity of the equalization tank.

4.3.5 Mixing and Flocculation

Flocculation is the 'snowballing' of smaller particles to enable easy settling. The process takes place in a tank equipped with mechanical paddles to encourage interparticle contact

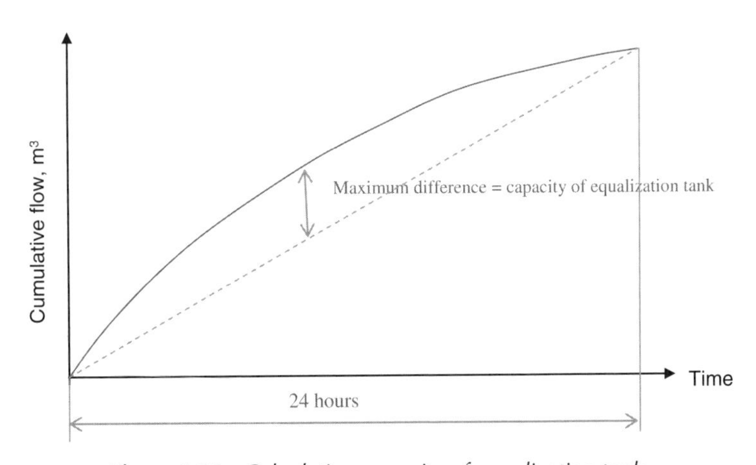

Figure 4.14 *Calculating capacity of equalization tank.*

and gentle enough to avoid disintegration of flocculated particles. Chemically, coagulant chemicals are either metallic salts or polymers. Polymers are long chains of smaller molecules, which are either cationic or anionic or nonionic. Improper coagulation may result from the use of old chemicals, the wrong coagulant and wrong concentration of coagulant.

Colloidal suspensions are regarded as 'stable' when they do not agglomerate naturally. The most important reason for the stability of colloidal suspensions is an excessively large surface-to-volume ratio due to their small size making surface phenomena predominate over mass transfer phenomena. The important surface phenomenon contributing to 'stability' is the accumulation of electrical charges on the particle surface. Electrical charges are formed mainly due to the molecular arrangement within crystals and loss of atoms due to abrasion of the surfaces. Ions in the water near the colloid will be affected by charges on surface of colloids. A possible configuration of ions around a charged particle is shown in Figure 4.15. The first layer of ions attracted to the charged surface is 'bound' to the colloid. This arrangement produces a net charge that decreases exponentially with distance from the colloid.

When two colloids come close to each other, there are two forces acting on them. The electrostatic potential that repels the particles and *Van der Waals' force* that make particles attract each other. Electrocoagulation and electroflocculation involving the coagulating metal ions from electrodes are considered to be more sustainable than chemical coagulation, which involves alum, ferric chloride or polyelectrolyte. The ions from sacrificial electrode in electroflocculation act in a similar way to coagulating chemicals such as alum and ferric chloride. The electroflocculation has advantage that there is no enhancement of the salinity

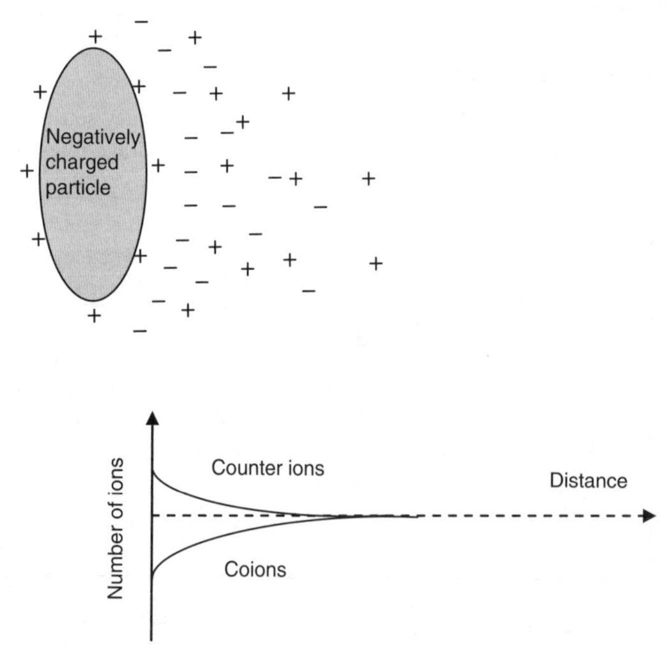

Figure 4.15 *Configuration of ions around a charged particle.*

of the treated water. Electroflocculation produces 30% to 50% of the sludge generated in chemical flocculation. In electroflocculation the pollutants are removed by the bubbles that float to the surface, capturing the coagulated pollutants and achieving more than 98% efficiency in a single-stage process.

The process involves the passage of an electric current, using aluminium electrodes that generate hydrogen. The aluminium coagulates the pollutants, which can be removed by either settling or filtration.

Electrocoagulation has been associated with high operating costs. A variation of electrocoagulation in which the gas in the process flocculates the coagulated pollutants is known as electroflocculation. Electroflocculation can be used for removal of clay, suspended solids and FOGs.

Chemical flocculation is used to prepare the solids for subsequent removal. Four mechanisms predominate during flocculation with chemicals: (i) ionic layer compression, (ii) adsorption and charge neutralization, (iii) sweep coagulation, (iv) interparticle bridging. It is necessary to achieve uniform mixing conditions in chemical flocculation to optimize floc size and to avoid breakup of floc. A wide variety of flocculation mixing mechanisms is in use. Usually the flocculation process is achieved with flow having relatively fast velocity gradient values of 60 to 70 s^{-1} followed by values of 10 to 30 s^{-1}.

Iron and aluminium salts are commonly used for flocculation. Table 4.2 shows commonly used flocculants. These salts hydrolyse in water and form insoluble hydroxides with calcium and manganese hydrogen carbonates, which are present in water. If the carbonates are present in adequate concentrations, sodium carbonate or hydrated lime needs to be added. Aluminium sulfate coagulates best between pH of 4.4 and 6. Sodium aluminate is usually used at pH values of 6.5 to 8. Iron salts are effective over a range of pH except for values between 7 and 8.5.

Flocculation aids may be used to improve the performance of flocculants and to decrease the quantity of coagulants like diatome, activated silica, activated carbon powder, certain types of adsorptive clays, bentonite, cellulose derived materials and organic substances.

In developing countries, various naturally occurring materials like fluvial clays from rivers, clarifying rock material from desert and earth from termite hills are used. pH does not have an effect on clay, which is an advantage, but clays contain traces of heavy metals which may have toxic effects.

Coagulants of plant origin, like seeds of the trees of the Moringaceae family, seeds from *Strychnos potatorum*, sap from the tuna cactus (*Opuntia ficus-indica*), the bark of the *Schinopsis quebracho-colorado* and potato starch, can be used where available.

Other natural coagulants include algae-derived substances, chitosan produced from the shells of lobster/shrimp, dough from millet bread or curds.

Coagulation and flocculation depend on a multitude of factors: turbidity, temperature, colour, alkalinity, pH, coagulant, intensity and duration of stirring. The higher the temperature, the more effective is the coagulation due to faster reaction. The higher velocity causes the shearing/breaking of floc particles and hence velocity around approximately 0.3 m/s in the flocculation tanks should be maintained. The higher the zeta potential (the charge at the boundary of the colloidal particle and the surrounding water) the greater is the repulsion between the particles. Higher zeta potential needs a higher coagulant dose. Hence optimum dosage needs to be assessed at field before finalizing the quantity of coagulants to be added.

Table 4.2 Commonly used flocculants.

Flocculants	Primary coagulant	Coagulant aid
Activated carbon powder	√	
Algae-derived substances	√	
Aluminium sulfate (alum)	√	
Anionic polymer		√
Bentonite		√
Calcium carbonate		√
Calcium hydroxide (lime)	√[a]	√
Calcium oxide (quicklime)	√[a]	√
Cationic polymer	√	√
Chitosan produced from the shells of shrimp/lobster	√	
Curds	√	
Datome (kieselgur)	√	
Dough from millet bread	√	
Drumstick seed powder (*Moringa oleifera*)	√	
Ferric chloride	√	
Ferric sulfate	√	
Ferrous sulfate	√	
Nonionic polymer		√
Potato starch	√	
Sap from the tuna cactus (*Opuntia ficus-indica*)	√	
Sodium aluminate	√[a]	√
Sodium silicate		√
Tamarind seed kernel powder	√	
The bark of the *Schinopsis quebracho-colorado*.	√	

Note: [a]Used as a primary coagulant only in the water-softening processes.

Typically all coagulant aids are expensive, so care should be taken while using these chemicals. Lime used to enhance the alkalinity of the water results in an increase in ions in the water, some of which are positively charged. These charged particles attract the colloidal particles, forming floc. Bentonite joins with the small floc, making it heavier and therefore making it settle more quickly.

Enhanced particle flocculation uses addition of inert ballasting agent (like silica sand or recycled chemically conditioned sludge) and a polymer.

4.3.6 Sedimentation

Sedimentation/clarification is used for the separation of suspended and colloidal materials that are heavier than water by gravity. Settling can be either discrete or hindered (Table 4.3). Table 4.4 shows densities of sludges separated by sedimentation. The performance of ASP mainly depends on temperature, aeration time, food to micro-organisms ratio, toxicity of the water and pH (Table 4.5). Table 4.6 shows troubleshooting problems in ASP and remedies. Primary sedimentation aims at removing readily settleable solids. Chemically enhanced primary treatment (CEPT) uses chemicals, usually metal salts (like ferric chloride

Table 4.3 Difference between discrete settling and hindered settling.

Discrete settling	Hindered settling
Particle will retain its physical identity (retain shape and size).	The particle size may increase during settling due to formation of agglomerates.

Table 4.4 Densities of sludges.

Sludge from	Density
Grit	2650 kg/m^3
Primary sedimentation	1250 kg/m^3
Secondary sedimentation	1360 – 1400 kg/m^3

Table 4.5 Parameters that affect the activated sludge process.

Parameter	Too low	Too high
Microbes	Sewage is treated partially	Filamentous growth
Oxygen	Partial treatment, filamentous growth	Pinpoint flocs, poor settling
Food	Filamentous growth	Partial treatment
Temperature	Microbes will not be active	Bacteria will die
pH	Will not favour microbial growth	Will not favour microbial growth
Toxic	Favours microbes	Will not favour microbial growth
Salinity	May not have significant effect	Poor flocculation, decrease in substrate utilization rate, and high effluent solids

Table 4.6 Troubleshooting problems in ASP and remedies.

Sl. no.	Problem	Description	Remedy
1	Rising of sludge	Dead pockets are formed due to poor vortex formation or choking of pipe or poor functioning of aerators resulting in anaerobic condition. Such conditions will lead to formation of nitrogen, which creates buoyancy of sludge.	Check for worn out blade and replace. Check for blower and pores in aerators.
2	Foaming	Presence of detergents and grease often leads to formation of foam, which is transferred in air causing nuisance.	Provide grease removal unit or rectify defects in grease removal unit.
3	Bulking of sludge	Toxic shock loads encourage growth of filamentous microbes due to death of bacteria. Such growth of filamentous microbes results in bulking of sludge.	Chlorinate for 10 to 15 min to kill filamentous microbes. Stop toxic shock loads.

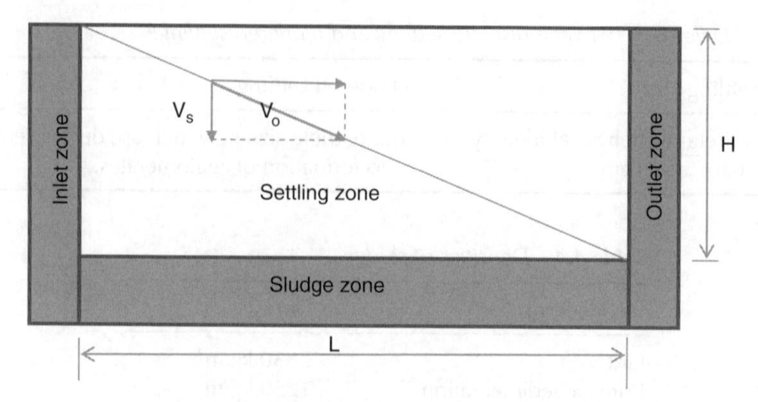

Figure 4.16 Different zones during sedimentation.

and alum) and/or polymers followed by primary sedimentation. These chemicals cause the suspended particles to join together through coagulation and flocculation. The flocs settle faster, enhancing treatment efficiency.

Any particle more than 1 micron in size will theoretically settle down. Colloidal particles (greater than one micron) take 4–5 days for discrete settling. Hence the colloidal particles are flocculated using flocculating chemicals/substances when colloidal particles need to be removed.

Sedimentation tanks can be divided into (i) inlet zone, (ii) settling zone, (iii) sludge zone, (iv) outlet zone. Even though they are in an actual sedimentation tank, these zones are not separated by sharp, distinct boundaries; for theoretical purpose boundaries can be marked as shown in Figure 4.16.

Taking congruent sides of similar triangles formed due to velocity vectors and dimensions of the sedimentation tank (Figure 4.16)

$$L/H = V_o/V_s$$

$$V_o = LV_s/H$$

$$Q = V_oBH$$

where

Q = flow rate
B = width of sedimentation tank
$V_s = Q/BL = Q/A$

Hence

$$\text{settling velocity} = \text{hydraulic surface loading}$$

Further Renauld's number R_e which is measure of laminar flow or turbulent flow, can be defined as

$$R_e = V_oR/\nu$$

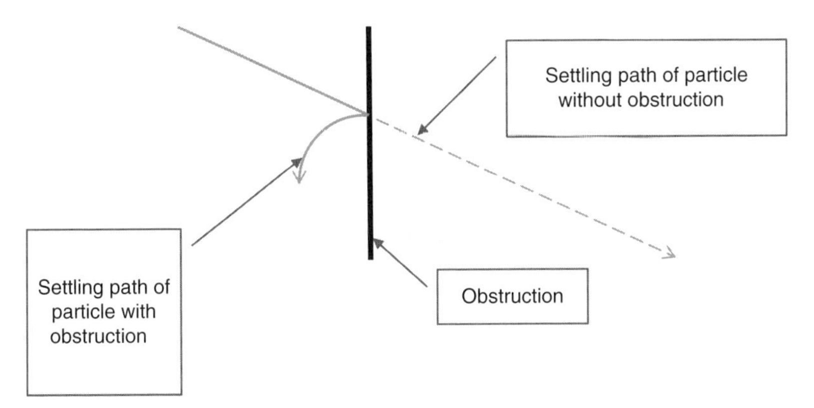

Figure 4.17 *Schematic diagram depicting movement of particle with obstruction.*

where

R = hydraulic mean radius
ν = kinematic viscosity of fluid = 1×10^{-6} m^2/s for water

(For rectangular channel R = BH/(B + 2H).)

For proper settling, R_e should be maintained below 1000.

If R_e is not within permissible limits, it is essential that height of the settling tank should increase.

In the case of rectangular sedimentation tanks, L: 2B has to be maintained for proper settling.

Wind causes waves in water if the sedimentation tank is exposed to air, making particles enter the outgoing water from the sedimentation tank.

$$\text{Frouds number} = V_o^2/gH \text{ should be maintained less than } 10^{-5} \text{ to avoid short circuiting,}$$

where g = gravitational acceleration = 9.81 m/sec^2.

The long path traversed by particles during settling can be cut short by introducing obstructions, as shown in Figure 4.17. This forms the basis for the lamella clarifier. A series of inclined surfaces or tubes is placed in the path of particles to increase efficiency. Increased efficiency is attained by a series of inclined plates (lamellae) or tubes occupying nearly 70% of the tank depth. A typical distance between the plates (or internal width/diameter of the tube) is 25–50 mm. Lamellae/tubes are set at an angle greater than 40° so that settled particles fall to the base of the tank. Sludge is removed by conventional methods.

The lamella clarifier (Figure 4.18) results in the use of a smaller tank, saving capital costs and space. Some drawbacks of tube/lamella settlers are: (i) a tendency to clog, due to a buildup of fats and grease and (ii) the growth of plants/biofilms on the plates/tubes.

Tube and lamella settlement tanks needs frequent draining and cleaning if they are used for wastewater streams with biodegradable pollutants to clear biofilm.

4.3.7 Flotation

Flotation, or floatation (Figure 4.19), is a separation process using very fine gas bubbles. Gas bubbles stick to the solid particles and move them toward the surface of the liquid

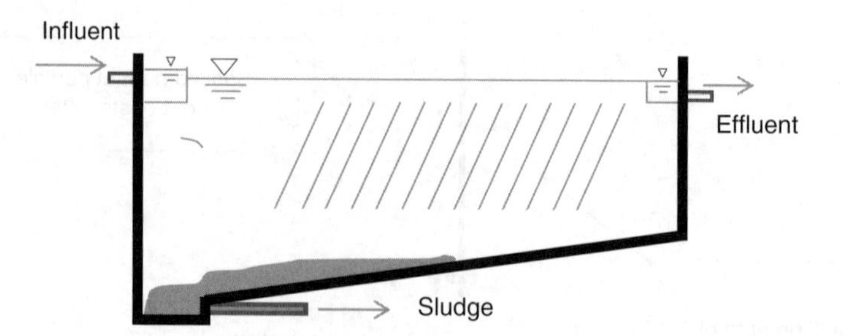

Figure 4.18 Schematic diagram of lamella flarifier.

due to buoyancy. This process is used for very light and small particles with low settling velocities.

Chemical additives that promote flotation are employed to enhance flotation. Commonly used flotation additives are: (i) activated silica, (ii) aluminium/ferric salts, and (iii) organic polymers.

Flotation in wastewater is achieved by: (i) dispersing air into the wastewater, (ii) applying a vacuum to the wastewater, (iii) dissolving air into pressurized effluent followed by releasing the pressure.

Although a theoretical analysis is possible, floating thickeners are generally designed using experimental tests. Design of the floating thickeners is based on: (i) wastewater flow rate and solid loading, (ii) overflow rate, (iii) final solid loading in sludge and (iv) air-to-solid ratio (A/S ratio).

The air-to-solid ratio is determined and is used to calculate the quantity of air required. A typical air-to-solid ratio is in the range of 0.005 to 0.060 ml air per mg solids.

The flow rate of air for flotation is:

$$F_{air} = Q(S_{in} - S_{out})$$

At equilibrium

$$S_{in}/S_{out} = P/P_o$$

The theoretical amount of air available for flotation is:

$$F_{air} = QS_{out}[(P/P_o) - 1] = QS_{out}[(P - P_o)/P_o]$$

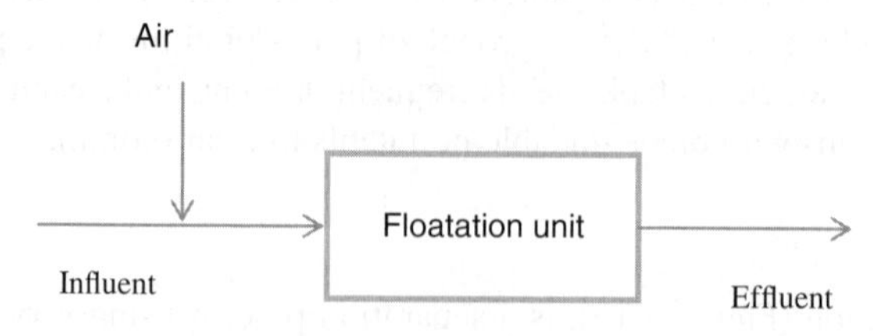

Figure 4.19 Scematic diagram of floatation unit.

F_{air} = net mass flow rate of air available for flotation (mg/h);
P = pressure of pressurized wastewater (atm);
P_o = pressure after depressurization (atm), usually equal to the atmospheric pressure;
Q = wastewater volumetric flow rate (l/h);
S_{in} = concentration of dissolved air in incoming wastewater (mg/L);
S_{out} = concentration of dissolved air in outgoing wastewater (mg/L).

4.4 Secondary Treatment

Secondary wastewater treatment usually follows the primary treatment process. The process is designed to remove or reduce contaminants left during the primary treatment process. The chemical/biological process dominates the secondary treatment process.

Secondary treatment can eliminate up to 90% of the organic substance in effluent. The aim of secondary treatment is to remove the organic and inorganic parameters remaining in effluent.

4.4.1 Biological Treatment

The objective of biological treatment in wastewater is to remove the dissolved and nonsettleable colloidal solids.

The biological process can be divided into aerobic and anaerobic treatments. These treatment processes can be further categorized into attached growth and suspended growth.

Micro-organisms are commonly classified as eukaryotes, and prokaryotes. The eucaryotic group includes (i) fungi (ii) protozoa and rotifers and (iii) algae.

4.4.1.1 Bacterial Growth

Effective biological wastewater treatment design and operation needs knowledge of the growth of micro-organisms.

4.4.1.1.1 General Growth Patterns in Pure Cultures

The usual growth pattern of microbes in a batch culture is shown in Figure 4.20. The growth pattern has four distinct phases:

1. *The lag phase:* in this phase microbes acclimatize to their new environment.
2. *The log-growth phase:* during this phase the microbes divide logarithmically.
3. *The stationary phase:* in this phase the population of microbes remains stationary due to limitations in food.
4. *The log-death phase:* in this phase, the microbial death rate exceeds the reproduction of new cells.

4.4.1.1.2 Growth in Mixed Cultures

It is important to note that the preceding discussion concerned a single population of micro-organisms. Most biological treatment processes consist of interrelated, complex, mixed biological populations, with each particular micro-organism having its own growth curve. The shape of a particular growth curve for such system depends on the food and nutrients

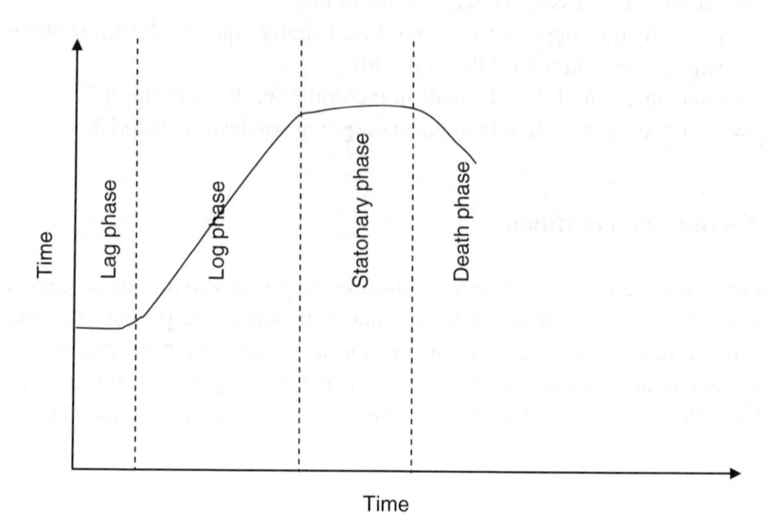

Figure 4.20 *Typical bacterial growth curve in terms of numbers.*

available and on environmental factors such as temperature, pH, type of micro-organism, available oxygen, presence of toxic substances.

4.4.1.2 Attached Growth Biological Treatment

Wastewater treatment processes in which microbes' biomass are attached to the media is called attached growth biological treatment. The effluent flows over the media, during which time the microbe will act upon large biodegradable molecules and degrade into molecules of smaller molecular weight. Attached growth biological treatment can be aerobic or anaerobic in nature. The advantage of the attached growth treatment is that the biomass will not be washed away from process during shock load.

4.4.1.2.1 Aerobic Treatment

In the attached growth aerobic treatment the microbes are aerobic in nature and the process is designed to operate in the presence of atmospheric oxygen.

4.4.1.2.1.1 Trickling Filters A trickling filter is a reactor in which packed material provides a surface on which to form biofilm growth. The system will have a provision to distribute wastewater over the medium and removing the effluent. Packing materials include stone, slag and plastic media. Stones/slag sizes range from 50 mm to 100 mm. Stones and slags at this size are capable of providing specific surface area of 50 to 65 m^2/m^3 with porosity of 40% to 50%. Plastic media designed specifically for a trickling filter can provide a specific surface area up to 200 m^2/m^3 and porosity of 90% (Peavy *et al.*, 1986).

A rotating distribution system is used. Under a hydraulic head of around 1.0 m, jet action through the nozzles is adequate to power the rotor. Electrical motors may be required where it is not possible to maintain the hydraulic head necessary for jet action. Proper plastic media can eliminate plugging problems due to the clogging of sloughed biomass. An under-drain system is provided below the trickling filter to collect treated wastewater and sloughed biomass.

Historically, trickling filters have played an important role but they have become almost obsolete. Several operational modes can be attained. Standard-rate filters have low hydraulic loading without recycling treated wastewater. High-rate trickling filters recirculate part of the treated effluent.

High organic loading results in fast growth of biomass and may result in plugging of biomass followed by the flooding of part of the trickling filter. At temperatures less than 3 °C to 4 °C, relatively little air movement occurs, which prevents good circulation of air. Excessive cold will result in the formation of ice and the destruction of biomass. If the rotating arm of the trickling filter does not move the biofilms it will dry and it may need 30 to 45 days for maturation.

4.4.1.2.1.2 Rotating Biological Contactors Rotating biological contactors (RBCs) are a type of attached growth reactor in which the biomass is attached to slowly rotating discs that are partly submerged in the flowing wastewater of the reactor. Oxygen is transferred to the biofilm when the biofilm comes in contact with air. Oxygen is also transferred to the wastewater by turbulence during the discs' rotation. Sloughed pieces of biomass are removed in the secondary sedimentation tank.

Advantages of RBCs are: (i) high contact time and high effluent quality, (ii) high process stability, (iii) resistance to shock hydraulic/organic loading, (iv) short contact periods, (v) low space requirements and (vi) low sludge production. Disadvantages of RBC are (i) a continuous electricity supply is required, (ii) high investment is needed, (iii) high operation cost, (iv) they need to be protected against wind, rain and freezing temperatures, (v) they require skilled technical labour and (vi) shaft failure would make entire plant cease to function.

4.4.1.2.2 Anaerobic Treatment

In attached growth aerobic treatment the microbes are anaerobic in nature and the process is designed to operate in the absence of oxygen.

4.4.1.2.2.1 Anaerobic Filter An anaerobic filter is an attached biological reactor with a fixed bed through which wastewater flows. Filter material used includes rocks, gravel or specially formed plastic. Ideally, the material should provide 90–300 m^2 of surface area for 1 m^3 of reactor volume for efficient treatment. The efficiency of the reactor will increase with higher surface area for increased contact between the microbes and organic matter. Effluent enters the anaerobic filter upward or downward. A hydraulic retention time of 0.5 to 1.5 days and surface-loading rate of 2.8 m/d are suitable. Biochemical oxygen demand and suspended solid removal is usually between 50% and 80%. Nitrogen removal will normally be 15% of total nitrogen.

An advantage of an anaerobic filter is its resistance to organic and hydraulic shock loads. Disadvantages include the requirement for further treatment to bring it to statutory standards, low reduction of pathogens/nutrients and long startup time.

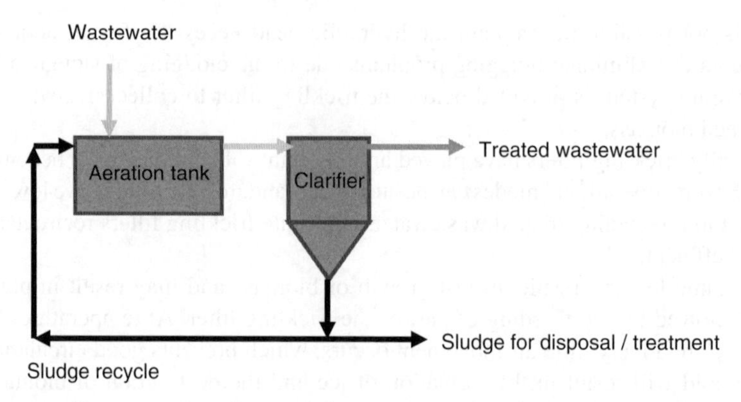

Figure 4.21 *Schematic diagram of activated sludge process.*

4.4.1.3 *Suspended Growth Biological Treatment*

Suspended growth biological treatment is a type of treatment in which the microbes are kept in suspension. The suspended microbes act on the organic material and degrade it to smaller molecules. In the process microbes also consume some of the organic matter, which will become part of the micro-organisms. As these organisms mature, they flock together, making it easy to remove them. Ultimately the sludge, which consists of microbial biomass, is separated, leaving behind treated water for disposal or reuse.

4.4.1.3.1 Aerobic Treatment

In suspended growth aerobic treatment the microbes are aerobic in nature – that is, they need oxygen to survive. Oxygen is supplied by aerating the wastewater by diffused aeration or surface aerator or maintaining depth of water required to maintain oxygen in treatment tank. The overall process can be explained in the simplified equation given below.

$$\text{organic matter} + \text{oxygen} \xrightarrow{\text{Microbes}} \text{water} + \text{carbon dioxide} + \text{biomass}$$

4.4.1.3.1.1 Activated Sludge Process The activated sludge process (ASP) (Figure 4.21) is an aerobic treatment process where the organic matter is degraded by microbes in the presence of oxygen, forming a microbial mass that can easily be separated by conventional sedimentation. Usually primary sedimentation precedes the activated sludge process, wherein the wastewater is allowed to stagnate so that suspended solids can settle but it is common to see wastewater from small industries being fed into aeration tanks immediately after screening. The breakdown of organic matter depends on: (i) oxidation of organic matter and (ii) syntheses producing new biomass.

The aeration tank (Figures 4.22 and 4.23) together with the sedimentation tank (referred to as secondary clarifier or secondary sedimentation tank) forms an activated sludge process. The principal function of the aeration tank is to maintain micro-organisms' population. The microbes are settled in a sedimentation tank and recycled back to the aeration tank.

Adequate mixing and aeration are important to sustain the suspended sludge, which consists of microbes and organic matter. The organic matter is converted into new microbial-cell material and other metabolic end products. The recycling of a proportion of the biomass

Figure 4.22 *Aeration process with submerged aerators.*

is done to maintain the required number of acclimatized microbes in the aeration tank. This makes the mean cell residence time more than the hydraulic retention time.

The quantity of diffused air required for respiration of the micro-organisms is always more than the quantity of air required for keeping the tank contents fully mixed. The rule of thumb is that 50–60 m^3/h of air is required per kg of BOD removed. On the other hand, a surface aerator can transfer 1 kg/hp/h of oxygen at optimum immersion (Figure 4.24). Hence, the power requirement for the surface aerator is calculated as below.

$$\text{Power required for surface aerator in hp} = \text{total BOD to be removed per day}/24$$

The purpose of the secondary clarifier is: (i) to allow settling of biomass, (ii) to thicken the settled biomass, (iii) to produce clear supernatant water. A membrane can be used for liquid-solid separation replacing the secondary clarifier to minimize space requirements in conventional activated-sludge processes. Usually 1/30th of the sludge (Figure 4.25) is removed every day in the activated sludge process, to ensure mean cell residence time of 30 days for microbes and rest of the sludge is recycled back. To achieve operational

Figure 4.23 *Piping arrangement for submerged aerators.*

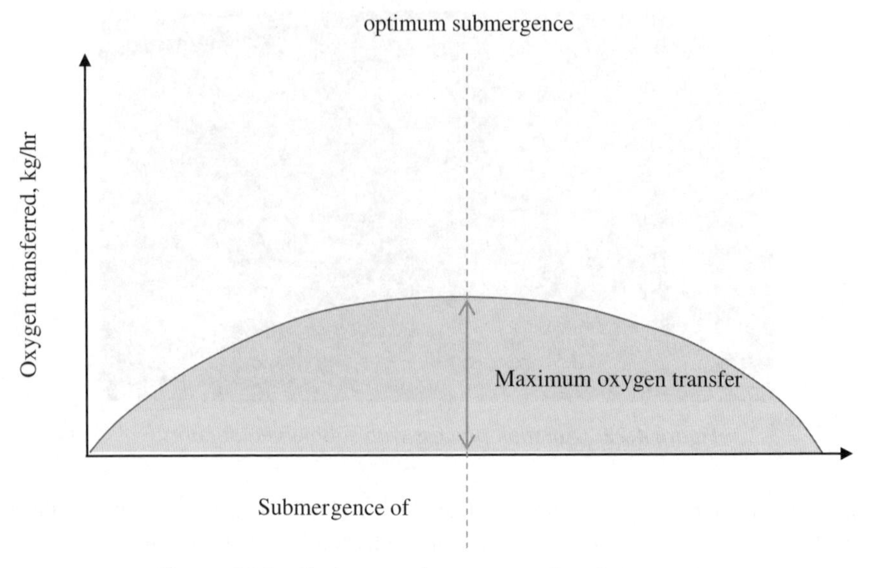

Figure 4.24 Optimum submergence of surface aerator.

simplicity in small ETPs, the aeration tank is divided into 30 parts vertically and an opening is provided at the bottom of top part by providing a pipe and valve so that the valve is opened every day to direct the sludge to be sent to the sludge-thickening disposal unit.

Nitrification in activated sludge processes is determined by the growth rate of nitrifiers, which is lower than heterotrophs in sewerage. Hence, most of the effluent treated in ASP will have nitrogen even though it looks clear and is devoid of carbonaceous BOD. Nitrification can be achieved along with carbonaceous BOD removal in secondary treatment systems. In other cases carbonaceous BOD removal and nitrification are done separately. In the next

Figure 4.25 Sludge from ASP.

Figure 4.26 *Treated effluent from ASP being further treated in anoxic tank to achieve denitrification.*

step, denitrification (removal of nitrogen by conversion to nitrogen gas) is achieved under anoxic conditions (Figure 4.26).

Nitrification and denitrification are optional treatment processes in many treatment plants. They are used only in large-scale treatment plants. The small treatment plants adopt extended aeration where an hydraulic retention time of one to one-and-a-half days is maintained as against 4–6 h in the conventional activated sludge process. Extended aeration produces high density flocs, which do not require sludge digestion to stabilize sludge.

4.4.1.3.1.1.1 Optimum Operational Conditions The performance of ASP mainly depends on temperature, aeration time, food to micro-organism ratio, toxicity of the water and pH. Other factors include the flow regime within the reactor, radiation, mixing efficiency and oxygen level. The improper design of the tank allows short-circuiting of wastewater bypassing the 'active zone'. Radiation/toxic substances that enter the reactor, which are detrimental to microbes, may also impede the reaction. Acclimatized microbes in aeration tank increase the treatment efficiency (Grady *et al.*, 1988) and improve the physical/biological characteristics of flocs (Clauss *et al.*, 1998; Tchobanoglous *et al.*, 2003). Activated sludge is a heterogenic matter, and a colloidal matter consisting of micro-organisms, organic polymers and mineral particles. Density, porosity and size have important roles in settling and Sludge Volume Index (volume in millilitres occupied by one gram of a sludge after 30 min settling) (Ghanizadeh and Sarrafpour, 2001). Performance of enzymes generated by bacteria, and the activity of the bacteria, are optimum in a pH range of 6 to 9 and a temperature range of 10 °C to 25 °C. Hence, the performance of ASP will be optimum at these pH and temperature values.

4.4.1.3.1.1.2 Optimal Operational Requirement for Aeration Tank

Aeration time: 16–18 h
Food to micro-organism (F/M) ratio: 1.12
MLSS: 3500 mg/l
Air: 50–60 m^3/h/kg BOD
Diffusers: flux rate 8–12 m^3/running metre/h

4.4.1.3.1.1.3 Optimal Operational Requirement for Secondary Clarifier

Overflow rate: 12–18 m^3/m^2/day
Detention time: 2.5–3.5 h
Solids loading: 2–3 kg/m^2/h
Weir loading: less than 50 m^3/running m/day

ASP has been modified for better performance and energy efficiency. The common variations (Figure 4.27) are:

- Step aeration: influent addition is done at intermediate points to provide more uniform BOD removal.
- Tapered aeration: air is added in proportion to the organic load exerted.

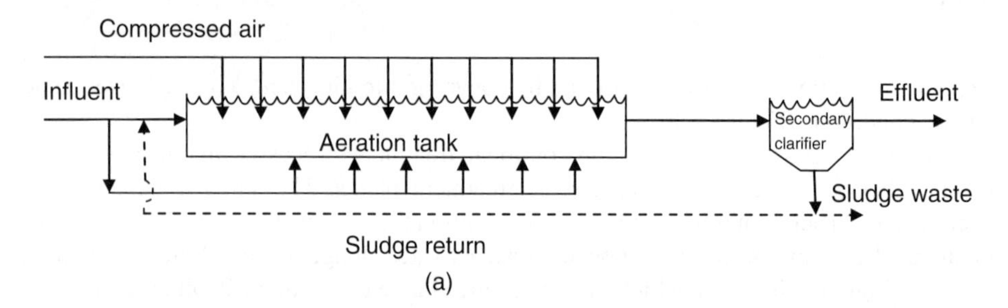

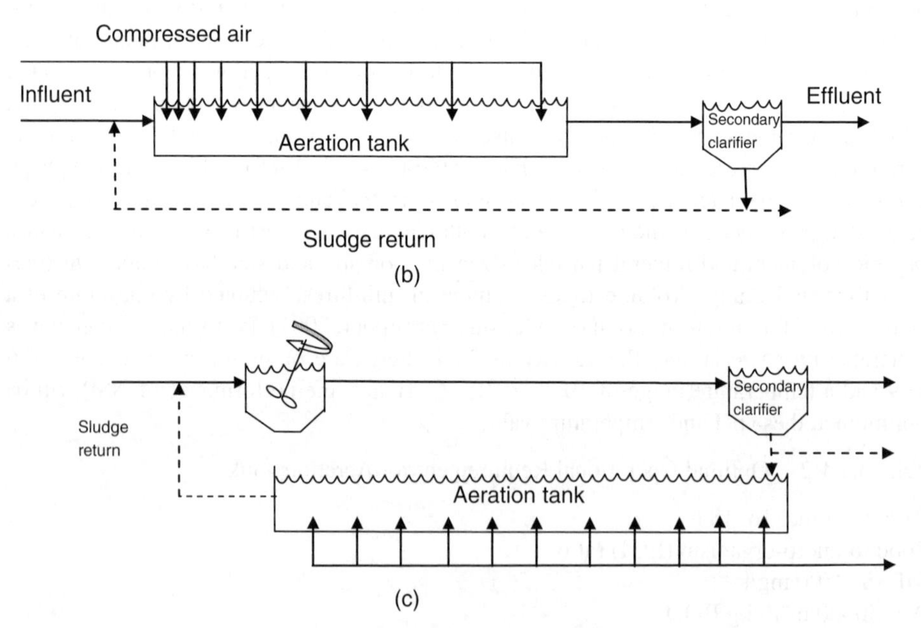

Figure 4.27 Variations in ASP: (a) step aeration, (b) tapered aeration, (c) contact stabilization.

- Contact stabilization: organic matter is brought in contact with biomass whereas inorganic matter adsorbs to biomass. The biomass with adsorbed organic matter is fed to an aeration tank.
- Pure oxygen ASP: pure oxygen is supplied instead of aeration.
- High rate: aerated for short detention time and high food/mass ratio is maintained.
- Extended aeration: a long detention time and low food/mass ratio are maintained to keep the culture in the endogenous phase.

4.4.1.3.1.2 Sequential Batch Reactor Sequential batch reactors (SBRs) or sequencing batch reactors are aerobic wastewater treatment unit operations that combine reaction and settling in one unit, thereby decreasing foot space.

Sequential batch reactors operate a single biological reactor that performs repetitive cycles of aeration, settlement and discharge of the treated effluent. Single batch reactor systems are filled and then operated as a batch reactor followed by settling and drawing clarified supernatant. The cycle in a typical SBR consists of fill, react, settle, draw and idle (Figure 4.28).

The time of 'fill' depends on the volume of the tank and quality/quantity of effluent. The fill operation is followed by a read operation where the biological reactions initiated during the fill are completed. The biological flocs are separated in a settling operation followed by a decanting operation to separate clear water. The idle period is the period between draw and fill where time is used effectively to remove excessive settled sludge. Sequential batch reactor technologies are much more flexible than conventional activated sludge processes.

4.4.1.3.1.3 Constructed Wetlands Wastewater can also be treated sustainably by natural systems, which take advantage of the chemical, physical and biological processes that occur in nature. Natural treatment involves floating aquatic plants, land treatment and constructed wetlands. Mechanical pretreatment usually precedes natural treatment systems for solid removal (ESCWA, 2003).

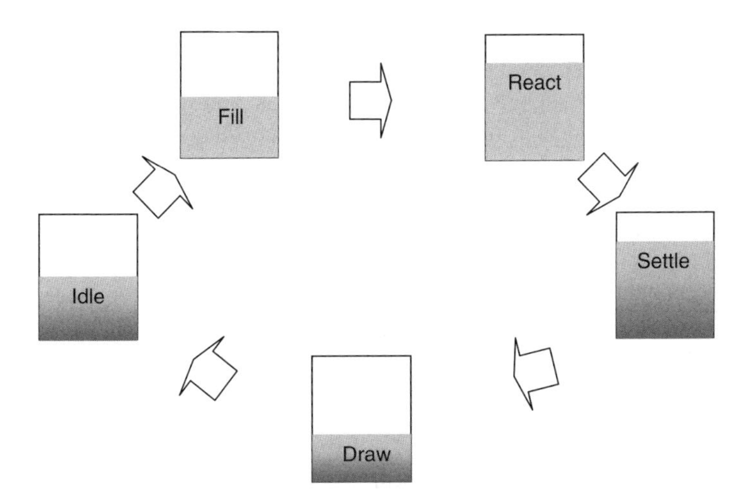

Figure 4.28 *Schematic diagram of sequential biological reactor (SBR).*

Land treatment is the application of wastewater on the land at rates that are compatible with the natural chemical, physical and biological processes that occur in the soil. Major types of land-treatment systems based on flow rates are slow overflow (OF) rate (SR) and rapid infiltration (RI) systems.

4.4.1.3.1.3.1 Slow Rate Constructed Wetlands This involves the application of wastewater to vegetated land. The applied water is lost through evapotranspiration or percolated into soil system. Wastewater percolating can be recovered by means of recovery wells or under drains.

4.4.1.3.1.3.2 Rapid Infiltration This involves high hydraulic and organic loadings applied to shallow infiltration/spreading basins intermittently. The system uses the soil for physical, chemical, and biological treatment. Chemical precipitation, physical straining, biological oxidation ion exchange and adsorption occur when the water infiltrate through the soil.

4.4.1.3.1.3.3 Overland Flow This involves wastewater treatment as it flows over vegetated sloping terraces. Wastewater is applied intermittently on terraces and flows down the terrace. The effluent wastewater undergoes physical, chemical and biological treatment as it proceeds along the surface runoff path.

4.4.1.3.1.3.4 Free Water Surface Systems This type of system consists of shallow basins/channels ranging from 0.1 m to 0.6 m with relatively impermeable soil and emergent vegetation. Usually prequalified wastewater is applied continuously. The effluent is treated as it flows in the system.

In the ecosan concept, constructed wetlands are used to treat greywater. The design of constructed wetlands for greywater treatment is not much different than for domestic wastewater, except that nitrogen and phosphorus removal are not important (as their presence is much lower in greywater than in domestic wastewater) and pathogen removal is also not a big issue (due to low levels of pathogens in greywater).

Prerequisites for being able to use constructed wetlands are: (i) the wastewater should not be too toxic for bacteria and plants, (ii) there should be sufficient incident light to allow photosynthesis, (iii) the temperature should not be too low, (iv) there should be adequate quantities of nutrients to support growth, (v) detention time long enough, (vi) organic loading should not be too high (expressed as g BOD/m^2/day), (vii) there is a need for enough space because it is a low-rate system.

Plant species employed in constructed wetlands include: (i) floating plants (e.g. Lemna, Nymphaea, duck weed, water hyacinth), (ii) submerged plants (e.g. Elodea), (iii) emergent plants (e.g. Phragmites, papyrus, cattail, reed, rush, bulrush, sedge). The use of local species is very much encouraged as they are acclimatized to the local environment. Constructed wetlands are classified according to the type of plants and flow pattern as depicted in Figure 4.29. Any combination of these systems is called hybrid flow. Constructed wetlands (CWs) have been used routinely worldwide to treat domestic and industrial wastewaters (Figure 4.30).

Constructed wetlands can also be categorized, as shown in Figure 4.29, into: (i) free-water surface (FWS) wetlands, and (ii) subsurface flow wetlands. Subsurface flow wetland can be fed by: (i) horizontal flows or, (ii) vertical flows (from just below the soil).

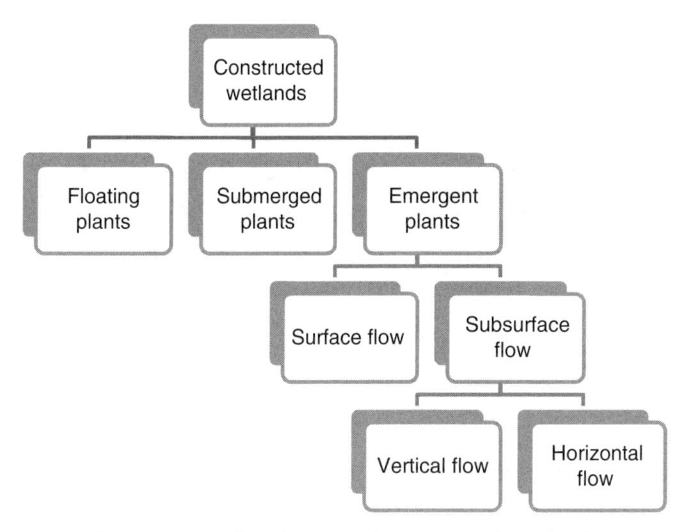

Figure 4.29 *Classification of constructed wetlands.*

Basic treatment mechanisms are: (i) sedimentation, (ii) chemical precipitation, (iii) adsorption, (iv) microbial action and (v) uptake by vegetation.

Horizontal flow beds (Figures 4.31 and 4.32) are fed continuously whereas vertical flow (Figure 4.33) is fed intermittently to allow drying of the bed for short a duration so that oxygen can pass into the bed. Constructed wetland systems can treat effluents with high BOD, SS, nitrogen, metals and pathogens efficiently. Phosphorus removal is usually low due to the limited contact period with the soil. Removal of pathogens in constructed wetland occurs due to (i) natural die off, (ii) predation, (iii) filtration, (iv) sedimentation, (v) adsorption, (vi) secretion of antibiotics from plants and (vii) solar radiation.

Space requirements for SF-wetlands are 0.5 to 5 m^2/person (Geldof *et al.*, 1997). Wetlands are natural systems and can be integrated into the landscape with no nuisance from smell/insects but regulatory authorities and consultants/architects in many parts of the

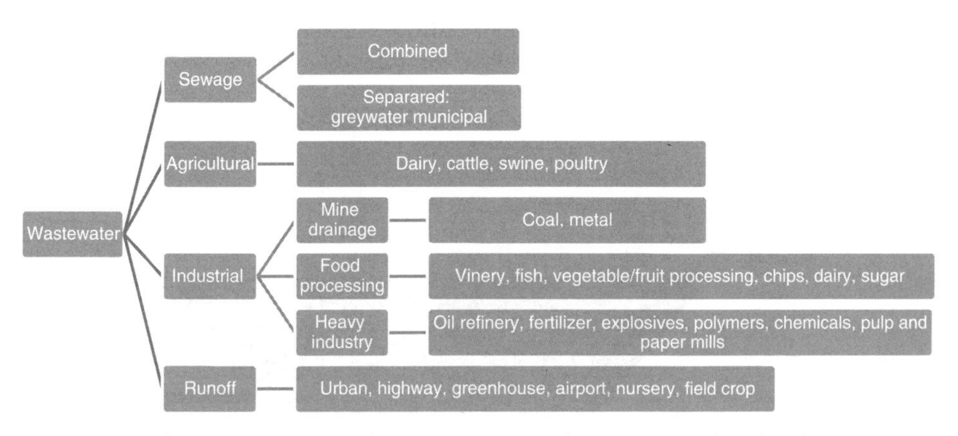

Figure 4.30 *Types of wastewater treated in constructed wetland.*

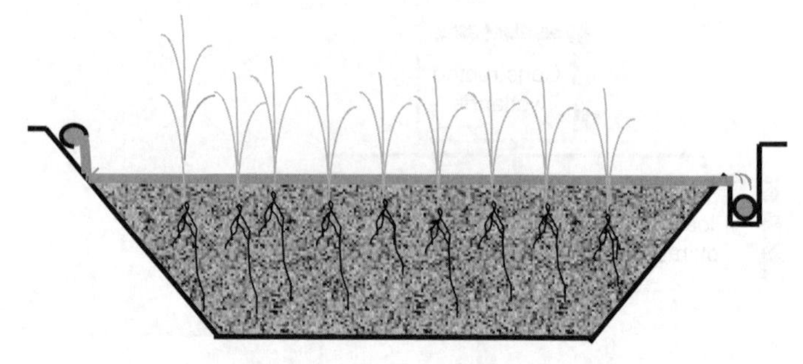

Figure 4.31 *Free water surface flow.*

world do insist on energy-intensive conventional physico-chemical and biological treatment, which is more costly both during construction and operation due to profit and commission associated with costly project.

Advantages of constructed wetland include: (i) site location flexibility (compared to natural wetlands), (ii) simple operation and maintenance, and (iii) it can be integrated attractively into landscaping.

Disadvantages of constructed wetland include: (i) mosquitoes (in free water surface systems), (ii) startup problems, (iii) space requirement, (iv) performance may vary depending on season and influent quality, and (v) designs still largely empirical.

The Guanting reservoir in the Yongding River is an important water sources for Beijing. The water reservoir was not used as a drinking water source after 1997 due to pollution. In order to restore the Yongding river watershed and use it for a water supply, Beijing's municipality and Brandenburg state government constructed the Heituwa wetland system at a place where the Yongding River flows into the Guanting Reservoir. A stabilization pond preceded the constructed wetland. The Guanting Reservoir usually freezes in early December or at the end of November and thaws in March and the ice depth reach around 48 cm. The wetland became fully functional in April 2005, with an inflow of about 1.5 m^3/s, with hydraulic loading of about 0.1–0.4 $m^3/m^2/d$ for the stabilization pond and about 0.4–0.8 $m^3/m^2/d$ for the constructed wetland. The removal rate of BOD_5 was 58–77%, and COD_{Mn} removal rate was higher than 40% (Bingbin *et al.*, 2006).

The North American Space Association fenced off sewage lagoons and introduced plants to successfully treat wastewater. For example, the town of Rio Hondo, Texas, dug lagoons

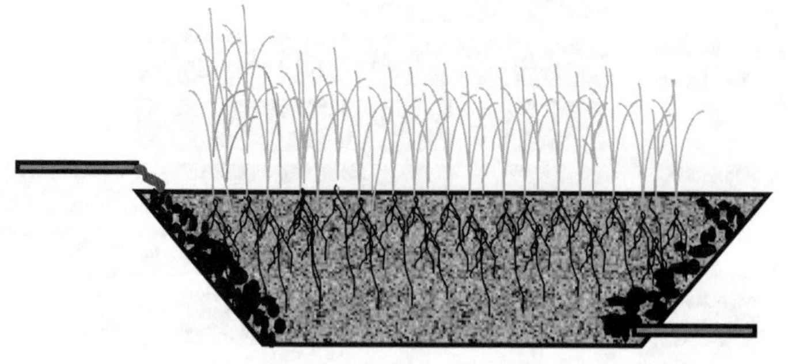

Figure 4.32 *Horizontal subsurface flow type constructed wetland.*

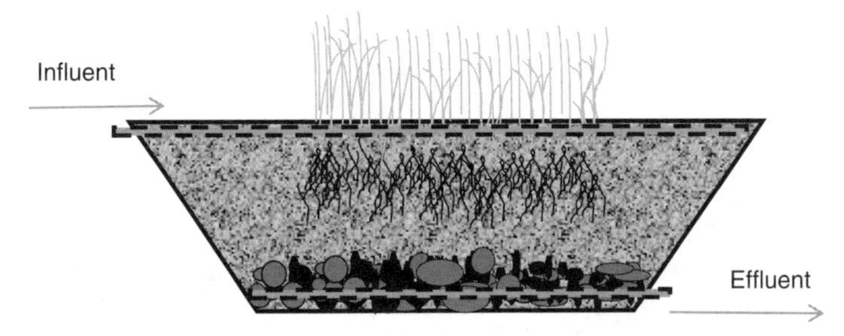

Figure 4.33 *Vertical subsurface flow constructed wetlands.*

and imported water hyacinths thereby reducing the initial cost of the facility by one-twentieth of the conventional system. Orange Grove, Mississippi, used hyacinths. Walt Disney, in the city of Hercules, Georgia, built a water-hyacinth wastewater treatment plant.

Figures 4.34 and 4.35 show constructed wetland. The majority of constructed wetland designers depend on the common reed, Typhas and Phragmites, but a sustainable solution should often consider local variety. In North America, cattails are common. Fishes are introduced to constructed wetlands to eliminate or reduce pests like mosquitoes.

4.4.1.3.1.4 Wastewater Treatment Ponds Effluent treatment using ponds is an economical method of treatment. The number and the type of ponds determine the degree of treatment. Wastewater treatment ponds are also called stabilization ponds.

The degradation in stabilization ponds occurs by aerobic, anaerobic and facultative bacteria. Unpleasant conditions connected with the anaerobic decomposition mean that it is essential to make sure that there is sufficient dissolved oxygen in the pond so that the aerobic and facultative bacteria can predominate.

Figure 4.34 *View of constructed wetlands.*

Figure 4.35 Constructed wetlands built between buildings. (For a colour version of this figure, see the colour plate section.)

One way to supply oxygen is by the use of algae. The algae produce the oxygen required by the bacteria and the bacteria in turn generate carbon dioxide needed by the algae. Algae will thrive in sunlight and cannot grow at night and in periods of cloudy weather. The water in the pond has to be clear to enable growth of algae throughout the pond depth.

If the water is not clear, DO levels in the pond will decrease, resulting in an increase in anaerobic activity. Shorter periods of daylight, snow or ice and cloudy weather prevent the sunlight from reaching the wastewater. Small quantities of oxygen will be produced by the algae and no aeration from wind occurs because of the ice cover. Lower temperatures will slow down the bacterial activity decreasing bacterial activity. In the spring, there will be an increase in bacterial action on the substances deposited during the winter resulting in odours from anaerobic conditions. As the pond warms in spring, the bottom layer of the wastewater starts to mix resulting in smelly gases and solids. Spring turnover odours usually last around seven days.

4.4.1.3.1.4.1 Types of Wastewater Treatment Ponds

- **Aerobic pond.** In the aerobic pond, oxygen will be present throughout the pond and therefore all decomposition occurs aerobically. These ponds are a maximum of 50–60 cm deep, so that the sunlight can reach the entire depth of the pond. The oxygen transferred at the surface and released by algae allows aerobic micro-organisms to live. These ponds are not used in areas predominated by colder climates as they will freeze completely in the winter.
- **Anaerobic pond.** Anaerobic ponds are usually used to treat industrial wastewater with high organic content like distilleries and sugar industry. All the biological activity occurs anaerobically. These ponds are 3 to 4 m deep and are anaerobic throughout. Scum on the top of most anaerobic ponds stops transfer of oxygen from the atmosphere.
- **Facultative pond.** Facultative ponds are commonly used to treat municipal wastewater. The ponds are normally 1 to 2 m deep and therefore sludge at the bottom will be anaerobic,

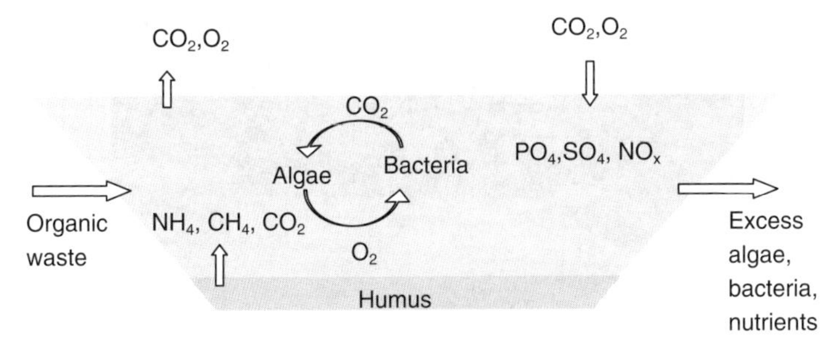

Figure 4.36 *Oxidation pond.*

whereas the 30 to 60 cm of the top of the pond is aerobic. The amount of dissolved oxygen in the middle varies, resulting in either aerobic or anaerobic decomposition.

- **Oxidation pond** (Figure 4.36)**.** This type of pond is designed to receive flows through a primary settling tank or stabilization pond. It is used to provide biological treatment and settling of particles.
- **Raw sewage stabilization pond.** This type of pond is designed to treat wastewater with no prior treatment except for shredding or screening. It is designed to provide a detention period of 45 days and to receive less than 25 kg of BOD_5 per day per acre. The usual operating depth is 1–2 m.
- **Polishing pond.** This type of pond is designed to treat partially treated effluents from the oxidation pond and other secondary treatment systems. They usually operate at a depth of 2–4 m and provide 1 to 3 day detention time. If the detention time is longer, then there will be a rise in the effluent suspended solids concentration.
- **Aerated pond.** In these ponds, oxygen is provided by artificial aeration using a mechanical or diffused air system. The artificial aeration also mixes wastewater in the pond keeping organics and bacteria in contact, thereby allowing the pond to have heavier loading and shorter detention times.

4.4.1.3.1.5 Lagoons When sufficient energy is provided by inserting aeration equipment to mix and aerate then the ponds are called *aerated lagoons*. When the energy is not sufficient to aerate the ponds completely, thereby allowing facultative bacteria to thrive, the lagoons are called facultative lagoon.

4.4.1.3.1.6 Subsurface Treatment Subsurface treatment (Figure 4.37) consists of pretreatment and final treatment. The subsurface treatment consists of single/series of compartments that reduces organic loading from wastewater on final treatment. While a septic tank is sufficient for small discrete dwellings, a series of tanks with packed beds is preferable as the quantity of wastewater increases. The treated water can finally be soaked into land over a large area with plants.

4.4.1.3.2 Anaerobic Treatment

Anaerobic treatment can be used for wastewater or wastewater sludge. Anaerobic treatment can be done in anaerobic ponds or anaerobic reactors where bio-gas can be collected. Some of the examples that use anaerobic treatment to recover energy are wastewater treatment

(a)

(b)

Figure 4.37 Subsurface treatment (a) pre-treatment (b) post-treatment.

plants located in Oakland, Millbrae, West Lafayette in United States, Delhi in India and Stockholm in Sweden (Box 4.3).

Box 4.3 Wastewater Treatment Plant in Bromma, Stockholm

The wastewater treatment plant (WWTP) at Bromma, Stockholm, is capable of treating about 10 000 tons of sewage sludge per annum. Before the 1970s this sludge was deposited in waste deposits. Since the 1970s, sludge has been treated using anaerobic digestion, producing biogas. The biogas was earlier used for heating at the WWTP, and sold for external use. The plant now separates methane from carbon dioxide, water and other contaminants and produces about 1.5 million m^3/year of gas for vehicle fuel of natural gas quality, which corresponds to about 1.5 million litres of petrol.

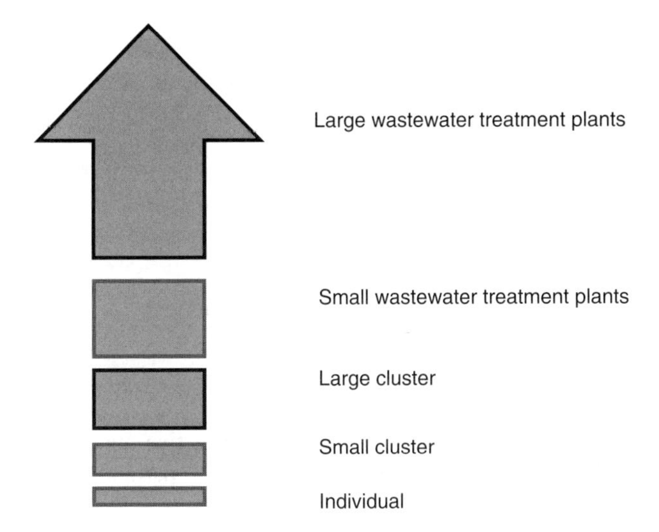

Figure 4.38 *Concept of choice of treatment for different needs.*

4.4.1.3.2.1 Upflow Sludge Blanket Reactor Various advances in anaerobic reactors have resulted in improvements in attached growth systems, suspended growth systems, or combinations thereof. One such process is upflow anaerobic sludge blanket process (UASB). The UASB reactor has a high loading capacity. It is a combination of physical and biological processes. This reactor does not need a separate settler to separate solids and reactor volume is not lost due to filter media (as in the case of anaerobic filters) (Bal and Dhagat, 2001).

4.4.1.3.2.2 DEWAT System Wastewater can be treated through centralized municipal-level systems or decentralized systems. Centralized systems normally discharge to surface waters. Decentralized systems can generate water for local reuse, or disposal. Conventionally, urban wastewater in developed countries is treated at centralized facilities. Industrial wastewater is usually treated on site, although a limited quantity is sent to centralized treatment facility (UNEP, 2010). Figure 4.38 shows the concept of choice of treatment for different needs and Figure 4.39 shows centralized treatment. A decentralized (Figure 4.40)

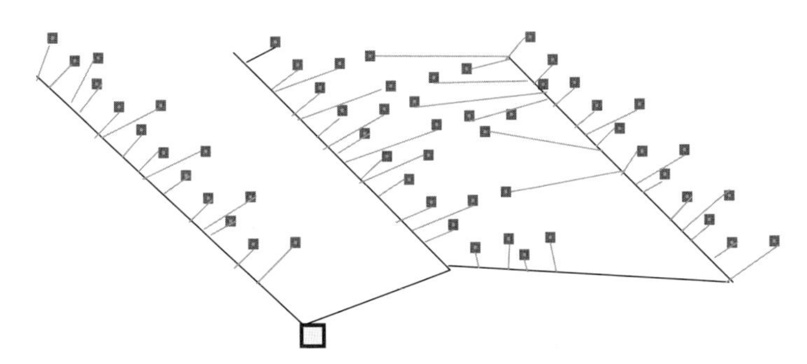

Figure 4.39 *Centralized treatment.*

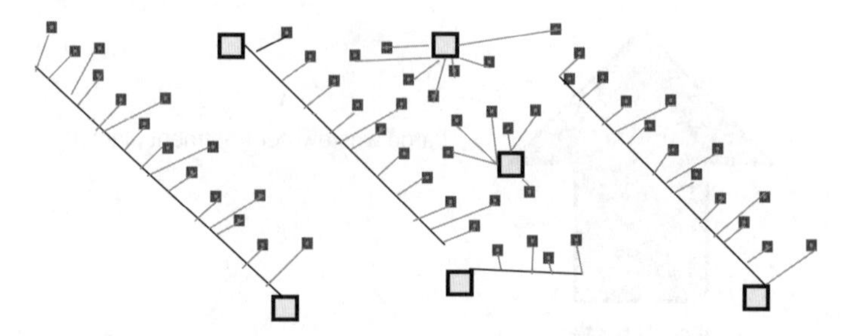

Figure 4.40 Decentralized treatment.

system is a cluster/individual wastewater system used to treat and dispose of small volumes of effluent.

Adequately managed DEWAT systems are a cost-effective and long-term option for meeting water-quality goals and promoting public health, particularly in less populated areas.

Box 4.4 The Kolding Pyramid, Denmark

The Pyramid 'green' sewage treatment plant was built on several stories housed in a pyramid made up of glass panels in the Hollændervej/Fredensgade block in Kolding, Denmark. The block contains 129 apartments with about 250 residents. The project also consists of (i) energy savings in the dwellings; (ii) passive solar heating, (iii) photovoltaics, (iv) water-saving installations, (v) use of rainwater for toilet flushing, (vi) use of sustainable materials, (vii) composting of organic waste, (viii) recycling of paper, glass and so on. The Pyramid was built as common space for the residents and later modified to treat all sewage in the block. The sewage collected was pretreated in a mechanical-biological process, sterilized in an ultraviolet-ozone filter treated by algae and plants. The total tank volume was 460 m^3. The sewage from the Pyramid was 'polished' in a reed-bed followed by infiltration into the ground (Cardiff, NA).

4.4.2 Vermifiltration

A typical conventional vermifilter will have a bed of soli/compost with earthworms over which water is passed. The organic matter is degraded due to a combination of activities involving microbes and earthworms but the earthworms cannot survive in stagnant water, high/low pH, and unfavourable temperatures. Toxic contaminants can also turn out to be unfavourable as they kill earthworms. Ghatnekar *et al.* (2010) have shown that a vermifilter was successful that had three filters with bedding material of leucaenaleucocephala foliage, sawdust, cow dung and bovine urine. The treatment units also employed microbial culture, enzymes, algal cultures, earthworm cultures and canna indica.

4.4.3 Chemical Treatment

Chemical treatment is a method in which chemical reaction is used to treat wastewater economically. *A chemical reaction* occurs when two or more chemically different materials

Plate 1.5 *Definition of blue, green and grey water footprint.*

Plate 2.34 *Green taxis in Stockholm.*

Plate 3.19 *Green roofs.*

Sustainable Water Engineering: Theory and Practice, First Edition. Ramesha Chandrappa and Diganta B. Das.

Plate 3.22 *(a) Colour of the water due to visibility of objects below the water; (b) Colour of the water due to reflection of sky; (c) Colour of the water due to colloidal soil particles; (d) Colour of the water in large forest is yellowish brown hues which is the colour formed during degradation of humus substance in nature. The water bodies in would look dark especially in forest with large pine trees due to colour released from pine needles on the ground.*

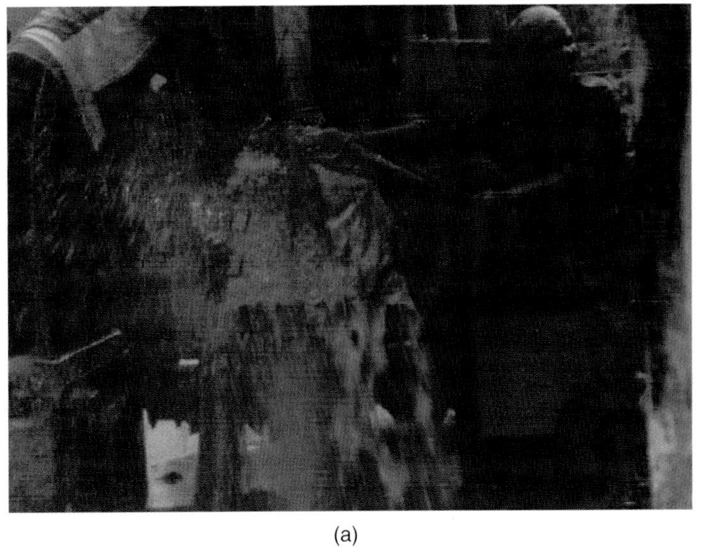

(a)

(b)

Plate 3.23 *Chromium-contaminated groundwater.*

Plate 3.31 *Colour of water that has turned green due to Eutrophication.*

Plate 4.35 *Constructed wetlands built between buildings.*

Plate 4.51 *Solar energy is an attractive idea but still not practical due to high coast associated with installation of potovoltaic cells.*

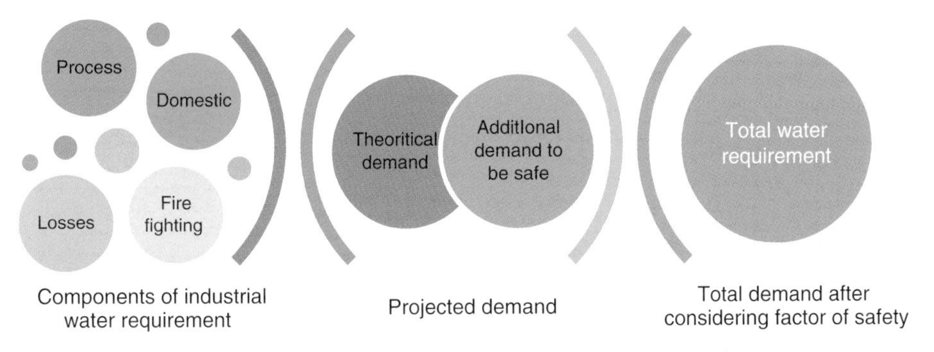

Plate 5.1 *Components of calculating water demand.*

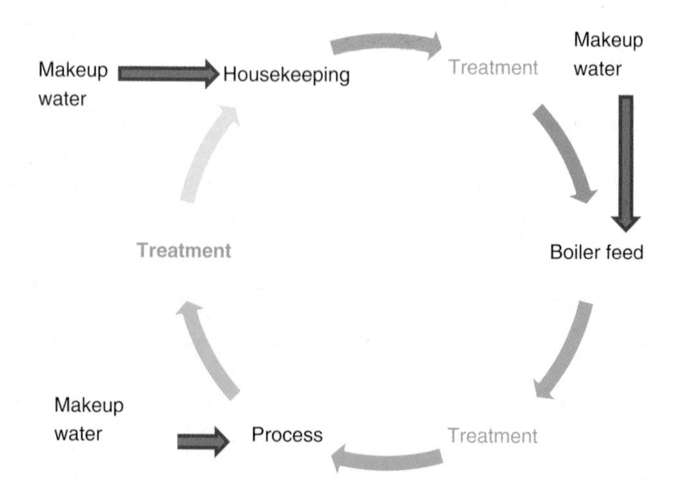

Plate 5.4 *Closed-loop system.*

Plate 5.8 *View of open-cast mining.*

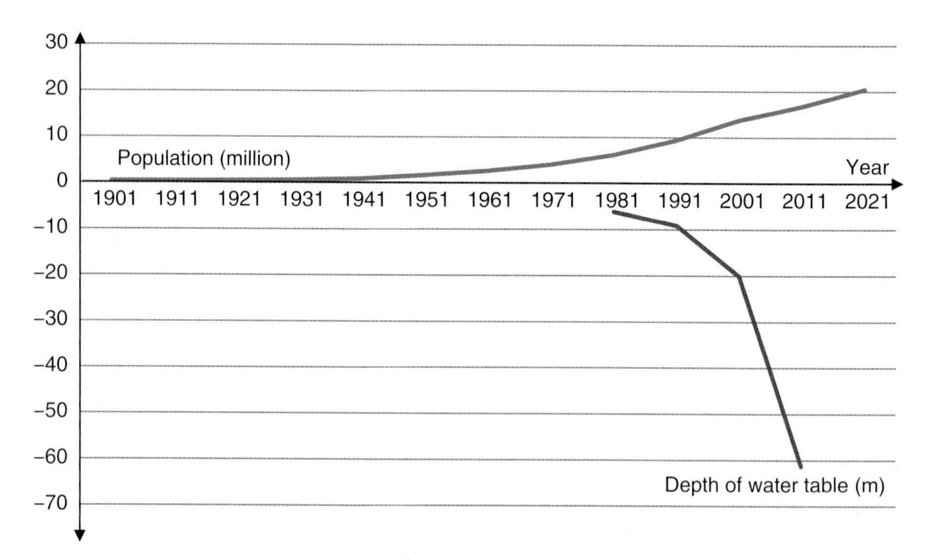

Plate 7.4 *Population growth versus groundwater depletion in Delhi (Source: based on India Today, 2012; India Online pages, 2013).*

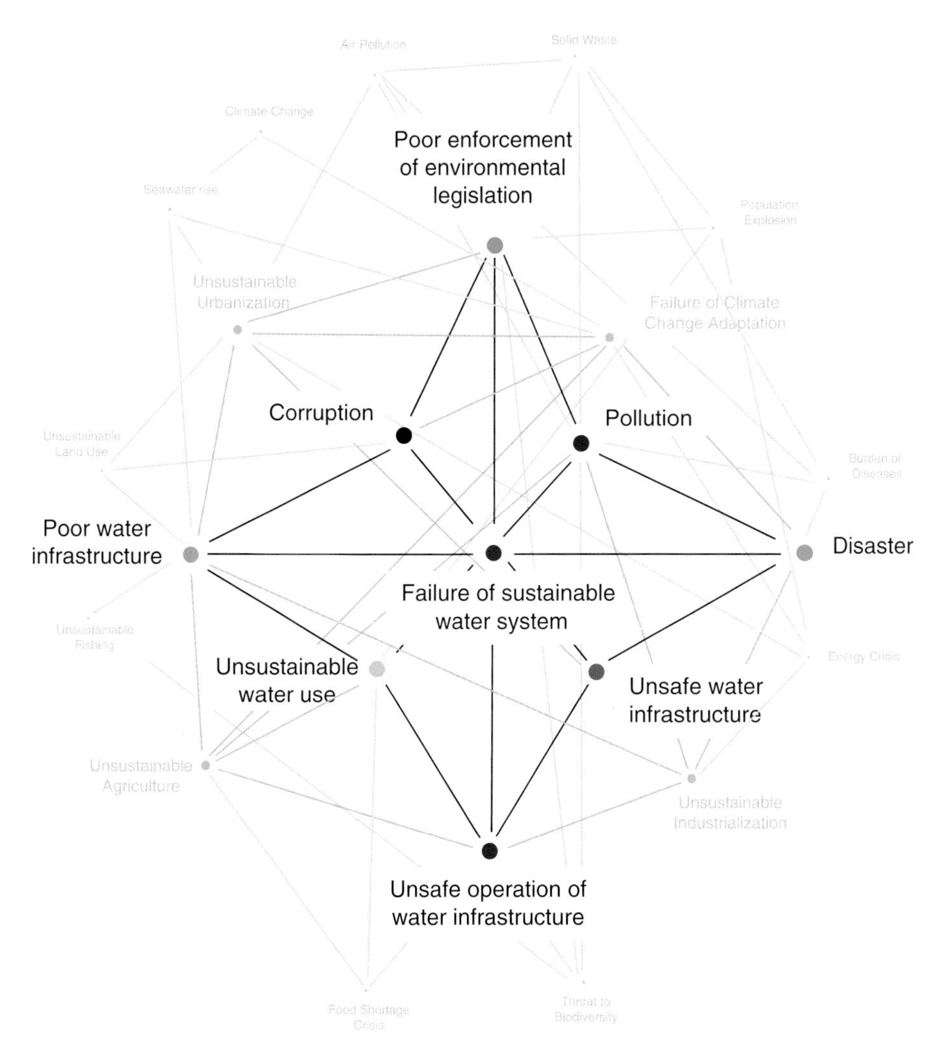

Plate 8.3 *Risk map for health, safety and sustainability.*

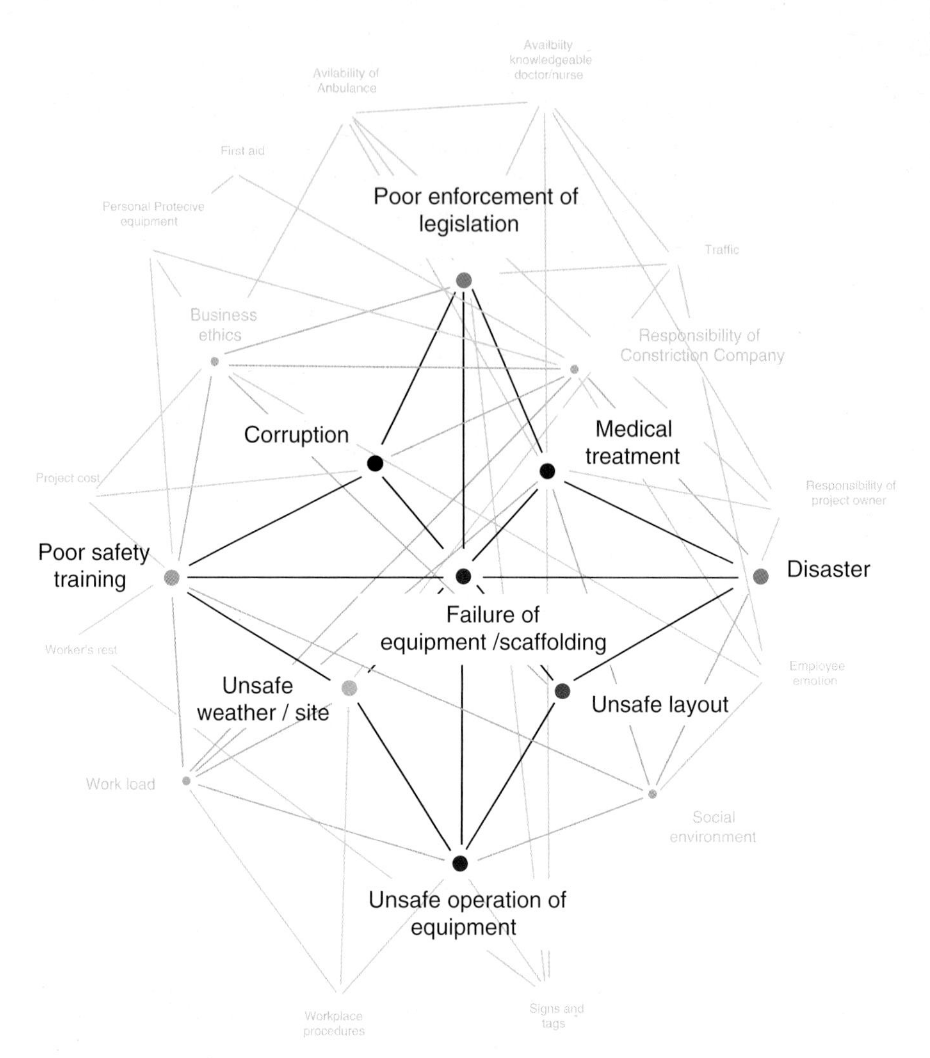

Plate 8.4 *Risk map of safety of construction workers.*

Plate 8.22 *Chlorine stored at wastewater treatment plant site.*

combine to form one. Atoms of different elements combine with each other only in particular ratios to form molecules. Wastewater treatment usually uses those chemical reactions that result in (i) physical change so that separation would be easy or (ii) reduction in harmful constituents.

4.4.3.1 Coagulation/Chemical Precipitation

A clumping of particles in water/wastewater to settle out impurities is called coagulation. Chemical precipitation in wastewater treatment involves the addition of chemicals to alter the physical state of suspended or dissolved solids to facilitate their removal by settling. In some cases removal is achieved by entrapment within flocs.

In the past, chemical precipitation was used to increase the suspended solids and BOD removal. Since about 1970, the need to provide complete removal of the nutrients in wastewater has increased interest in chemical precipitation.

There are two general types of colloidal solid particle dispersions in liquids. When water is the solvent, these are called hydrophobic, or 'water-hating' and hydrophilic, or 'water-loving' colloids. These two types are based on the attraction of the particle surface for water. Hydrophobic particles have relatively little attraction for water whereas hydrophilic particles have a great attraction for water. It should be noted, however, that water can interact to some extent with hydrophobic particles. Some water molecules will generally absorb on the typical hydrophobic surface but the reaction between water and hydrophilic colloids occurs to a much greater extent.

4.4.3.2 Chemical Oxidation

Chemical oxidation is appropriate for *in situ* treatment for environmental remediation used for soil/groundwater to reduce targeted environmental contaminants to acceptable levels.

4.5 Tertiary Treatment

The tertiary treatment of wastewater is the process that follows secondary treatment. It is a polishing process to improve the removal of suspended solids and the nutrient-removal process. The nutrient-removal process in tertiary treatment includes nitrification/denitrification, phosphorous precipitation, ammonia stripping and land application or overland flow.

4.5.1 Filtration

Even though filtration is one of the major unit operations used for drinking water, its use for effluent treatment processes is a much more recent practice. Filtration is now used extensively to remove suspended solids from wastewater effluents of chemical and biological treatment processes. Filtration is also used to remove chemically precipitated phosphorus.

The ability to design filters and to predict their performance must be based on:

- an understanding of the variables that control the process and
- a knowledge of the pertinent mechanism or mechanisms responsible for the removal of particulate matter from a wastewater.

The discussion in this section therefore covers the following topics: (i) description of the filtration operation, (ii) classification of filtration systems, (iii) filtration-process variables, (iv) particle-removal mechanisms, (v) a general analysis of filtration operation, (vi) analysis of wastewater filtration and (vii) the need for pilot plant studies. The literature dealing with filtration is so voluminous that the information presented in this section can only serve as an introduction to the subject. For additional details, the references in the text should be consulted.

In filters that operate continuously, such as the travelling-bridge filter and the upflow filter, the filtering and cleaning (backwashing) phases take place simultaneously. It should be noted that with filters that operate continuously there is no turbidity breakthrough or terminal head loss.

In the travelling-bridge filter, the incoming wastewater floods the filter bed, flows through the medium by gravity and exits to the clear well via effluent ports located under each cell. During the backwash cycle, the carriage and the attached hood move slowly over the filter bed, consecutively isolating and backwashing each cell. The backwash pump, located in the effluent channel, draws filtered wastewater from the effluent chamber and pumps it through the effluent port of each cell, forcing water to flow up through the cell and backwashing the filter medium of the cell. The wash water pump located above the draws water with suspended matter collected under the hood and transfers it to the backwash water through. During the backwash cycle, effluent is filtered through the cells not being backwashed.

In the upflow filter, the liquid to be filtered flows upward through the filter bed. At the same time the sand bed, moving in the counter-current direction, is being cleaned continuously. An airlift is used to pump the sand from the bottom of the filter up through a central pipe to a washer assembly located at the top of the filter. As the sand is being pumped up to the top of the filter, the individual sand grains are cleaned of accumulated material by abrasion (sand against sand) and fluid shear forces. In the sand washer, the accumulated material removed from the sand is removed over a weir. Additional washing of the sand occurs as it passes through the zig-zag flow channel in the lower portion of the sand washer and before it falls back on the surface of the sand bed. Because the effluent water level is higher than the water level in the sand washer, there is a positive upward flow of filtered effluent through the sand washer.

A number of individual filtration-system designs have been proposed and built. The principal types of granular-medium filters may be classified according to (i) the type of operation, (ii) the type of filtering medium used, (iii) the direction of flow during filtration, (iv) the backwashing process and (v) the method of flow rate control.

Flow Control. The rate of flow through a filter may be expresses as follows:

$$\text{Rate of flow} = \text{driving force}/\text{filter resistance}$$

In this equation, the driving force represents the pressure drop across the filter. At the start of the filter run, the driving force must overcome only the resistance offered by the clean filter bed and the under-drain system. As solids start to accumulate within the filter, the driving force must overcome the resistance offered by the clogged filter bed and the under-drain system. The principal methods now used to control the rate of flow through gravity filters may be classified as (i) constant-rate filtration and (ii) variable-declining-rate filtration.

In effluent control systems, at the beginning of the run, a large portion of the available driving force is dissipated at the valve, which is almost closed. The valve is opened as the head loss builds up within the filter during the run. The required control valves are expensive and because they have malfunctioned on a number of occasions alternative methods of flow rate control involving pumps and weirs have been developed and are coming into wider use.

In the application of filtration for the removal of residual suspended solids it has been found that the nature of the particulate matter in the influent to be filtered, the size of the filter material or materials and the filtration flow rate are perhaps the most important of the process variables.

The sand filter is used for the tertiary treatment of wastewater. Rapid sand filters are preferred for large treatment plants whereas pressure sand filters are preferred for small treatment plants.

The most important influent characteristics are the suspended-solids concentration, particle size and distribution and floc strength. Typically, the suspended-solids concentration in the effluent from activated-sludge and trickling-filter plants varies between 6 and 30 mg/l. This concentration is usually the principal parameter of concern, so turbidity is often used as a means of monitoring the filtration process. Within limits, it has been shown that the suspended-solids concentrations found in treated wastewater can be correlated to turbidity measurements. A typical relationship for the effluent from a complete-mix, activated-sludge process is:

$$\text{suspended solids, SS, mg/L} = (2.3 \text{ to } 2.4) \times (\text{turbidity, NTU})$$

Other mechanisms are probably also operative even though their effects are small and, for the most part, masked by the straining action. These other mechanisms include interception, impaction and adhesion.

It is reasonable to assume that the removal of some of the smaller particles must be accomplished in two steps involving (i) transportation of the particles to the surface where they will be removed and (ii) the removal of particles by one or more of the operative removal mechanisms.

The removal of suspended material by straining can be identified by noting (i) the variation in the normalized concentration-removal curves through the filter as a function of time and (ii) the shape of the head loss curve for the entire filter or an individual layer within the filter.

4.5.2 Activated Carbon Treatment

Activated carbon treatment is normally used for polishing residual organic/inorganic pollutants. The nature of activated carbon, the use of granular carbon and powdered carbon for wastewater treatment and carbon regeneration are discussed below.

Activated carbon (Figure 4.41) is produced from carbonaceous materials such as nutshells, lignite, coal, peat, wood, coir and petroleum pitch by physical/chemical means:

- **Physical reactivation**. This is generally done by pyrolysis of carbonaceous material at 600–900 °C, in the absence of oxygen followed by exposure to oxygen, carbon dioxide, or steam at temperatures above 250 °C. Typically the temperature range used is from 600 to 1200 °C.
- **Chemical activation**. In this procedure the raw material is impregnated with chemicals like an acid (phosphoric acid), strong base (sodium hydroxide, potassium hydroxide), or a

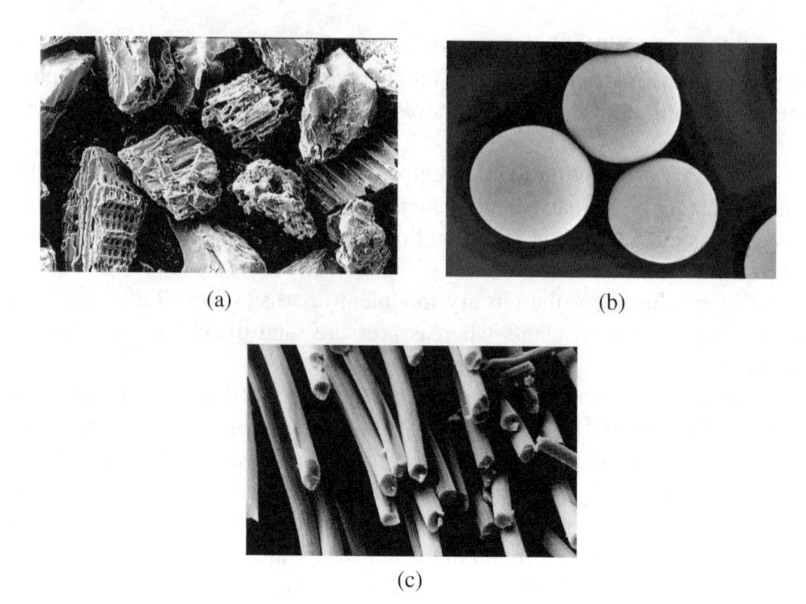

Figure 4.41 Electron microscope picture of activated carbon (a) spherical: polymer pyrolized carbons (b) irregular granular: wood-based active carbon (c) polyacrylonitrile derived carbon fibre. (Source: Dr Danish Malik, Loughborough University, UK).

salt(calcium chloride, and zinc chloride) followed by carbonization at lower temperatures (450–900 °C).

Activated carbon is mainly made up of coconut shells. It is normally used in water filtration systems. It has numerous applications such as: (i) groundwater remediation, (ii) spill cleanup, (iii) drinking water filtration.

4.5.3 Ion Exchange

A variety of dissolved solids can be removed by ion exchange. In water softening (Figure 4.42) calcium and magnesium in the water are exchanged with non-hardness-causing ions, usually sodium. In the case of demineralization, hydrogen is exchanged for cations and hydroxide for the dissolved anions. Ion exchange is done in two or three steps. Generally cations are removed first and anions are removed later.

4.5.4 Forward and Reverse Osmosis, Membrane Filtration, Membrane Bioreactor, Membrane Distillation, and Electro Dialysis

The major desalination technologies in use are based on reverse osmosis and thermal distillation with RO accounting for more than 50% of the installed capacity (Zhou and Tol, 2005; Veerapaneni *et al.*, 2007). Conventional thermal desalination processes suffer from corrosion and are inefficient in their use of energy. High capital and operational costs of thermal desalination limit adoption. However, small solar thermal distillation and humid air desalination technologies are finding an increasing role in remote parts of the developing

Figure 4.42 *View of pressure sand filter and softener.*

world, mainly in inland semi-arid places with access to saline lakes and aquifers (Bourounia *et al.*, 2001; Shannon *et al.*, 2008).

Despite possessing yearly renewable water resources of 2.4 billion m^3 (FAO, 2008), Saudi Arabia's water extraction was more than 20 billion m^3 in 2010 (FAO, 2008; SAMC, 2010), making it the third biggest per capita water user worldwide (SAMC, 2010). Large-scale thermal desalination in Saudi Arabia is an energy- and capital-intensive practice. Plant operation results in GHG emissions of 10 to 20 kg CO_2/m^3 of desalinated water generated (AFED, 2010) and the environmental impacts of the ocean disposal of the remaining brine remain are yet to be properly ascertained. Desalination processes using multi-effect distillation (MED) and multistage flash distillation (MSF) use steam from power-generation plants (ESCWA, 2001) and electricity.

Desalination technologies currently under investigation are forward osmosis (McCutcheon *et al.*, 2005) and membrane (Figures 4.44 and 4.45) distillation (Mathioulakis *et al.*, 2007), may be used alone, or as hybrid systems with reverse osmosis (RO) (Figure 4.43). The specific energy of desalination has reduced from over 10 kWh/m^3 in the 1980s to below 4 kWh/m^3 (Alonitis *et al.*, 2003; Veerapaneni *et al.*, 2007).

Major challenges of RO desalination are membrane fouling, poor removal of low molecular weight contaminants and poor recovery for seawater desalination ($< 55\%$).

4.5.5 Air Stripping

Transferring of volatile compounds of a liquid into an air is called air stripping. The process, which is a chemical engineering technology, is now widely used for wastewater treatment and purification of ground waters containing volatile compounds.

The contaminants stripped include benzene, toluene, ethylbenzene, xylene and ammonia. Some of the compounds stripped are toxic in nature, so the stripped air is passed thorough air-pollution-control equipment like carbon adsorption or bio towers.

Soil venting is an *in situ* air-stripping method used to remove volatile pollutants/contaminants from the vadose zone, where air is forced into the soil subsurface, followed by venting.

Figure 4.43 Reverse osmosis system.

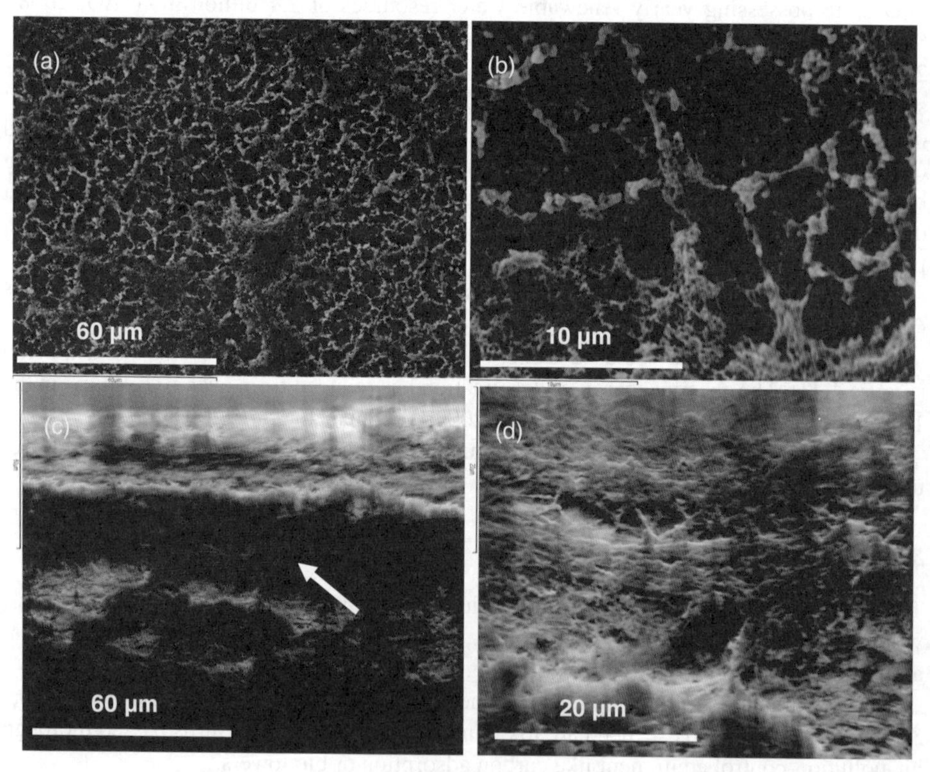

Figure 4.44 SEM micrographs showing morphology of cellulose nitrate membrane: (a) top view of membrane; (b) zoom-in top view of membrane; (c) cross section of membrane; (d) a zoom-in SEM image corresponding to the arrowhead pointed area in (c).

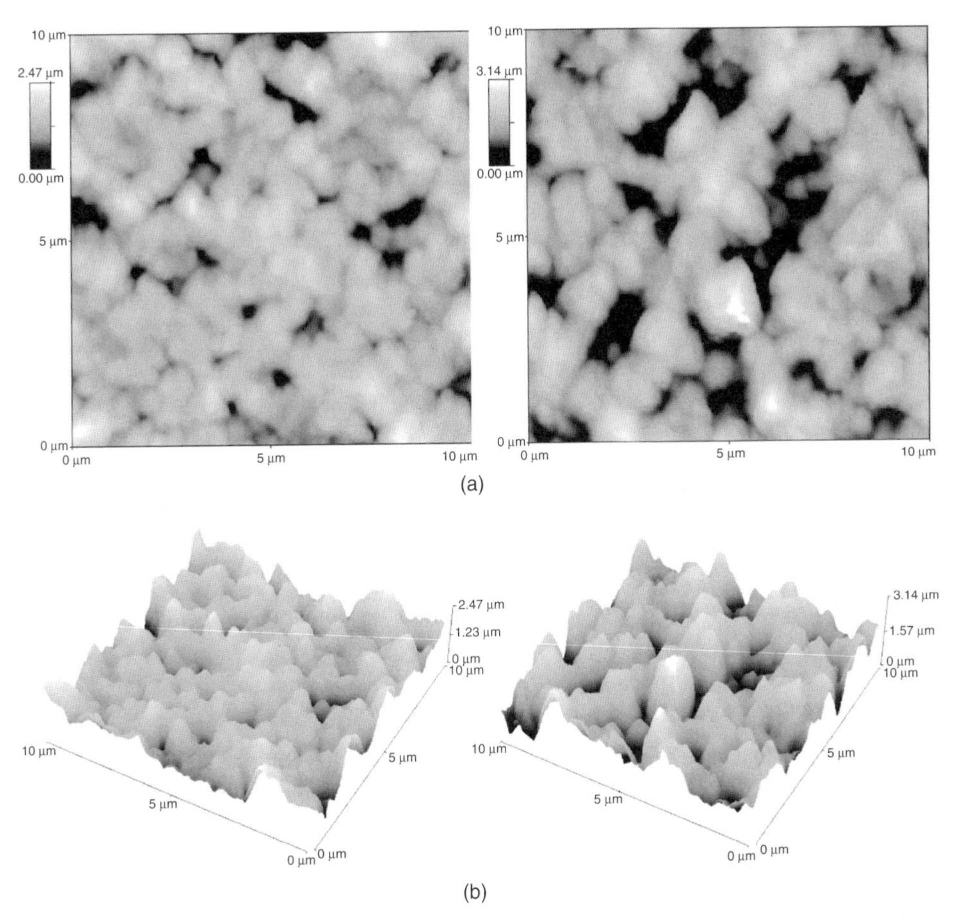

Figure 4.45 *Atomic force microscopy (AFM) topographic images of cellulose nitrate membrane: (a) 2D AFM image; (b) 3D AFM image.*

4.5.6 Disinfection and Fluoridation

Disinfection is selective destruction of disease-causing microbes. Not all the microbes are killed during the process, as in the case of sterilization.

In the context of wastewater disinfection, viruses, bacteria and amoebic cysts are important. Diseases caused by bacteria include cholera, typhoid, bacillary dysentery and paratyphoid; diseases caused by viruses include infectious hepatitis and poliomyelitis. The purpose of this section is to introduce the reader to the general concepts involved in the disinfection of micro-organisms.

Disinfection is usually accomplished by the use of (i) physical agents, (ii) chemical agents, (iii) radiation and (iv) mechanical means. Each of these techniques is considered in the following discussion.

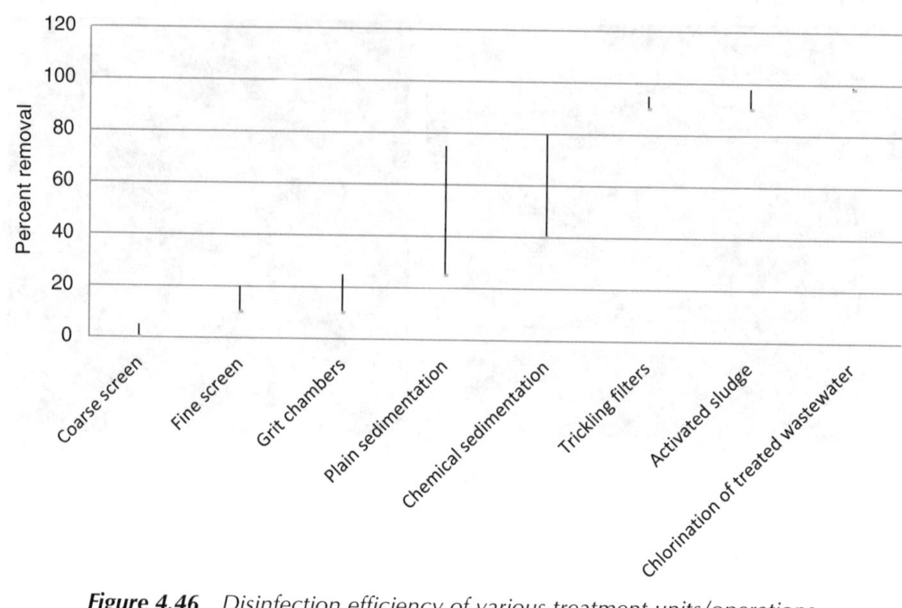

Figure 4.46 *Disinfection efficiency of various treatment units/operations.*

4.5.6.1 Chemical Agents

Chemical agents that have been used as disinfectants include (i) chlorine and its compounds, (ii) bromine, (iii) iodine, (iv) ozone, (v) phenol and phenolic compounds, (vi) alcohols, metals and related compounds, (vii) dyes, (viii) soaps and synthetic detergents, (ix) quaternary ammonium compounds, (x) hydrogen peroxide and (xi) various alkalies and acids.

The most common disinfectants is chlorine. Iodine and bromine have also been used for wastewater disinfection. Ozone is an effective disinfectant and it does not leave a residue.

4.5.6.2 Physical Agents

Physical disinfectants commonly used are heat and light. Heating water to boiling point will destroy most of the disease-causing non-spore-forming bacteria. Heat is not a feasible means of disinfecting large quantities of wastewater because of the high cost but pasteurization of sludge is done extensively in Europe. Special lamps that emit ultraviolet rays are used to disinfect small quantities of water.

4.5.6.3 Mechanical Means

Micro-organisms are removed by mechanical means reported in Figure 4.46. The first four operations shown in figure may be considered to be physical.

4.5.6.4 Radiation

The main types of radiation used are acoustic, electromagnetic and particle. Gamma rays have been used to disinfect both wastewater and water due to penetration power.

Ultraviolet radiation can be done by UV lamp powered from an available electricity source or through solar UV radiation technology. There are two mechanisms used for solar UV radiation: (i) using the natural wavelengths of sun rays during sunshine hours; (ii) using photo voltaic (PV) panels, to light UV lamps.

4.5.6.5 Mechanisms of Disinfectants

Mechanisms proposed for disinfection are: (i) alteration of cell permeability, (ii) damage to the cell wall, (iii) inhibition of enzyme activity and (iv) alteration of the colloidal nature of the protoplasm.

Agent such as phenolic compounds and detergents destroy the selective permeability of the membrane. Radiation, heat and highly acidic/alkaline agents alter the protoplasm. Heat will also coagulate the cell protein; acids or bases will denature proteins, resulting in a lethal effect. Oxidizing agents alter the chemical arrangement of enzymes, deactivating them.

4.5.6.5.1 Solar Pasteurization

Pasteurization has been widely used in purifying milk but has only recently been used in water treatment. Water is pasteurized by heating it to the pasteurization temperature (65 °C). The volume purified by solar pasteurization per unit area is about 0.1 m^3/m^2. Solar pasteurization can be done with (i) a flat panel collector or (ii) compound parabolic collectors. In a flat panel collector, water is moved through a flat structure so that sunlight passing through the transparent side heats the water to the pasteurization temperature. In compound parabolic collectors, pasteurization is done in a parabolic-shaped container.

4.5.7 Removal of Specific Constituents

Removal of specific constituents needs a target-based approach. If an industry or activity uses radioactive material it is essential that radioactive material be expected in wastewater and that arrangements be made well in advance, even prior to trial productions. Specific constituents are determined based on the mass balance and water balance of the industry rather than analysing wastewater for all the known pollutants after the industry is set up. An industry has to know the characteristics of effluents at the planning stage itself and arrangement must be made to construct the effluent treatment plants before trial production. The possible ways in which the characteristics of wastewater are determined are: (i) literature survey, (ii) mass balance and (iii) effluent analysis from pilot plants in case of new products. The characteristics calculated theoretically do vary during production but calculation should aim to predict characteristics within a manageable range as a slight deviation of 5–10% is manageable during commissioning stages but major deviation from predicted characteristics needs major alteration of the effluent-treatment plants.

4.6 Emerging Technologies

Most of the treatment methods have been evolved using fundamental or applied science. The natural degradation of organic matter has led to the invention of basic treatments like the septic tank and the Imhoff tank. The advanced water/wastewater treatment options used knowledge of fluid mechanics, electrical science, electronics, mechanical engineering, chemical engineering and civil engineering. The invention of the computer and its use in environmental engineering has led to its application in modelling, simulation and operation control. Over the years, scientists across the world have continuously attempted to bring knowledge from other areas of pollution control to: (i) cut construction/operational cost, (ii) reduce the requirements for cumbersome operations, (iii) reduce energy/chemicals, (iv) decrease space requirements and (v) enhance safety. Some of the promising emerging technologies are given in Table 4.7.

4.6.1 Nanotechnology applied for Water Purification

Nanotechnology is applicable in the detection and removal of various pollutants. Methods such as photocatalysis, nanofiltration, adsorption and electrochemical oxidation use TiO_2, ZnO, nanowire membranes, polymer membranes, ceramic membranes, carbon nanotubes, submicron nanopowder, magnetic nanoparticles and metal (oxides). Nanostructured boron-doped diamond is used to diminish problems related to water quality in natural environment (Mueller and Nowack, 2009; Mamadou and Savage, 2005).

Nanosized zero-valent ions or nanofiltration membranes are used for pollutant removal/separation from water. Nanoparticles can also be used as adsorbents, catalysts for chemical/photochemical oxidation for the destruction of contaminants. Nanoscale materials used in water treatment can be classified into dendrimers, zeolites, metal-containing nanoparticles and carbonaceous nanomaterials. Dendrite polymers include dendrigraft polymers, random hyperbranchedpolymers, dendrons and dendrimers. Metal-oxide nanoparticles include zinc oxide (ZnO), titanium dioxide (TiO_2) and cerium oxide (CeO_2) and are good adsorbents for water purification. MgO nanoparticles and magnesium (Mg) nanoparticles act as biocides against Gram- negative and Gram-positive bacteria and bacterial spores. Nano TiO_2 and Cu_2O electrodes can be used for electrocatalytic oxidation of organic components. Silver-loaded nanoSiO_2coated with crosslinked chitosan has biocidal activity against *Escherichia coli* and *Staphylococcus aureus* (Mei *et al.*, 2009). Zinc-oxide nanoparticles can be used to remove arsenic. Zeolite nanoparticles can be used for ion-exchange media for metal ions and effective sorbents for metal ions. Carbon-based nanoparticles like carbon nanotubes, nanodiamonds, fullerenes/buckyballs (Carbon 60, Carbon 20, Carbon 70) and nanowires act as sorbents. Nanosilver has strong antimicrobial uses and is used for a wide variety of products (Tyagi *et al.*, 2012).

4.6.2 Photocatalysis

Photocatalysis is an acceleration of a photoreaction in the existence of a catalyst. Organometallic complexes (pigments) and semiconductors are photocatalytic substances. Their general functions of photocatalysis include deodorizing, antibacterial effect, sterilizing, antifouling and removal of toxic substances. Compared to other semiconductor photocatalysts, nanocrystalline titanium dioxide (NTO) is unique in terms of biological inertness, photostability and low cost of production (Mills and LeHunte, 1997). Photocatalytic air/water purification using NTO is predominant in an advanced oxidation process (AOP) due to its efficiency and eco-friendliness. Nanocrystalline titanium dioxide photocatalysis can be used for pharmaceuticals, cosmetics, phenolic compounds, dyes, pesticides, herbicides and toxins.

Table 4.7 *Emerging technology in wastewater treatment.*

Sl. no.	Technology	Description
1	Adsorption process using macroporous resin	Macroporous adsorption resin is a polymer without ion exchange that does not dissolve in organic solvent, acid and alkali. Macroporous adsorption is used in adsorbing, recovery and purifying of chlorobenzene, benzene nitrobenzene, fluorinbenzene, phenol and other organic compounds.
2	Nanotechnology	When modified at nanoscale, materials can exhibit certain useful properties that were not observed before. Nanotechnology is used for desalination, as a catalyst for photochemical destruction, biocides, adsorption, removal/destruction of contaminants.
3	Advance Oxidation Process	Advanced oxidation processes (AOPs) are treatment processes based on highly reactive radicals. The process includes ozone, ozone-hydrogen peroxide, ozone-ultraviolet, hydrogen peroxide-ultraviolet, ltraviolet-TiO_2. Application of ozone include coagulation aid, Fe and Mn removal, colour removal, taste and odour control, algae removal, synthetic organic compound (SOC) oxidation, disinfection, disinfection dy-products (DBP) control, bromate formation, assimilable organic carbon (AOC) formation. Application of O_3-H_2O_2 technology is used for SOC oxidation, Taste and odour reduction, DBP control and disinfection. Application of O_3-UV technology is used for micro-pollutant destruction. H_2O_2-UV photolysis is used for SOC oxidation. UV-TiO_2 technology is used for hazardous compound oxidation. Ultrasonic technology may be used for wastewater treatment and water. This technology acts as an AOP. The process leads to decomposition of complex organic compounds to simpler compounds during cavitation process (Mahvi, 2009).
4	Photocatalytic degradation	Semiconductor photocatalytic process is a low-cost, environmentally friendly and sustainable treatment option. This process can be used to remove persistent organic compounds and micro-organisms in water. Currently, the main constraint is the postrecovery of the catalyst after wastewater and water treatment.
5	Electrochemical oxidation	In electrochemical oxidation, an electrochemical cell is used to generate an oxidizing species (metal ion) at the anode. Electrochemical technology can be used to treat toxic effluents with high concentrations of organic compounds. The disadvantages of this technology include: passivation (building of layers of blocking films leading to the inefficiency) and destabilization of electrode material.

4.6.3 Evaporation

Evaporation can be carried out either with solar energy or external energy. Solar evaporation by stagnating water in open shallow ponds of 5–10 cm is a solution that has been recorded in many developing countries but in reality the solution fails to deliver results during rainy days due to overflow from tanks and an increase in volume of the wastewater that is being evaporated. Solar distillation will be a sustainable solution in such cases. A normal solar still structure used for solar distillation will have a slanted glass cover over a black basin. Solar energy absorbed by the basin is transferred to the water. Evaporated water collects on the glass cover and condensate beads trickle down the slanted glass to be collected in a gutter designed to collect pure water. Solar distillation requires solar energy for longer periods.

Due to the slow rate of evaporation, the production of pure water per square metre of solar still is low. Hence solar distillation is a technology that can be used in a small business that generates a small quantity of wastewater that can be evaporated by the sun followed by sludge disposal along with other solid waste.

4.6.4 Incineration

Incinerators have been a sustainable option where all other treatments failed to deliver the desired results. Incinerators can be used for almost all types of liquids. High-COD liquids, liquid chemical waste, low CV (calorific value) liquids, dye waste, acidic liquid waste, insecticides, pesticides, fungicide waste, herbicides, mixed liquor wastewater, agricultural liquid chemical waste, waste oil refinery waste, ethanolamine, waste of chemical plants, alcoholic waste, inorganic chemical waste, organic chemical waste, black liquor waste, grey water waste, chemical laboratory liquid waste, petrochemical waste, winery wastewater, oil refinery waste, coffee wastewater waste, petrochemical waste, slurry, sludge liquid waste, slug mix liquid, biomedical lab liquid, restaurant and hotel liquid waste, clinical liquid waste, bulk drug liquid waste, oil-mixed water waste and many more. Taking into account the economics of operation and safety requirements, the incinerator is used where there is sufficient skilled man power available. The choice has proven very practical in case of spent wash distillery effluent, which is a high-COD effluent with high colour, which cannot be treated with conventional biological processes to acceptable standards.

4.6.5 Sono-Photo-Fenton Process

An AOP is an attractive option for the degradation of nonbiodegradable organic pollutants. AOPs have the advantage that they generate chemical or biological sludge and almost totally demineralize organics but the processes are associated with high costs for energy and reagents.

Fenton's reagent is a solution of an iron catalyst and hydrogen peroxide used to oxidize contaminants. It can be used for destroying organic compounds. Oxidation of an organic chemical by Fenton's reagent is fast and exothermic resulting in the oxidation of chemicals primarily to carbon dioxide and water.

Iron (III) is then reduced to an iron (II) peroxide radical and a proton by the hydrogen peroxide. The hydroxyl free radical produced by Fenton's reagent is a nonselective and powerful oxidant.

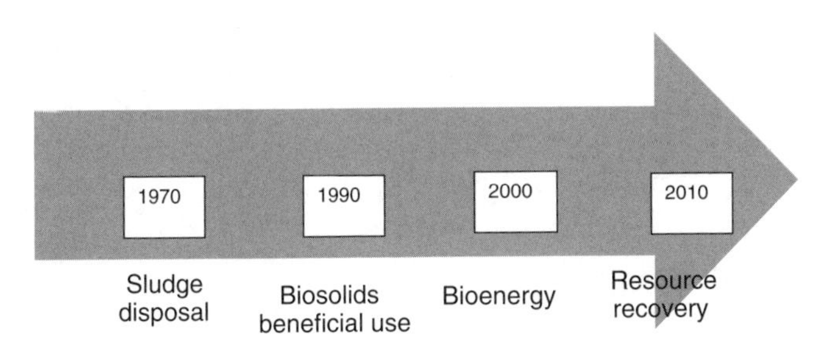

Figure 4.47 *Sludge management in developed countries.*

Sono- and photo-Fenton processes are applied separately and can be combined to provide improved results for wastewater and potable water treatment (Segura *et al.*, 2009).

4.7 Residual Management

Sewage sludge consists of the organic and inorganic solids present in the effluent. The generated sludge will be in the form of a liquid/semisolid, containing 0.25 to 12% solids by weight depending on wastewater and the treatment processes. Sludge is treated by a variety of processes in various combinations. Thickening, conditioning, dewatering and drying are used to remove moisture. Digestion, incineration, composting, wet-air oxidation and vertical tube reactors are employed to stabilize the organic matter in the sludge.

Residual management can be made highly profitable by recovering energy and resources. A focus on solids as a resource has increased in developed countries (Figure 4.47) and is expected to continue and expand. Major operational themes likely to affect the future of residual management technology are sustainability, funding and a shifting and/or unclear regulatory landscape (CDM, 2011).

The most common method is sludge-to-biogas processes by anaerobic digestion. In addition, new technologies like thermal hydrolysis and cell-destruction methods like pulse electrical fields, ultrasonic treatment, ozone, and mechanical disintegration are also used. The pyrolysis and gasification of sludge are also employed in many treatment facilities (in Germany, Australia, Canada, Japan and the United States) to recover energy. A sludge-to-liquid process called supercritical water oxidation, involving heating at temperatures of about 374 °C, at high pressure and using pure oxygen, is emerging as promising solution o residual management.

Biosolids are organic matter resulting from the treatment of sludge generated from wastewater treatment facilities. Sixteen Bay Area agencies in the San Francisco Bay Area created the Bay Area Biosolids to Energy coalition in order to create a sustainable solution to biosolids by management. The Encina Wastewater Authority, in San Diego manages biosolids by (i) prioritizing short-term uses like biofuel generation and landfill application, (ii) using biosolids in other markets like recreational fields and/or orchards (USEPA, 2013).

Many technologies are available to recover nitrogen, phosphorus, sculpture, volatile acids and inert building materials (cement, lightweight aggregates, brick and glass) (USEPA, 2013).

Figure 4.48 Sludge drying beds.

4.7.1 Thickening

Thickening is the process of increasing the solid content in sludge by removing a portion of its liquid. A rise in solids from 3% to 6% can reduce sludge volume by 50%. Sludge thickening methods include gravity flotation, centrifugation, settling and gravity belts.

4.7.2 Drying

Sludge drying varies from simple solar drying to energy-intensive drying. Sludge-drying beds (Figure 4.48) are used most widely where solar energy is available sufficiently. The key advantages of drying beds are low cost and high solids content after drying. The principal disadvantages are the large amount of space required and the effects of climatic changes, insects and potential odour. Table 4.8 gives types of sludge dryers commonly employed across the globe.

4.7.3 Stabilization

Sludges are stabilized to remove offensive odours, pathogen content and reduce/remove the potential for putrefaction. Methods used for sludge stabilization include heat treatment, aerobic digestion, lime stabilization, composting and anaerobic digestion.

Lime is added to sludge in lime stabilization to increase the pH to 12 or more. Hydrated lime and quicklime are usually used for lime stabilization. The high pH environment hinders the survival of microbes and therefore eliminates the risk of putrefaction and odour formation. Lime is added either before dewatering or after dewatering.

In heat stabilization, sludge is heated in a pressure vessel to 260°C at a pressure of up to 2760 kN/m^2 for about 30 s, leading to the hydrolysis of proteinaceous compounds and

Table 4.8 *Types of sludge dryers.*

Sl. no.	Type of sludge dryer	Description
1	Sand drying beds	Sand beds comprises of a layer of coarse sand on a graded gravel bed with underdrains. Sludge is allowed to dry by evaporation and drainage after which sludge cake is removed manually.
2	Paved drying beds	These are two types of paved drying beds: (i) decanting type and (ii) drainage type. The drainage type involves agitation and uses a front-end loader. The decanting type uses impermeable paved beds. It relies on mixing of the supernatant and drying sludge for enhanced evaporation.
3	Wedge-wire drying beds	Wedge-wire beds consist of beds made up of high-density polyurethane or stainless steel wedge wire and the process is controlled by an outlet valve.
4	Vacuum-assisted drying beds	In this system, drying is accelerated by application of vacuum from underside of filter plates.
5	Drying lagoons	Sludge-drying lagoons consist of shallow earthen basins where sludge is placed and allowed to dry. The supernatant is decanted to the plant while the liquid evaporates.
6	Flash dryer	Sludge is dried rapidly in the presence of hot air.
7	Spray dryer	Liquid sludge is sprayed into a drying chamber.
8	Rotary dryer	Involves physical contact of sludge with heat in a rotary drum. Direct heating involves contact with hot gases, while in case of indirect heat a drying chamber is surrounded with steam.
9	Multiple-hearth dryer	Heated air and combustion products are passed over sludge, which is continuously raked to expose fresh surfaces.

resulting in cell destruction and the release of soluble organic compounds and nitrogen. The process also releases bound water and causes the coagulation of solids.

4.7.4 Digestion

Sludge digestion can be anaerobic or aerobic. Anaerobic digestion (Figure 4.49) involves the reduction of organic compounds in the sludge by anaerobic biological activity. The methane generated can be recovered for reuse.

Aerobic sludge digestion involves the oxidation of biodegradable matter and microbial cellular material. Aeration is achieved naturally or by means of mechanical aerators and diffusers.

4.7.5 Composting

Composting is the decay of organic matter by micro-organisms in a controlled facility. Waste composting is a method of treating organic fractions and involves three steps: (i) preparation, (ii) decomposition and (iii) product preparation. Composting is used for sludge stabilization and final disposal. During composting, enteric micro-organisms will die due to the increase

Figure 4.49 *View of digesters.*

in temperature of the composted organic matter and will undergo biological degradation, resulting in a 20% to 30% decrease in volatile solids.

The composting processes used are in-vessel composting and windrow-based composting. In the windrow method, waste is formed into windrows of 1 to 2 m high, which are turned from time to time, mechanically or manually, to maintain a steady temperature. Water is added to sustain optimum moisture content. After the required decomposition has occurred, the composted material can be used as manure. Aeration of windrows can also be achieved by passing air into the windrows through pipes. Plastic sheets are used in some places to protect the windrows from the weather.

In in-vessel composting the waste is kept in a series of waste vessels with a provision to maintain the required humidity and temperature.

Vermicomposting is a process of composting using earthworms. The vermicomposting is carried out at 10–32 °C. A disadvantage of vermicomposting is that the process requires more labour and space as worms operate only up to about 1 m in depth. The worms are susceptible to changes in temperature, toxic substances, pH, and other inhibiting factors such as waterlogging.

4.7.6 Dewatering

A number of methods are used for dewatering. When there is a limited amount of space available and the amount of sludge is small, drying beds and drying lagoons are usually adopted. Mechanical dewatering methods like the centrifuge, filter press, belt filter press and vacuum filter systems are used when the amount of sludge to be handled is high and sufficient energy is available.

4.7.7 Incineration

Incineration as a technology is used for treating wastewater when it has high calorific value, as in the case of distillery waste. The wastewater is passed through multiple evaporators to concentrate the waste and it is finally fed into the incinerator with supplementary fuel. For example, the wastewater treatment facility at the Erie Wastewater Treatment Plant,

Pennsylvania, in the United States, used to grow algae in a bioreactor. A photo-bioreactor is fed with carbon dioxide generated during the incineration of flue gas.

4.7.8 Remediation of Contaminants in Subsurface

Approximately 40% of the drinking water in the United States comes from groundwater and nearly 97% of the rural population depends on groundwater for drinking water (Sharma and Reddy, 2004). The percentage utilization of ground water varies from country to country and the importance given to ground water varies hugely. The USEPA has estimated that more than 217 000 contaminated sites need urgent remedial action (Reddy, 2008).

4.7.8.1 Site Characterization

Site characterization consists of the collection and assessment of data about pollutant types and distribution at the site and forms the basis for decisions relating to the requirements of remedial action. It includes data collection pertaining to site contamination, site geology and site conditions.

4.7.8.2 Risk Assessment

Risk assessment is evaluation to determine the risk due to contamination. Risk assessment consists of four steps: (i) exposure assessment, (ii) hazard identification, (iii) risk characterization and (iv) toxicity assessment.

4.7.8.3 Remedial Action

Typically, remediation methods can be divided into (i) *ex situ* remediation methods and (ii) *in situ* remediation. *Ex situ* methods are employed to treat extracted groundwater. *In situ* methods treat contaminated groundwater in place.

4.7.8.4 In Situ *Air Sparging*

Air sparging or biosparging, is employed for the remediation of sites contaminated with VOCs. During air sparging, a gas, normally air, is sparged into the saturated soil. As the air comes into contact with the contaminants it will strip the contaminant away. Ultimately, the contaminant-laden air is collected using a vapour extraction system followed by on-site treatment.

4.7.8.5 In Situ *Flushing*

In situ flushing is done by pumping a solution into groundwater through injection wells where it solubilizes, desorbs and/or flushes the contaminants from the groundwater and/or soil. After the contaminants are dissolved, the solution is pumped out and is treated using an effluent treatment process and recycled by pumping treated back.

4.7.8.5.1 Permeable Reactive Barriers

In this method the contaminated groundwater is made to flow through a barrier of reactive substance, which will decontaminate the contaminated ground water. The process is discussed in detail in Chapter 8.

Figure 4.50 View of portable water purifier.

4.8 Portable Water Purification Kit

Portable water-purification devices (Figure 4.50) are used by military personnel, survivalists, recreational enthusiasts and others to purify water from untreated sources. Many portable water-purification systems are available commercially for camping, hiking and for use at home. These kits are becoming increasingly popular in developing countries like India, where people may not have faith in bottled water and the municipal water supply. Technology used in these kits includes membrane technology, filter cartridge, solar disinfection, UV disinfection, chemical disinfection and ion exchange. These kits become most essential during disasters to curb epidemics.

4.9 Requirements of Electrical, Instrumentation and Mechanical Equipment in Water and Wastewater Treatment to Achieve Sustainability

Water and wastewater facilities have numerous options to save energy ranging from upgrading pumps and motors to changing light bulbs and installing co-generation systems and renewable energy technologies.

Solar energy (Figure 4.51) can be a reliable energy source for water and wastewater facilities but the cost currently associated with it makes it impractical in developing countries. Even many developed countries have not fully shifted to the use of solar energy. However demand for photovoltaic cells is increasing year after year, making it viable option for the future. The wastewater treatment facility in Oroville, California, was powered by solar energy (USEPA, 2013). Wind can be another sustainable source of energy. Effluent treatment facilities in Browning, Montana and Farmington, Maine have installed wind turbines to generate electricity.

Figure 4.51 *Solar energy is an attractive idea but still not practical due to high coast associated with installation of potovoltaic cells. (For a colour version of this figure, see the colour plate section.)*

The Clearwater Cogeneration Wastewater Treatment Facility, Corona, California, installed a co-generation system that provides heat back to the digesters and provides power to the local power utility. The Ina Road Water Pollution Control Facility, Tucson, Arizona, achieved a savings of $1.26 M /annum by installing a co-generation system. The Burlingame Wastewater Treatment Facility, Burlingame, California saves over $100 000/ year with co-generation. The East Bay Municipal Utility District, Oakland, California, saves $200 000–$300 000 per year (USEPA, 2013).

Apart from a solar and hydroelectric power supply, in-conduit hydropower is another option for sustainable electric generation. In-conduit hydropower is the production of hydroelectric power in existing man-made water conveyances like canals, tunnels, pipelines, aqueducts, ditches and flumes. Hydroelectricity is generated by installing electric generating equipment in these water-conveyance systems. The San Diego County Water Authority and Eastern Municipal Water District in California and City of Prescott, Arizona are some of the examples that have benefitted from in-conduit hydropower systems.

4.9.1 Electrical Equipment and Energy Requirement

Water is a widely consumed product and it is a strategic resource. It needs to be stewarded carefully and managed in order to meet the challenges of (i) access to clean drinking water, (ii) water pollution and sustainable development, (iii) regulatory and environmental issues. Labour and energy costs are rising at faster rate. A WWTP electrical network usually accounts for 5% to 10% of the total plant cost. Table 4.9 shows energy requirements for conventional effluent treatment plants and Table 4.10 shows power requirements for some of the machines/equipment used in effluent treatment plants. The average lifespan of a WWTP is 25 years to 50 years and energy costs up to 30% of the WWTP's operating costs.

Table 4.9 Energy requirements for conventional effluent treatment plants.

Size of WWTP	m3/day	Power demand	Design principle
Small autonomous	1000–5000	25–125 kVA	Single feeder configuration with low-voltage power supply supplied from electric utility. All process units are supplied power from a single low-voltage switchboard. Motor control is performed using variable speed drives and soft starters.
Small	5000–50 000	125–1250 kVA	Design principles are the same as above and these plants have more motors.
Medium-sized	50 000–200 000	1.25–5 mVA	Plants like these comprise many motors, which use variable speed drives and thus require intelligent motor control systems. Hence it is recommended to have a ring-main service from the electric utility to feed the medium voltage (MV) system. Installing several MV and/or low voltage (LV) transformers – of equal power – as close as possible can reduce the low-voltage connection distances.
Large to very large	200 000–1 000 000	5 – 25 mVA	Plants of this type require a large power base. These plants contain many motors, which usually use variable speed drives (VSDs), and hence require intelligent motor-control systems. These plants should have automated service recovery for improved personnel safety and maximum service continuity. Installing several MV/LV transformers of equal power as close as possible can reduce the LV connection distances. Each LV and MV switchboard bus bar should be split into two half bus bars connected with a bus tie. Such configuration improves power availability and allows maintenance without stopping the process.

Table 4.10 Power requirement for some of the machines/equipment used in effluent treatment plants.

Sl. no	Electrical equipment	Power requirement in kWh
1	Pump	$P_h = q\ \rho\ g\ h/(3.6\ 10^6)$ where P_h = hydraulic power (kW) ρ = density of fluid (kg/m^3) q = flow capacity (m^3/h) h = differential head (m) g = gravity (9.81 m/s^2) $P_s = P_h/\eta$ where η = pump efficiency P_s = shaft power (kW)
2	Surface aerator	1.7 to 2.2 kg oxygen transferred per kWh
3	Diffused aerator	2–8 kg oxygen transferred per kWh
4	Sludge scrappers	1 kw per 1250 m^2 of tank area

Any chemical and biological process requires optimum conditions. Power failure affects performance of the treatment system operated with electric power. The power from a backup generator is required to avoid a drop in the efficiency of a water/wastewater treatment plant during power failures. A generator's backup time depends on the quantity of available fuel. A backup generator produces power only after it has reached its rated speed and hence these generators are usually coupled with a UPS system to ensure a continuous supply of power, especially for critical processes. Power monitoring and control systems should be used to ensure control of MV and LV devices on an electrical network.

4.9.2 Piping and Instrumentation

The correct selection, application and maintenance of flow measurement devices is important for the efficient operation of effluent treatment facilities but this does not mean that all WWTPs should have sophisticated piping and instrumentation. Low-cost solutions always avoid instruments. A flow-measurement system consists of (i) a sensor/detector and (ii) a converter device.

A number of devices can be used to measure flow in the pipes/channels of WWTP. In open channels, the cross-sectional wetted area or flume or weir plate is used to measure the head and corresponding velocity is used to find out the flow rate.

Techniques commonly used for flow measurement in closed conduits are: (i) insertion of an obstruction to generate a head loss or pressure difference, (ii) measuring the effect of the moving fluid (e.g. sonic wave transmittance, magnetic field shift, momentum change), (iii) measuring the incremental units of fluid volume.

Rotameters, pitot tubes, flow tubes, orifices and venturi tubes can be used to generate pressure differentials that can be changed into flow rate readings. Ultrasonic, magnetic and vortex measuring devices, turbine and propeller meters can be correlated with velocity and flow rate.

A piping and instrumentation diagram is a type of engineering drawing that describes the process design aspects of a plant, which include:

- major and minor equipment;
- valves;
- instrumentation;
- stand-alone controllers;
- buttons;
- motors and drives;
- limit and point devices;
- piping;
- virtual devices (graphical representations that interact with the plant from the control room).

4.9.3 Mechanical Equipment Requirements and Related Issues

Conventional water treatment involves several items of mechanical equipment for water/sludge/chemical handling. Figure 4.52 compares mechanical equipment requirements for different treatment units/processes. Mechanical equipment is usually attached to electrical motors. Such equipment includes sludge scrapers, chemical dosers, mixers, flocculators, filter press, surface aerators rotating discs and scrapers. This equipment uses gears, belts, chains and shafts to produce motion.

4.9.4 Systems and Operational Issues

Daily and seasonal variations pose serious operational issues. The load on water and wastewater treatment varies depending on the rainfall. With a changing climate it is very

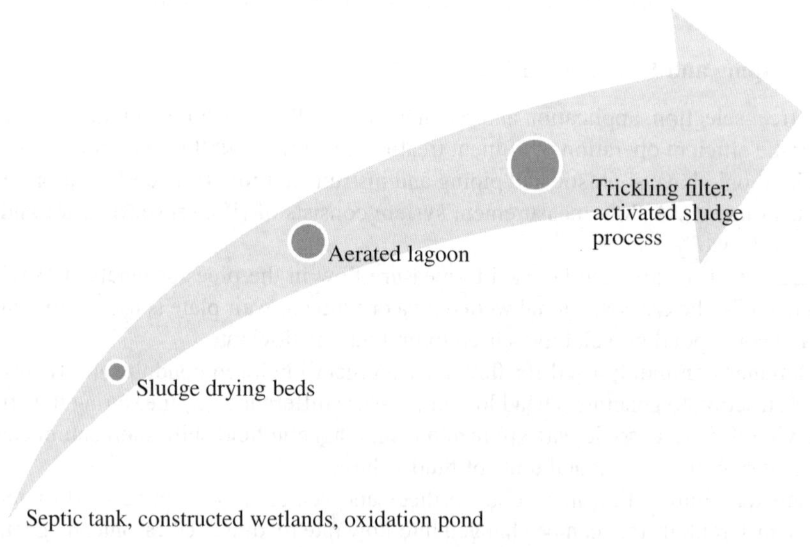

Figure 4.52 *Comparison of mechanical equipment requirement for different treatment unit/process.*

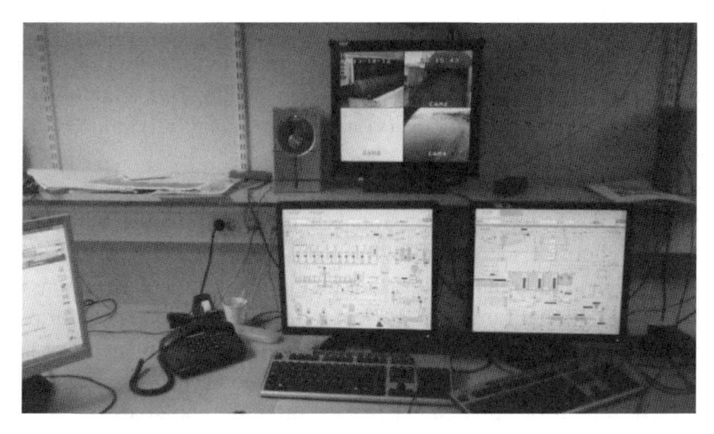

Figure 4.53 *Sophisticated real time monitoring setup.*

difficult to predict water and wastewater variation. Changes in population growth will also makes the system unsustainable if it cannot adapt to them quickly enough. Development planning should consider climate change as disasters cause disruption of the water system. Hence to make the system sustainable and reliable, planners should consider probable hazards due to a changing climate. The impact of disasters includes load on treatment plants and pump house and damage to overall infrastructure including the water distribution and collection network. Disruption of power will also pose challenge during disaster and recovery periods.

4.9.5 Real-Time Control

Small effluent treatment plants have limited controls and are operated manually but, as the size of treatment plant increases, real-time monitoring and control become essential to obtain the optimum results and reduce operational costs. Monitoring of water flow at various points, with auto controls and emergency release, is essential to avoid inefficiency and to ensure the safety of the investment and the people at the site.

Figure 4.53 shows sophisticated real-time monitoring setup of STP. The information monitoring includes water level, water flow, critical water quality parameters and a real-time view of critical working equipment. Such systems are capable of changing flow rate and operating machines/valves from the control room. The systems are also designed to show messages in case of failure of machines/valves/devices, thereby enabling staff to attend to the problem manually.

4.9.6 Indicators of Sustainable Performance; Systems Approach for Sustainability Assessment of Water Infrastructure

Water policies and laws have aimed to create infrastructure along with checks and balances all over the world but they need to improve in many developing countries. monitoring of water resources in Cameroon was abandoned since the economic crises of the 1980s, resulting in barely 10% of 408 rain gauges and 36% of 74 flow measurement stations being functional throughout the national territory (UNEP, 2012). Even though some countries have management instruments in place, they lack the capacity to implement them properly. The indicators used in the study consisted of 42 indicators in following categories: (i) water resources governance; (ii) ecosystems, (iii) human health, (iv) state of the resources, (v) food, agriculture and rural livelihoods, (vi) risk assessment, (vii) industry and (viii) valuing and charging for resources.

Irrespective of international efforts, many countries carry inbuilt lacunas, which extend to the water sector as well. The lack of professionals in the area is coupled with poor leadership. Many governments choose cricketers, movie/sport stars, political workers and relatives of people in power to lead the organizations that include wildlife conservation, environment and water. Huge international funds for the sector are often siphoned off for personal gain. The lack of carrier development, poor human training, poor recruitment procedure and poor-quality education add to unsustainability. Influential people from politics/sports/glamour are more involved in their original career than in attaining the goals of the organizations they are leading. The indicators adopted often fail to address the inbuilt inherited lacunas. Many countries that reserve about 30–40% of their revenue for military expenditure neglect the water sector, thereby causing more death due to poor water quality and inadequate water quantity.

4.9.7 Troubleshooting

Lack of knowledge about troubleshooting is often the cause of investment failure and of failure to achieve end results. Many effluent treatment plants have failed because buyers often think that WWTPs and water treatment plants are *plug-and-play* equipment. The local bodies that receive funding from international agencies often fail to operate plants because they do not have any intention of operating them or do not have the knowledge to operate them. There have been innumerable examples where water-treatment plants donated by donor agencies will stop functioning within a few months of withdrawal of support, especially in disaster-affected areas. On the other hand, a WWTP is the first thing that industries look at when they want to cut expenses. Many industries often think it is easier and more profitable to manage enforcing officers rather than wastewater treatment plants. This section discusses some of the troubleshooting that can be necessary.

4.9.7.1 Aerobic Wastewater Treatment

Water in the aeration chamber should be brownish white (the colour of chocolate milk). Blue or grey/blue water is an indication of the heavy use of detergents/chemicals. Water in the clarifier should be clear. Foamy wastewater is an indication of too much detergent in the system. Grease/oil forms white balls floating in the system. A sludge blanket in the aeration/clarifier is an indication of improper pumping into the chamber. Shock chemicals that are toxic to acclimatized microbes will eventually lead to the failure of the system and necessitate the process of reacclimatization. Raw sewage may enter the clarifier because of loose/improperly installed seals in the piping that goes through the clarifier wall. Odour problems could arise from inadequate aeration or chemicals/detergents killing healthy bacteria.

All tanks should be level. If ground settling occurs, action should be taken to correct it. Improper shape and location of aerators can lead to the formation of dead zones in aeration tanks, leading to the accumulation of sludge, which eventually leads to anaerobic activity and sludge shooting.

Foaming in activated sludge can occur in the aeration tank and secondary clarifier. Foam in WWTP is usually sticky, viscous and brown. It floats and accumulates on the surface of water and can take up a large proportion of the reactor volume, thus decreasing the effluent quality. The foam can overflow onto walkways and surrounding areas, leading to operational difficulties and causing a risk to the environment.

Foaming in an aeration tank can occur due to slowly biodegradable surfactants, excess production of extracellular polymeric substances by activated sludge microbes under nutrient-limited conditions, propagation of filamentous organisms and gas generated in the anoxic zones of aeration tanks and secondary clarifiers.

Growth of filamentous bacteria also causes foaming. Common strategies for foaming control are: (i) reduction of sludge-retention time, to wash out filamentous bacteria; (ii) removal of hydrophobic substances/substrates that enhance foaming or favour the increase of filamentous bacteria and (iii) addition of oxidizing agents such as chlorine to reduce the population of filamentous bacteria. Such practices need field investigation as the addition of excess chlorine would lead to the death of other microbes as well.

4.9.7.2 Disinfection

Disinfection may fail due to (i) low disinfectant residual, (ii) instrumentation malfunctions, (iii) insufficient contact time (iv) sampling techniques and (v) water quality. Disinfection though UV lamps may not provide the desired result due to the low intensity of the UV lamp or improper penetration of UV wave due to the presence of colloidal substances or due to the formation of microbial floc.

4.9.7.3 Mixing

Improper design and improper erection of mixers can often lead to improper mixing of chemicals in WWTPs.

4.9.7.4 Pumps

Pumps are an important part of most of the WWTPs. Except in a few situations, like DEWAT and low-cost wastewater-treatment plants, pumps are used in almost all WWTPs. Common reasons for pumps (and any electrical equipment) not starting are (i) blown circuit breakers or fuses due to incorrect rating of fuses or circuit breakers; (ii) nonfunctioning of the automatic control mechanism; (iii) terminal connections are loose/broken somewhere in the circuit; (iv) switch contacts are corroded or shorted; (v) the motor is shorted or burnt out; (vi) incorrect wiring hookup or service; (vii) warm fuses or thermal units; (viii) contacts of the control relays are dirty and arc; (ix) switches are not set for operation and (x) wiring is short circuited.

Common causes for reduced pump efficiency are: (i) the pump is not primed, (ii) the speed of the motor is too slow, (iii) incorrect wiring, (iv) there is air in the wastewater, (v) there is a defective motor, (vi) the discharge line is clogged, (vii) suction lifts higher than expected, (viii) the impeller is clogged, (ix) the discharge head is too high, (x) the pump is rotating in the wrong direction, (xi) there are air leaks in the packing box or suction line, (xii) the inlet to suction is too high, allowing air to enter, (xiii) valves are partially or entirely closed, (xiv) packing is worn or defective, (xv) incorrect impeller adjustment, (xvi) the impeller is damaged or worn, (xvii) check valves are stuck or clogged, (xviii) the flexible coupling is broken, (xix) belts slip and (x) a worn wearing ring.

Common causes for high energy requirements are (i) the speed of rotation is too high, (ii) the specific gravity or viscosity of liquid being pumped is high, (iii) check valves are open, (iv) operating heads are lower than the design specification, (v) clogged pump, (vi) sheaves on the belt drive are maladjusted/misaligned, (vii) packing is too tight, (viii) rotating elements are binding, (ix) the pump shaft is bent, (x) wearing rings worn/binding, (xi) impeller rubbing.

Causes for noisy pump operation are: (i) improper priming of pump, (ii) inlet is clogged, (iii) inlet is not submerged, (iv) pump is not lubricated properly, (v) worn impellers, (vi) strain on pumps due to unsupported piping secured to the pump, (vii) insecure foundation, (viii) mechanical defects in pump, (ix) misalignment of motor/pump and (x) rags or sticks bound around impeller.

4.9.8 Operational Checks for the STP

The success of sustainable operations depends on the skill and knowledge of the operator. In the absence of sophisticated instrumentation and operational control it is essential that the operator should do the checks manually. Some of the important checks are listed in Table 4.11.

4.9.9 Design, Construction and Engineering Checks for the WWTP

4.9.10 Odour Management

Apart from the skill and knowledge of the operator, the success of WWTPs depends on design, construction and engineering checks made during the design phase (Tables 4.12 and 4.13)

Odours emanating from water treatment are due to gases or fumes emitted by chemicals in wastewater or compounds formed during treatment. Table 4.14 shows the origins of odours in various industrial activities. The list of pollutants responsible for odour nuisance includes many compounds of nitrogen, sulfur, hydrocarbons and so on.

Long-term exposure to high-strength odours affects human health (leading to headaches, nausea, and respiratory problems). Hence, the minimization and abatement of unpleasant odours is a major challenge for WWTP utilities.

Odour control in wastewater treatment facilities includes prevention of odorant formation, impact minimization, emission and end-of-the-pipe odour-abatement technologies. Odour prevention includes (i) design and operation practices and (ii) odour containment by the installation of process covers.

Production of sulfides can be prevented by addition of O_2, NO_{3-}, ozone or by precipitating sulfides by the addition of iron salts. Good design includes minimizing heights in weirs and process covering. Good operation practices such as regular cleaning of screening units, grit chambers, minimizing sludge retention time in sedimentation tanks, sludge-handling units and operation under sufficient aeration and mixing.

Table 4.11 *Some of the important operational checks.*

Sl. no.	Treatment unit / operation	Checks
1	Bar screen chamber	• Check and clean at frequent intervals • Do not allow solids to escape from the screen • Ensure no gaps are formed due to corrosion • Replace corroded bar screen immediately
2	Oil and grease/grit trap	• Check and clean at frequent intervals • Remove settled solids and the floating grease
3	Equalization tank	• Keep air and mixing on at all times • Ensure uniform air flow/mixing • Keep the tank nearly empty before the peak load hours • Check and clean diffusers at regular intervals • Evacuate settled muck/sediments frequently
4	Pumps	• Switch between the standby and main pump approximately every four hours. • Check alignment of motor to pump after every dismantling operation • Check oil in the pump every day • Check condition of coupling • Replace damaged parts immediately • Check for vibrations frequently • Tighten the anchor bolts and other fasteners • Completely drain out oil and replace as per manufacturer's recommendation • Check condition of oil seals, bearings, mechanical seal • Keep safety guard in proper position • Follow the lockout tagout (LOTO) safety principles during maintenance • Maintain the flow rate at designed level
5	Aeration tank	• Maintain the correct MLSS • Dead zones on the wastewater surface indicate blockage of membranes. Both conditions call for cleaning or replacing • Local violent bubbling is indicates ruptured membranes. • Foaming may be caused by excessive foaming chemicals in the wastewater and may need addition of antifoaming agent.
6	Secondary clarifier	• Motor and gearbox of mechanical rake need routine maintenance. • The rubber squeegees sweeping the clarifier need to be checked and replaced
7	Clarified water sump	• Inspect the tank periodically for sediments and clean
8	Pressure sand filter	• Perform backwash when the pressure drop across the pressure sand filter exceeds 0.5 kg/cm^2
9	Activated carbon filter	• Perform backwash when the pressure drop across the activated carbon filter exceeds 0.5 kg/cm^2. When the carbon gets exhausted refill with fresh carbon.
10	Plate-and-frame filterpress	• Avoid storage of sludge for longer durations in the holding tank. • Add desired quantity of polymer 15–30 min before dewatering. • Filter cloths must be cleaned after every dewatering operation. • Replace the filter cloth with a new set when the filtration process becomes very slow.

Table 4.12 *Design, construction and engineering checks for WWTP.*

Sl. no.	Treatment unit / process	Design, construction and engineering checks
1	Bar screen	The screen chamber should have sufficient cross-sectional opening to allow passage of wastewater at peak flow rate at a velocity of 0.8 to 1.0 m/s. The screen must extend a minimum of 0.3 m from the floor of the chamber above the maximum design level of wastewater in the chamber during peak flow conditions.
2	Oil and grease/grit trap	The length of trap must be about 2 times its depth. Residence time should be 5–20 min at peak flow. Plan area of the trap in m^2 shall be around 1.5 to 2 times the depth of chamber in metres.
3	Equalization tank	The equalization tank should be designed to hold wastewater during peak time.
4	Aeration tank	Should not have any dead zones. Should be designed in such way that there should not be any short circuit. Aerator should be optically positioned.
5	Secondary clarifier/settling tank	Clarifier cross-sectional area is usually computed between 12 and 18 $m^3/h/m^2$ of throughput flow of wastewater. The depth of clarifier is usually kept between 2.5 and 3.0 m. In order to avoid localized turbulence, clarifiers need to be provided with an adequate length of 'weir' over which treated wastewater flows.
6	Pressure sand filter (SPF)	12 $m^3/ m^2/h$ filter cross-sectional area, is considered as a good filtration rate. Depth and distribution of filter media are equally important.
7	Activated carbon filter (ACF)	It is recommended that the diameter of the ACF be 25% larger than the (SPF).
8	Disinfection	Efficiency of disinfection depends on residual concentration of the chemical and contact time. Usually a contact time of 20 to 30 min is used to achieve about 99% germicidal efficiencies.

The use of buffer zones or masking/neutralizing agents can contribute to the mitigation of odour nuisance in the areas around WWTPs. Passive impact reduction can also be achieved by providing fences, trees, or buffer zones. The neutralizing agents or masking, composed of aliphatic/aromatic aldehydes, terpenes and alcohols, are applied by air spraying, mixing and direct surface application.

Chemical scrubbers are a commonly used abatement technique in wastewater treatment facilities where odorous gases are transferred into an aqueous solution/activated carbon, zeolite or silica gel-based odour-adsorption systems.

Table 4.13 *Engineering checks.*

Sl. no.	Treatment unit/process	Design, construction and engineering checks
1	Bar screen	Bar screen racks are usually made of 25 mm × 6 mm bars. The bars are epoxy-coated mild steel or stainless steel. The screen frame is inclined at 60° away from the incoming side. The platform must have weep holes, to drain liquid in debris.
2	Oil and grease/grit trap	An oil and grease/grit trap should not be too deep. Compressed air can be forced into the tank to avoid septicity and keep solids in suspension. Membrane diffusers fail frequently due to repeated cycles of expansion and contraction.
3	Secondary sedimentation tank	An equalization tank with a steep slope in the hopper-bottom settling tank is provided for easy collection of sludge. A weir at uniform level is provided to reduce turbulence of incoming mixed liquor. A minimum difference of 0.2 m is provided in water level between the aeration tank and the clarifier.
4	Aeration tank	The aeration tank is usually made of waterproof RCC. Operating platforms must be provided so that all the diffusers are easily accessible. Wastewater depth is usually maintained between 2.5–4.0 m.
	Pressure sand filter (PSF)	The filter vessel is designed to withstand a pressure of 5 kg/cm^2. Small diameter vessels are provided for maintenance. The PSF are provided with valves, backwash waste line, bypass line and pressure gauges at the inlet and outlet. The shell height usually varies between 1.2 m and 1.5 m in small plants. The PSF is provided with graded pebbles ranging from 1 to 3 cm at bottom layers up to a depth of about 0.5 to 0.6 m. The top layers consist of coarse and fine sand to a depth of 0.6 to 0.7 m with freeboard of about 0.3 m above the sand. Essential appurtenances are provided at the top and bottom to ensure distribution and collection.
5	Activated carbon filter (ACF)	Construction and engineering of the ACF is similar to the PSF. Epoxy paint coating is recommended inside the filter shell to preserve the shell from abrasiveness and corrosiveness of activated carbon.
6	Disinfection	The chlorine disinfection system with a hypo-holding tank and an electronically metered dosing pump is set to have a chlorine dose rate, of around 3 to 5 PPM.
7	Filter Press	The plate-and-frame filter press consists of (i) a sludge-holding tank with mixing/aeration, (ii) a high-pressure filter-press feed pump, (iii) a polymer solution-preparation tank and dosing, (iv) plate-and-frame filter press. Usually the polymer requirement is about 1 to 2% of the surplus sludge on dry weight basis. The sludge-holding tank should have sufficient capacity to accommodate (i) the excess sludge to be dewatered and (ii) the polymer solution added.

Table 4.14 Origin of odour.

Industries	Origin of odour
Pharmaceutical industries	Fermentation products
Food industries	Fermentation products
Fish processing	Amines, mercaptans, sulfides
Rubber industries	Mercaptans, sulfides
Textile industries	Phenolic compounds
Pulp and paper industries	Hydrogen sulfide, sulfur dioxide
Organic compost	Ammonia, sulfur compounds

Biologically based odour technologies like biotrickling filters, biofilters and bioscrubbers can also be used. Biofilters are commonly employed for odour treatment. Odours are sparged in the aeration tank; activated sludge diffusion systems' odorous emissions are diffused along with fresh air into the activated sludge process. Biological treatments based on organic media do not allow complete odour removal. Microbial acclimation can take many weeks, hence a polishing treatment needs to be implemented (Lebrero *et al.*, 2011).

References

AFED (2010) Arab Environment: Water. Sustainable Management of a Scarce Resource, Arab Forum for Environment and Development, Beirut.

Alonitis, S.A., Kouroumbas, K. and Vlachakis, N. (2003) Energy consumption and membrane replacement cost for seawater RO desalination plants. *Desalination* **157**, 151–158.

Bal, A.S. and Dhagat, N.N. (2001) Upflow anaerobic sludge blanket reactor–a review. *Indian Journal of Environmental Health* **43** (2), 1–82.

Bingbin, H., Chunli, H. and Yubin, Y. (2006) Demonstration Study on Water Quality Improvement by Constructed Wetlands in Low Temperature Areas of Northern China. Sustainable Water Management: Problems and Solutions under Water Scarcity International Conference, Beijing, P.R. China, 6–8 November 2006.

Bourounia, K., Chaibib, M.T. and Tadrist, L. (2001) Water desalination by humidification and dehumidification of air: state of the art. *Desalination* **137**, 167–176.

CDM (Camp Dresser and McKee Inc.) (2011) Charting the Future of Biosolids Management. Final report. Water Environment Federation, Alexandria.

Clauss, F., Helaine, D., Balavoine, C. and Bidault, A. (1998) Improving activated sludge floc structure and aggregation for enhanced settling and thickening performance. *Journal of Water Science and Technology* **38** (8–9), 35–44.

Daw, J., Hallett, K., DeWolfe, J. and Venner, I. (2012) *Energy Efficiency Strategies for Municipal Wastewater Treatment Facilities*, National Renewable Energy Laboratory, Denver CO.

Electric Power Research Institute (EPRI) (2002) Water and Sustainability (Volume 4): US Electricity Consumption for the Water Supply and Treatment–The Next Half Century. Topical Report 1006787. EPRI, Palo Alto, CA.

ESCWA (Economic and Social Commission for Western Asia) (2001) Energy Options for Water Desalination in Selected ESCWA Member Countries, United Nations, New York.

ESCWA (Economic and Social Commission for Western Asia) (2003) Waste-Water Treatment Technologies: A General Review, United Nations, New York.

FAO (2008) Irrigation in the Middle East region in figures: AQUASTAT Survey – 2008, Food and Agriculture Organization of the United Nations, Rome.

Galbraith, K. (2011) How Energy Drains Water Supplies. New York Times, http://www.nytimes.com/2011/09/19/business/global/19iht-green19.html?_r=1 (accessed 30 December 2013).

Geldof, G.D., Jong de, S.P., Braal de, A.J. and Marsman, E.H. (1997) Water in de stad, behandlingstechnieken (Water in the City, Treatment Technologies), RIZA rapport 97.091, SPA rapport 97004, ISBN 9036951380, Deventer, 11 April 1997.

Ghanizadeh, Gh. and Sarrafpour, R. (2001) The effects of temperature and PH on settlability of activated sludge flocs. *Iranian Journal of Public Health* **30**, (3–4), 139–142.

Ghatnekar, S.D., Kavin, M.F., Sharma, S.M. *et al.* (2010) Application of vermi-filter-based effluent treatment plant (pilot scale) for biomanagement of liquid effluents from the gelatine industry. *Dynamic Soil, Dynamic Plant* **4** (1), 83–88.

Grady, C.P.L., Daigger, G. and Lim, H.C. (1988) *Activated Sludge Process: Theory and Application*, Marcel Decker Inc., New York.

Green, C. (2013) *Case Study Brief – Sustainable Urban Water Management in London*, http://www.switchurbanwater.eu/outputs/pdfs/w6-1_gen_dem_d6.1.6_case_study_-_london.pdf (accessed 26 December 2013).

Khataee, A.R. and Kasiri, M.B. (2011) Modeling of biological water and wastewater treatment processes using artificial neural networks. *CLEAN – Soil, Air, Water* **39** (8), 742–749.

Kodavasal, A.S. (2011) The STP Guide – Design, Operation and Maintenance, Karnataka State Pollution Control Board, Bangalore.

Lebrero, R., Bouchy, L., Stuetz, R. and Muñoz, R. (2011) Odor assessment and management in wastewater treatment plants: a review. *Critical Reviews in Environmental Science and Technology* **41**, 915–950.

Lee, P.O. (2005) Water Management Issues in Singapore. Paper presented at the Water in Mainland Southeast Asia, Siem Reap, Cambodia.

Mahvi, A.H. (2009) Application of ultrasonic technology for water and wastewater treatment. *Iranian Journal of Public Health* **38**(2), 1–17.

Mamadou, S.D. and Savage, N. (2005) Nanoparticles and water quality. *Journal of Nanoparticle Research* **7**, 325–330.

Mathioulakis, E., Belessiotis, V. and Delyannis, E. (2007) Desalination by using alternative energy: review and state-of-the-art. *Desalination* **203**, 346–365.

McCutcheon, J.R., McGinnis, R.L. and Elimelech, M. (2005) A novel ammonia-carbon dioxide forward (direct) osmosis desalination process. *Desalination* **174**, 1–11.

McGhee, T.J., Mojgani, P. and Viicidomina, F. (1983) Use of EPA's CAPDET program for evaluation of wastewater treatment alternatives. *Journal of Water Pollution Control Federation* **55** (1), 35–43.

Mei, N., Xuguang, L., Jinming, D. *et al.* (2009) Antibacterial activity of chitosan coated Ag-loaded nano SiO2 composites. *Carbohydrate Polymers* **78** (1), 54–59.

Mills, A. and LeHunte, S. (1997) An overview of semiconductor photocatalysis. *Journal of Photochemistry and Photobiology A: Chemistry* **108**, 1–35.

Morgenroth, E., Arvin, E. and Vanrolleghem, P. (2002) The use of mathematical models in teaching wastewater treatment engineering. *Water Science and Technology* **45** (6), 229–233.

Mueller, N.C. and Nowack, B. (2009) Nanotechnology Developments for the Environment Sector, Report of the Observatory NANO, European Commission. http://www.nanopinion.eu/sites/default/files/ts_environment_vs_1.pdf (accessed 5 January 2014).

Peavy, H.S., Rowe, D.R. and Tchobanoglous, G. (1986) *Environmental Engineering*, McGraw-Hill International, New York.

Reddy, K.R. (2008) Physical and chemical groundwater remediation technologies, in *Overexploitation and Contamination of Shared Groundwater Resources* (ed. C.J.G. Darnault), Springer Science + Business Media B.V., Dordrecht.

SAMC (2010) Forty Sixth Annual Report: The Latest Economic Developments 1431H (2010G), Saudi Arabian Monetary Agency: Research and Statistics Department, Riyadh.

Seeger, H. (1999) The history of German wastewater treatment. *European Water Management* **2** (5), 51–56.

Segura, Y., Molina, R., Martínez, F. and Melero, J.A. (2009) Integrated heterogeneous sonophoto Fenton processes for the degradation of phenolic aqueous solutions. *Ultrasonics Sonochemistry* **16** (3), 417–424. Epub 2008 Oct 17. doi: 0.1016/j.ultsonch.2008.10.004

Shannon, M.A., Bohn, P.W., Elimelech, M. *et al.* (2008) Science and technology for water purification in the coming decades. *Nature* **452**, 20 March, doi: 10.1038/nature06599

Sharma, H.D. and Reddy, K.R. (2004) *Geoenvironmental Engineering: Site Remediation, Waste Containment, and Emerging Waste Management Technologies*, John Wiley & Sons, Inc., Hoboken, NJ.

Slaughter, S. (2010) Improving the sustainability of water treatment systems: opportunities for innovation. *The Solutions Journal* **1** (3), 42–49.

Spearing, B.W. (1987) Sewage treatment optimization model - STOM - the sewage works in a personal computer. *Proceedings of Institutions Civil Engineers, Part 1* **82**, 1145–1164.

Sustainable Cities (2012) *Sustainable Cities: Building Cities for the Future.* http://www.sustainablecities2013.com/images/uploads/documents/SC2012.pdf (accessed 5 January 2014).

Swedish EPA (2011) *Information Facts, Best Practice Examples – Biogas, Klimp – Climate Investment Programme*, http://www.naturvardsverket.se/Documents/publikationer6400/978-91-620-8535-3.pdf (accessed 5 January 2014).

Tchobanoglous, G., Burton, F.L. and Stensel, H.D. (Eds) (2003) *Wastewater Engineering, Treatment, and Reuse*, McGraw Hill, Inc., New York.

Tyagi, P.K., Singh, R., Vats, S. *et al.* (2012) Nanomaterials Use in Wastewater Treatment. International Conference on Nanotechnology and Chemical Engineering (ICNCS'2012), 21–22 December 2012, Bangkok.

UNEP (2010) Clearing the Waters: A Focus on Water Quality Solutions, UNEP, Nairobi.

UNEP (2012) Status Report on The Application of Integrated Approaches to Water Resources Management, UNEP, Nairobi.

UN ESCAP (Economic and Social Commission for Asia And The Pacific) (2007) Sustainable Infrastructure in Asia Overview and Proceedings, Seoul Initiative Policy Forum on Sustainable Infrastructure Seoul, Republic of Korea, 6–8 September 2006.

USEPA (2000) *The History of Drinking Water Treatment*, Fact sheet No. EPA-816-F-00-006, February 2000. http://www.epa.gov/safewater/consumer/pdf/hist.pdf (accessed 5 January 2014).

USEPA (2013) *Water and Energy Efficiency in Water and Wastewater Facilities*, http://www.epa.gov/region9/waterinfrastructure/technology.html (accessed 27 December 2013).

van der Vleuten-Balkema, A.J. (2003) Sustainable wastewater treatment, developing a methodology and selecting promising systems. Technische Universiteit Eindhoven.

Veerapaneni, S., Long, B., Freeman, S. and Bond, R. (2007) Reducing energy consumption for seawater desalination. *Journal of the American Water Works Association* **99**, 95–106.

Zhou, Y. and Tol, R.S.J. (2005) Evaluating the costs of desalination and water transport. *Water Resources Research* **41**, 1–10.

5

Sustainable Industrial Water Use and Wastewater Treatment

Conventionally all engineering practices assume a *factor of safety* while designing an infrastructure, machine or engineering solution. Most of the time the assumed factor ranges between two and three. The water industry is not an exception. Hence, to avoid losses it forecasts greater demand than is really required. The industry usually assumes greater losses and multiplies the final water requirement by the factor of safety to arrive at the total water demand (Figure 5.1).

As water is a limited resource, the quantity available usually diminishes as the number of competitors increases (Figure 5.2). Furthermore, the amount of water within a given locality may also diminish due to climate change and diminishing quality.

Precise information on effluent characteristics is an essential factor in the design of suitable treatment facilities. In theory, design and operation is governed by the relevant microbial/chemical kinetics but, in reality, design and operation are mainly based on rules of thumb and the experience of the people involved. Wastewater composition varies depending on the raw materials, machinery installed, skills of staff, enforcement efficiency, leakages, production procedures, cleaning procedures and recovery of byproducts. Water use can be made sustainable by following procedures like using condensate and blowdowns for dust suppression and cleaning the floor.

5.1 Sustainable Principles in Industrial Water Use and Wastewater Treatment

Even though the industrial sector consumes less water than agriculture and domestic users, it has high pollution potential as it releases complex pollutants, many of which are yet to be understood. In many countries, significant emphasis is given to the prevention of industrial pollution as the industrial sector is owned by entrepreneurs and, hence, it is easy to impose

Sustainable Water Engineering: Theory and Practice, First Edition. Ramesha Chandrappa and Diganta B. Das.

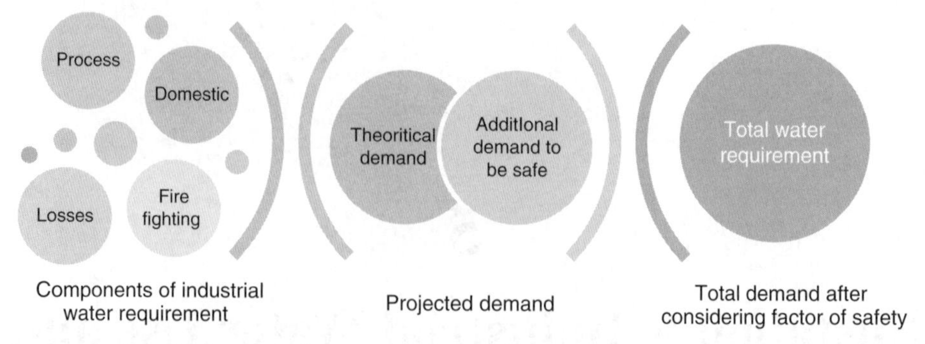

Figure 5.1 *Components of calculating water demand. (For a colour version of this figure, see the colour plate section.)*

laws and take action against defaulters compared to many local bodies that are managed by officers of the government itself.

- **Reuse to maximum extent.** Water was used and wasted when it was abundant. Instead of keeping water in a *closed loop system* (Figure 5.4) it is often used in a *linear system* (Figure 5.5).
- **Use treatment options that require large quantities of chemicals and energy only if low chemical-/energy-use options are not available.** Options requiring large quantities of chemicals or large amounts of energy are often recommended by consultants or suppliers to increase their profit. Construction costs are often increased by unnecessary use of steel and cement. Instead of costly reinforced cement concrete tanks, one can use earthen bund with low-cost lining.
- **Avoid water wherever possible.** Avoiding water is a major principle adopted by many industries. Water cooling is being replaced by air cooling (Figure 5.3) in thermal power plants, cement plants and other industries.
- **Segregate wastewater with different characteristics.** Different processes from industry are likely to have different characteristics and, consequently, they have different significance during water treatment. For example, mixing of organic wastes with inorganic wastes will lead to the inhibition of microbial activity during the treatment process. Similarly, mixing hot water, like boiler blowdown, with wastewater streams at ambient temperature may kill microbes in the treatment plant, which may affect the

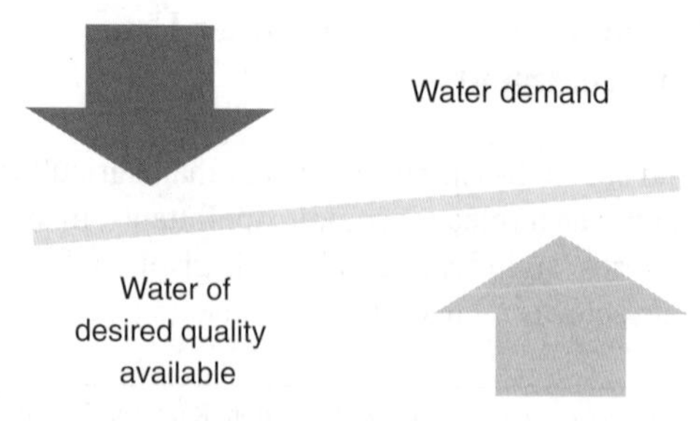

Figure 5.2 *Imbalance between water availability and demand.*

Figure 5.3 Air-cooling arrangement made to cement kiln.

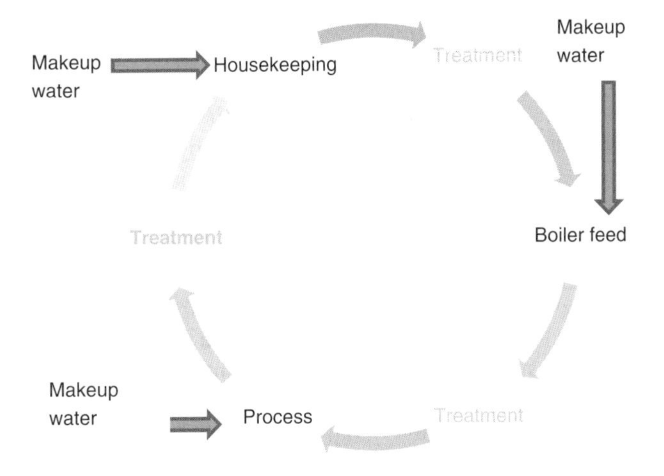

Figure 5.4 Closed-loop system. (For a colour version of this figure, see the colour plate section.)

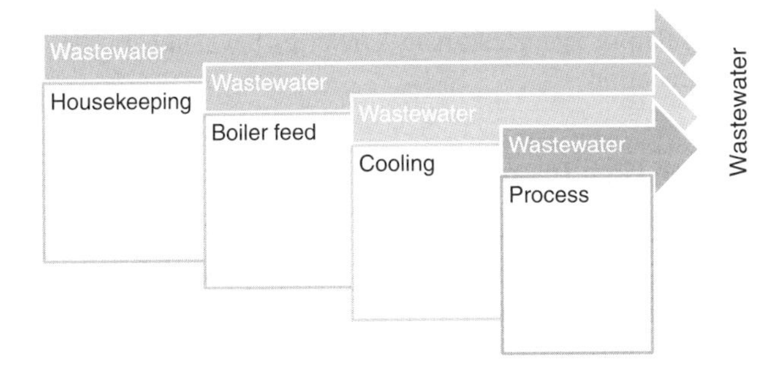

Figure 5.5 Leaner system.

performance of the effluent treatment plant (ETP). Hence, it is important to segregate the waste streams into a number of streams in complicated processes such as oil refinery or integrated steel plants and then treat the individual streams as necessary. In the case of a single-product industry one might prefer to decide the wastewater treatment process after knowing the effluent characteristics. For example, if the BOD/COD ratio is more than 0.7, biological processes would be a cheaper option. If the BOD/COD ratio is less than 0.4, then physico-chemical treatment would be essential as living things cannot remove refractory substances. If the ratio of BOD/COD is between 0.4 and 0.7 then treatment can be carried out by acclimatizing/choosing microbes to treat the these effluents by inoculating microbes that are readily available for that purpose or by acclimatizing microbes by slowly increasing refractory organics. If process effluents contain chemicals that are toxic to microbes then the wastewater is treated by physico-chemical methods. Most of the industries that have manufactured products over many years have evolved cost-effective methods based on practical experience. One example is the spentwash distillery where process wastewater can be treated, theoretically, by anaerobic digesters followed by aerobic treatment. However this treatment does not remove colour. Hence incineration of concentrated spentwash distillery waste has been adopted in many parts of the world.

- **Minimize the carbon footprint.** Systems are often designed with unnecessary pumps and equipment. Technology providers often market the product using impressive presentations or by paying bribes to decision makers or regulators. Sometime costly solutions are chosen to syphon out investors' money. On other occasions the design process of a water treatment plant and the solution provided by designer are not questioned. Components that are responsible for a high carbon footprint during construction are overdesigning and the use of large quantities of cement and metal. Components that are responsible for a high carbon footprint during operation are electromechanical devices such as pumps, scrapers, aerators and heaters.
- **Research and design the manufacturing unit to attain sustainability.** Most industries are designed and erected with the help of consultants. In many cases existing plants from other countries/places are transported to the developing countries. Most such plants and equipment may not be efficient with respect to energy and water utilization. If a manufacturing unit is not designed in the initial stage of the project to attain environmental sustainability, its incorporation in the latter stage of the life of a manufacturing unit would be costly and may not be feasible.
- **Investment in human resources.** Most industries do not plan human-resource investments well. As a result of a poorly trained and unknowledgeable workforce, both in industry and in regulatory agencies, many countries have ended up paying in terms of environmental degradation.

5.1.1 Industries with High Dissolved Solids

Industrial wastewater can be classified based on biodegradability and dissolved solids. Figure 5.6 shows the industrial wastewater classification and Figure 5.7 shows the classification of dissolved solids based on biodegradability.

The pharmaceutical, sugar, distillery, chemical, dye, electroplating, electropolishing and surface coating industries often discharge effluents containing a large amount of dissolved

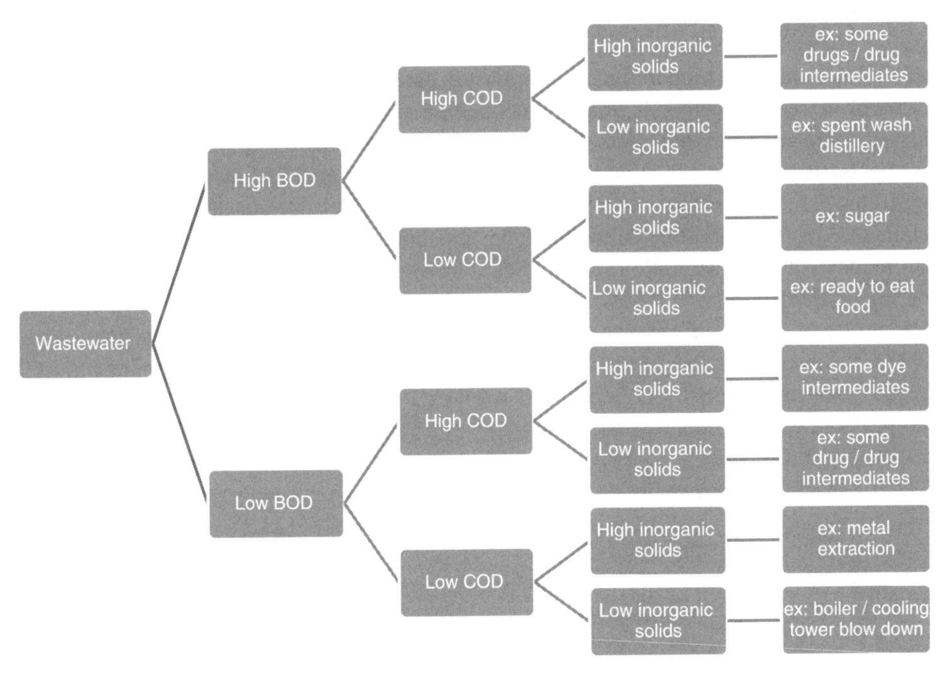

Figure 5.6 *Industrial wastewater classification.*

solids. These effluents have been problematic due to their associated colour and odour. The characteristics of effluents from bulk drug and drug intermediate industries usually do not vary from day to day. Pharmaceutical formulations are characterized by effluents that change every day as these industries manufacture different products by combining different raw material to make different drugs in accordance with market demand.

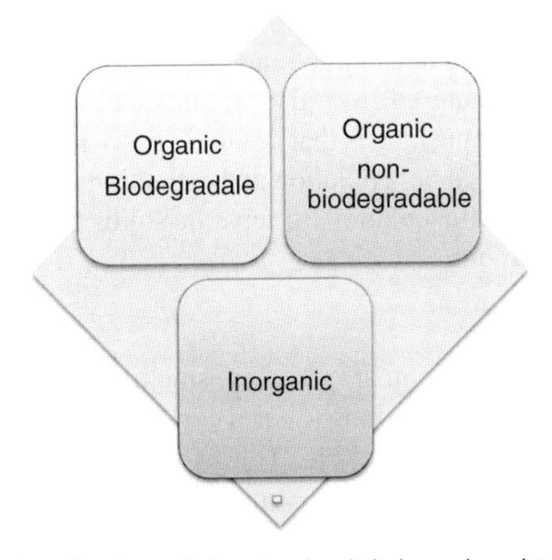

Figure 5.7 *Classification of dissolved solids based on biodegradability.*

5.1.2 Industries/Activities with High Inorganic Content

Major industries with high inorganic content are: electroplating, electropolishing, manufacturing of inorganic chemicals, metal beneficiation and extraction. Industrial activities that discharge effluent with inorganic content are (i) manufacturing processes, (ii) reverse osmosis rejects, (iii) regeneration of ion exchange resins, (iv) boiler blowdown and (v) housekeeping. Table 5.1 shows sources of chemicals from various industries. Sustainability cannot be attained by end-of-pipeline solutions alone. Changes in processes and efficient operation can reduce water consumption. In practice, however, the efforts may not become part of the operation/design of the system due to negligence, poor implementation of laws and availability of sufficient water to enable poor practices to continue. Such practices generally lead to economic and ecological unsustainability. Even if water were abundant during the early stages of industrial operation, it would soon become scarce, forcing industries to invest at a later stage.

5.1.2.1 Metal Beneficiating and Extraction

Most metals are extracted from the earth by mining. Mining (Figures 5.8–5.10) without affecting water resource is a big challenge. Management of water use and quality, in and around mine sites, is a noteworthy issue. Significant pollution of water sources occurs during the exploration stage due to population in-migration. Depletion of groundwater and available surface water is a concern. Hence, the quantity and quality of mine effluents, (leach pad drainage, stormwater and process effluents) should be managed to safeguard the environment. Efficient oil/grease traps should be installed and maintained at workshops, fuel storage depots, refuelling facilities and containment areas. Stormwater drains and stream channels should be protected against erosion using rip-rap and lining. Appropriate combinations of terracing, contouring techniques and slope reduction will help safeguard water resources (IFC, 2007).

Beneficiation eliminates unwanted clay and silica. Metal extraction separates the useful metallic content. Metal is beneficiated and extracted by a series of physical and chemical processes like grinding, smelting, flotation and electrorefining. The operations that are carried out depend on the metals to be extracted. Wastewater streams are generated as water is used in almost all beneficiating and extraction processes. Since the wastewater streams from metal beneficiating and extraction will have minute fractions of biodegradable material, which would have entered from the ore, this cannot be treated using biological processes due to presence of substances that are toxic to living organisms.

The waste streams from metal beneficiation and extraction are treated by (i) plain settling to remove heavy fraction; (ii) flocculation, to separate solids that do not settle, contributing to turbidity; (iii) pH adjustment.

5.1.2.1.1 Copper

The main steps in copper production are roasting, smelting and refining. Roasting is done to reduce sulfur content. Smelting is done to form copper sulfide matte and a siliceous slag.

Roaster gases are rich in sulfur dioxide. This is used for the manufacture of sulfuric acid. The major wastewater streams from copper smelting operations are (i) blowdown from the acid plant; (ii) overflow from the thickener; (iii) effluents from scrubbers; (iv) effluent from cooling of anodes and (v) washing. Solids from these effluents are separated by sedimentation and are dredged periodically.

Table 5.1 *Source of chemicals from various industries.*

Sl. no.	Chemical	Industry
1.	Acetic acid	Acetate rayon
2.	Acids	Acid manufacturing, chemical manufacturing involving acids, mines, textiles manufacture, plating
3.	Alcohol	Breweries, distilleries
4.	Alkalies	Wool scouring, cotton/straw kiering
5.	Ammonia	Coke/gas and chemical manufacture
6.	Arsenic	Sheep dipping
7.	Cadmium	Plating
8.	Chromium	Alum anodizing, chrome tanning, plating
9.	Chlorine	Health care establishments, bleaching powder
10.	Chloride	Fish processing, pickling
11.	Citric acid	Citrus fruit processing and soft drinks
12.	Copper	Copper pickling, copper plating
13.	Cyanides	Gas manufacture, metal cleaning, plating, steel hardening, electropolishing
14.	Fats, oils, grease	Wool scouring, laundries, textile industry, food industries using oil/fat/grease, vehicle servicing, petroleum industry
15.	Fluorides	Scrubbing of flue glass, gases, etching
16.	Formaldehyde	Synthetic penicillin and resins manufacture
17.	Free chlorine	Textile bleaching, paper mills, laundry
18.	Gold	Plating
19.	Hydrocarbons	Rubber factories and petrochemical
20.	Hydrogen sulfide	Petrochemical industry
21.	Lead	Plating
22.	Iron	Iron and steel
23.	Mercaptans	Oil refining, pulp
24.	Nickel	Plating
25.	Nitro compounds	Chemical works and explosives
26.	Organic acids	Fermentation plants and distilleries
27.	Pesticides	Pesticide manufacturing
28.	Phenols	Gas and coke manufacturing, chemical plants
29.	Phosphate	Soap and detergent
30.	Radioactive material	Atomic power station, radioactive processing industry
31.	Silver	Plating
32.	Sodium	Fish processing, pickling
33.	Starch	Food processing, textile industries
34.	Sugars	Breweries, dairies, sweet industry, confectionaries, fruit juice, soft drinks, sugar, jaggary
35.	Sulfides	Textile industry, tanneries, gas manufacture
36.	Tannic acid	Tanning, sawmills
37.	Tartaric acid	Wine, leather, chemical manufacture, dyeing
38.	Tin	Electroplating
39.	Zinc	Zinc plating, rubber process, galvanizing.

Figure 5.8 View of open-cast mining. (For a colour version of this figure, see the colour plate section.)

Figure 5.9 Afforestation efforts in mining area and surface water inundation in mining area.

Figure 5.10 Mining activity in progress and exposure of groundwater.

Copper refining is carried out in an electrolytic cell, from which discharges are made once the electrolyte is no longer suitable for refining.

5.1.2.1.2 Aluminium

Aluminium is mainly produced from bauxite, which contains 40–60% alumina with small quantities of titania, silica and iron oxide. Alumina is largely extracted using the Bayer process, in which the aluminium component of bauxite ore is dissolved in sodium hydroxide to remove impurities and precipitate alumina trihydrate, which is calcined to form aluminium oxide.

The electrolytic manufacturing of aluminium involves the reduction of anhydrous aluminium chloride with potassium using carbon electrodes. Sustainable techniques for disposal of red mud include thickening of the mud followed by solar drying. Wastewaters generated from primary aluminium processing are the clarification and precipitation activities in the Bayer process. A major portion of this water is reused.

5.1.2.1.3 Iron and Steel

Steel is manufactured by the electric arc furnace (EAF) and basic oxygen furnace (BOF). The input materials in BOF are molten iron, scrap and oxygen. The raw materials for EAF are electricity and scrap.

Pig iron is produced from sintering (heating without melting) and pelletizing iron ores using coke and limestone. Afterwards the iron is fed to the BOF in molten form, with fluxes, scrap metal and high-purity oxygen. In some plants sintering is done to agglomerate fines.

In the process from pig-iron production without recirculation the effluent is generated from cooling operations at an average rate of 80 m^3/t of steel manufactured. The main pollutants in untreated effluent from pig-iron manufacturing are total suspended solids (7000 mg/l), total organic carbon (100–200 mg/l), dissolved solids, COD (500 mg/l), and zinc (35 mg/l) cyanide (15 mg/l) and fluoride (1000 mg/l) (World Bank, 1998). The major pollutants in effluent from steel manufacturing using the BOF are total suspended solids (4000 mg/l), lead (8 mg/l), zinc (14 mg/l), fluoride (20 mg/l), chromium (5 mg/l), cadmium (0.4 mg/l) and oil and grease. The effluents are at a high temperature too (World Bank, 1998). Despite these characteristics, over 90% of the wastewater from iron and steel can be reused. The quantity of wastewater generated varies from 5 m^3/t of steel manufactured. Wastewater treatment systems include sedimentation, precipitation of heavy metals and filtration.

5.1.2.2 /Electroplating/Electropolishing

The earliest form of plating was done for the decorative gilding of domestic vessels during the Middle Ages. Figure 5.11 shows the usual metal deposition processes that are used for coating metals. Electroplating emerged as an industry in the 1740s after silver was plated onto copper by fusing silver plate on copper ingots, which were worked into finished products (DoE, 1995). In the 1800s the first commercial electroplating unit was established for gold and silver electroplating by deposition of metal from boiling alkaline carbonate solution. During 1840, the use of alkaline cyanide baths was started, which produced superior-quality deposition.

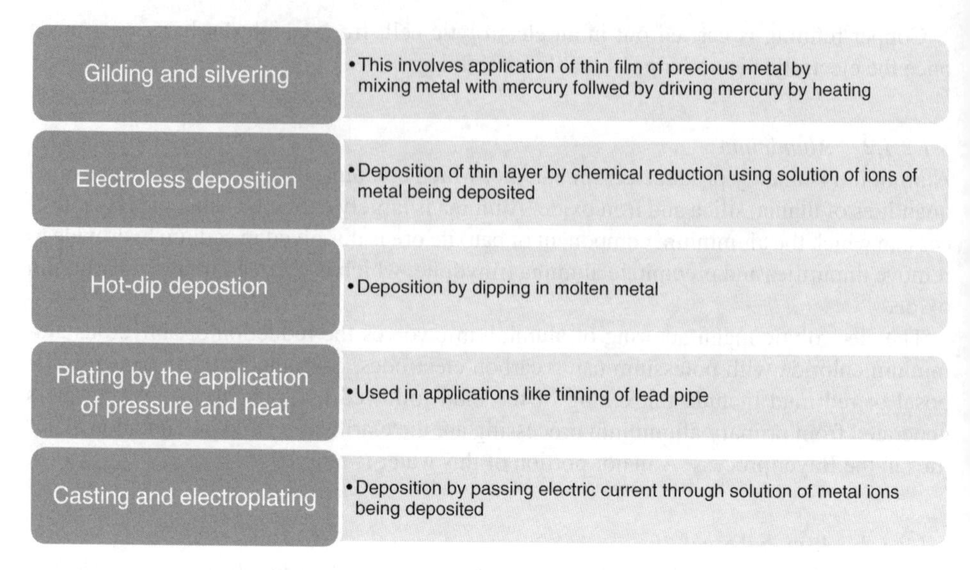

Figure 5.11 *Usual metal-deposition process.*

5.1.2.2.1 Process

Electroplating is the process of depositing a protective layer (usually metal) coating to an object by passing an electric current through electrolyte in contact with the object. Electroplated materials are used for a specific property such as decoration or corrosion resistance. The metals and alloys electroplated on a commercial scale are cobalt, copper, chromium, cadmium, indium, gold, silver, tin, zinc, lead, iron, platinum, nickel, brass, bronze, nickel-phosphorus, nickel-cobalt, tin-nickel, gold alloys, nickel-iron, lead-tin, zinc-nickel, zinc-cobalt and zinc-iron.

The crucial components of an electroplating process are: (i) the electrode to be plated (the cathode or negative electrode), (ii) the electrolyte comprising the metal ions to be deposited, (iii) the anode (positive electrode) and (iv) a direct current power source. The plating tank is made up of an inert material or lined with inert materials. Anodes are either soluble or insoluble.

The electroplating (Figure 5.12) process involves cleaning, degreasing, plating, rinsing, passivating and drying. Cleaning and pretreatment involve of the use of solvents such as chlorinated hydrocarbons and surface-stripping agents such as caustic soda and strong acids.

The postelectroplating steps include sealing and air drying. Sealing is usually performed to aluminium products by immersing the finished product in a water bath at 88 to 99 °C for about 15 minutes, resulting in the hydration of the aluminium oxide, which seals the pores in the aluminium surface. The electroplated articles are allowed to air dry.

The plating tanks are usually equipped with heat exchangers. Electroplating requires a constant temperature, plating time, current density and bath composition. The electroplating activities include preprocessing to prepare a surface ready for electroplating. Some typical pretreatment processes are given in Table 5.2 and the compositions of typical electroplating baths are given in Table 5.3.

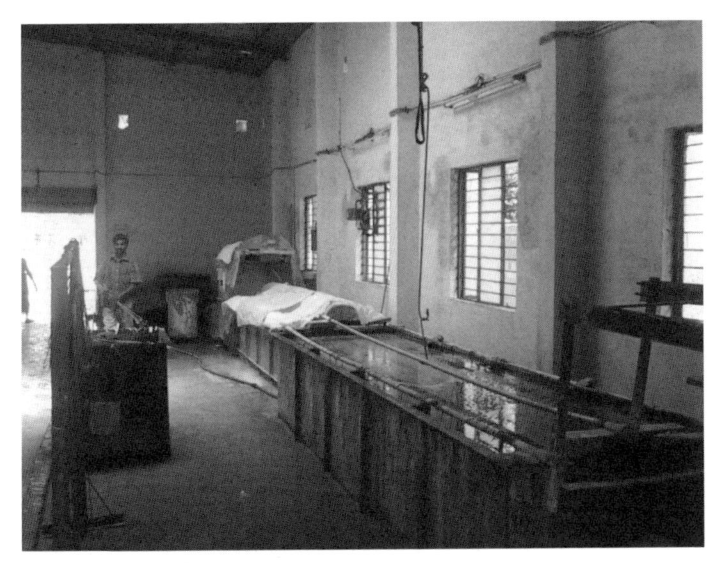

Figure 5.12 *View of an electroplating unit.*

Conventionally, cyanide solution is used widely in electroplating baths to improve the conductive properties of the base metal. Due to many hazards associated with cyanide, alternative chemicals/processes have been identified or are under research – for example, trivalent chromium process, which is a sustainable solution compared to hexavalent chromium.

Electroplating is conventionally a highly toxic activity so a large number of industries usually outsource the activities to small-scale units, which prefer to use time-tested procedures. Such units usually will not have the capability to do research and invest in pollution control and patented eco-friendly processes, thereby releasing pollutants in the ecosystem.

5.1.2.2.2 Waste Characteristics

The effluents from electroplating activity arise from the scrubber for air pollution control, spillage, housekeeping, the release of used chemicals and so on. The chemicals in preprocess and plating vats are replaced by fresh chemicals to ensure quality and the desired properties. The solvents/vapours from hot plating baths result in emissions of volatile organic compounds (VOCs) and, in some cases, volatile metal compounds. About

Table 5.2 *Typical pretreatment processes.*

Sl. no.	Preprocessing	Preprocessing baths
1	Polishing	—
2	Grinding	—
3	Degreasing	Trichloroethylene or perchloroethylene
4	Alkaline cleaning	Sodium carbonate, sodium hydroxide or sodium phosphate.
5	Acid dipping	10 to 30% sulfuric or hydrochloric acid.

Table 5.3 *Typical constituents of electroplating bath.*

Metal/alloy electroplated	Description	Uses	Process	Chemicals used
Hard chromium electroplating	Layer of chromium is plated on the base metal to provide a surface with a wear resistance, hardness, low coefficient of friction, corrosion resistance	Used for items such as industrial rolls, hydraulic cylinders and rods, castings, plastic moulds, zinc die, marine hardware and engine components	Polishing, degreasing, alkaline cleaning, acid dipping and chromium electroplating	Hexavalent chromium bath: sulfuric acid, chromic acid and water
Decorative chromium electroplating	Base material is deposited with layers of copper and nickel and a layer of chromium to give a bright surface and wear and stain resistance	Automotive parts, hand tools, metal furniture, plumbing fixtures and bicycles	Polishing, grinding and degreasing, alkaline cleaning, acid dipping, followed by strike plating of copper, followed by copper/nickel/chromium electroplating Chromium electroplating is similar to hard chromium plating except that decorative electroplating requires shorter plating times and lower current densities	Copper strike plating: copper cyanide solution Copper electroplating: copper cyanide or acid copper solution Semibright and bright nickel plating, Watts plating baths. Chromium electroplating bath: chromic acid, sulfuric acid (some plating units fluosilicate or fluoborate instead of sulfuric acid) and water

Chromic acid anodizing	Chromic acid anodizing is done to provide an oxide layer on aluminium for electrical insulation, corrosion protection, ease of colouring and to improve dielectric strength	Used on architectural structures and aircraft parts that are subject to high corrosion and stress	During anodizing, the voltage is applied step-wise and maintained at 40V for the remainder of the anodizing time. The aluminium substrate acts as anode; the tank is the cathode; and sidewall shields (like herculite glass, neoprene and wire safety glass and vinyl chloride polymers) are used instead of a liner to avoid short circuits and to lessen the effective cathode area Pretreatment include desmutting, alkaline soak, etching and vapour degreasing. The postanodizing steps involve sealing and air drying	Copper strike plating: copper cyanide solution Copper electroplating: copper cyanide or acid copper solution Semibright and bright nickel plating: Watts plating baths Chromium electroplating bath: chromic acid, sulfuric acid (some plating units fluosilicate or fluoborate instead of sulfuric acid) and water

(*continued*)

Table 5.3 *(Continued)*

Metal/alloy electroplated	Description	Uses	Process	Chemicals used
Trivalent electroplating	Base material is deposited with layers of copper and nickel and a thin layer of chromium to give a bright surface and wear and stain resistance. Trivalent chromium electroplating baths are used to replace decorative hexavalent chromium plating baths. Currently trivalent chromium processes has single-cell and double-cell methods. In a double-cell process the solution will have minimal to no chlorides and uses lead anodes that are positioned in anode boxes, which comprise a dilute sulfuric acid solution that are lined with a permeable membrane Single-cell process solution will have a high concentration of chlorides with carbon or graphite anodes that are positioned in direct contact with the plating	Automotive parts, metal hand tools, furniture, bicycles and plumbing fixtures	Polishing, grinding and degreasing, alkaline cleaning and acid dipping, which were described previously, followed by strike plating of copper, copper electroplating, nickel electroplating and chromium electroplating Chromium electroplating is similar to hard chromium plating except that decorative electroplating requires shorter plating times and lower current densities.	Copper strike plating: copper cyanide solution Copper electroplating: copper cyanide or acid copper solution Semibright and bright nickel plating: Watts plating baths Trichromium electroplating bath: proprietary trivalent electroplating chromium bath

Brass electroplating	Base material is deposited with layers of brass	Decorative and engineering applications.	Polishing, grinding and degreasing, alkaline cleaning, acid dipping and brass electroplating.	Brass plating bath: copper cyanide, zinc cyanide, sodium cyanide, soda ash, and ammonia
Cadmium electroplating		Engineering applications.	Polishing and degreasing, alkaline cleaning, acid dipping and brass electroplating	Alkaline cyanide bath: cadmium oxide in a sodium cyanide solution. OR acid fluoborate, neutral sulfate and acid sulfate (non-cyanide cadmium plating solutions are used to avoid hazards associated with cyanide)
Copper electroplating		Copper cyanide plating is used in numerous plating operations as a strike		Copper cyanide bath: cuprous cyanide, potassium cyanide (or sodium cyanide)
Copper pyrophosphate	Used for plating on plastics and printed circuits	Decorative and engineering applications		Copper pyrophosphate plating bath: copper pyrophosphate and potassium pyrophosphate
Copper sulfate	Used for decorative uses, plating printed circuits, rotogravure, electronics and plastics and for electroforming.	Decorative and engineering applications		Copper sulfate bath: copper sulfate and sulfuric acid

(continued)

Table 5.3 (Continued)

Metal/alloy electroplated	Description	Uses	Process	Chemicals used
Gold electroplating	Gold plating can be classified as: (i) neutral cyanide gold, for high purity gold plating; (ii) alkaline gold cyanide, for gold and gold alloy plating; (iii) noncyanide (generally sulfite), for gold and gold plating; (iv) acid gold cyanide, for bright hard gold and gold alloyplating; and (v) miscellaneous	Decorative and engineering applications		Alkaline gold cyanide plating bath: potassium gold cyanide and potassium cyanide Neutral gold cyanide plating bath: potassium gold cyanide Acid gold cyanide plating bath: potassium gold cyanide
Lead electroplating		Engineering applications.		Lead fluoborate plating bath: lead fluoborate, fluoboric acid and boric acid.
Nickel electroplating		Engineering, decorative and electroforming purposes.		Decorative nickel plating bath: benzene trisulfonic acid, contain organic agents, such as benzene disulfonic acids, naphthalene trisulfonic acid, benzene sulfonamide, formaldehyde, ethylene cyanohydrin and butynediol. Engineering nickel plating bath: nickel and boric acid

Palladium and palladium-nickel electroplating	Engineering applications.	Palladium plating solutions are classified as ammoniacal, chelated, or acid Ammoniacal palladium plating bath: palladium ammonium nitrate or palladium ammonium chloride Palladium acid plating baths: palladium chloride Palladium nickel electroplating bath: palladium metal and nickel metal
Platinum electroplating	Engineering applications.	Platinum plating bath: dinitroplatinite sulfate or chloroplatinic acid
Rhodium electroplating	Decorative plating in jewellery/silverware. Electronic and industrial application	Decorative rhodium plating bath: rhodium phosphate or rhodium sulfate concentrate and phosphoric or sulfuric acid. Industrial and electronic rhodium plating bath: rhodium metal as sulfate concentrate and sulfuric acid

(continued)

Table 5.3 (Continued)

Metal/alloy electroplated	Description	Uses	Process	Chemicals used
Ruthenium electroplating		Engineering applications		Ruthenium plating bath: ruthenium as sulfamate or nitrosylsulfamate and sulfamic acid
Silver electroplating		Decorative and engineering applications		Silver plating bath: silver as potassium silver cyanide and potassium cyanide
Tin-lead, electroplating		Engineering applications		Tin-lead plating bath: lead, stannous tin, boric acid and fluoboric acid
Tin electroplating	Tin plating typically is performed using (i) sodium or potassium stannate) or (ii) stannous fluoborate, stannous sulfate, or (iii) halogen tin process.	Engineering applications.		Stannous fluoborate plating bath: stannous fluoborate, fluoboric acid, boric acid. Stannous sulfate plating bath: stannous sulfate, stannous tin, sulfuric acid Sodium/potassium stannate plating bath: sodium stannate or potassium stannate and tin metal

Tin-nickel electroplating		Used in light engineering and electronic applications as a substitute for decorative chromium plating		Tin-nickel fluoride plating bath: stannous chloride anhydrous, nickel chloride and ammonium bifluoride Tin-nickel pyrophosphate plating bath: stannous chloride, nickel chloride and potassium pyrophosphate
Zinc electroplating	The typical zinc plating solutions are categorized as (i) acid chloride, (ii) alkaline noncyanide and (iii) cyanide.	Engineering applications.		Zinc plating bath: acid chloride zinc, cyanide zinc, alkaline noncyanide zinc
Zinc-nickel		Engineering applications		Acid zinc-nickel plating bath: zinc chloride and nickel chloride. Alkaline zinc-nickel plating bath: nickel metal and zinc metal.
Zinc-cobalt		Engineering applications		Acid zinc-cobalt plating bath: cobalt metal and zinc metal Alkaline zinc-cobalt plating bath: cobalt metal and zinc metal.
Zinc-iron		Engineering applications		Acid zinc-iron plating bath: zinc sulfate and ferric sulfate Alkaline zinc-iron plating bath: zinc metal and iron metal.

30% of the solvents and degreasing agents used may be released. Typical emission-control technology includes (i) the moisture extractor, (ii) the use of polypropylene balls, (iii) fume suppressant, (iv) mesh-pad mist eliminator, (v) chevron-blade mist eliminator and (vi) scrubber. Electroplating units in developed countries usually avoid emission control to cut production costs, thereby exposing their employees and the environment to toxic fumes The mixing of cyanide with acidic wastewaters can generate hydrogen cyanide gas. The quantity of effluent varies from 1 to 500 l/m^2 of surface plated.

5.1.2.2.3 Pollution Prevention and Control

Changes in processes to replace treatment at the end of the pipeline are highly advisable. Sustainable solutions to electroplating need constant research but the following practices have been practically tested in the fields: (i) replacing cadmium with zinc plating, (ii) using cyanide-free systems, (iii) giving preference to water-based surface cleaning instead of organic cleaning agents, (iv) using trivalent chrome instead of hexavalent chrome, (v) regenerating acids and other process ingredients whenever feasible, (vi) reduction in dragout and wastage, (vii) allowing dripping time of about 10 to 20 seconds before rinsing, (viii) minimizing dragout through effective draining of bath solutions from the plated articles, (ix) using fog spraying of parts while dripping, (x) maintaining the viscosity, density and temperature of the baths, (xi) placing recovery tanks before the rinse tanks, (xii) agitation of rinse water or articles being plated to increase rinsing efficiency, (xiii) using spray rinses, (xiv) using counter current rinses, (xv) recycle process baths after concentration and filtration, (xvi) recovering and regeneration of plating chemicals, (xvii) recycling rinse waters after filtration, (xviii) analysing and regenerating process solutions regularly, (xix) cleaning racks between baths to reduce contamination, (xx) covering baths (especially volatile chemicals) when not in operation.

5.1.2.2.4 Effluent Treatment

Waste characteristics are highly toxic and do not have any organic matter that acts as food for micro-organisms, so biological processes are not used. This leaves the option of physical and chemical treatment procedures. The costly (but highly efficient) choice of removing dissolved solids includes ion exchange, reverse osmosis, electrodialysis, membrane filtration, forced evaporation, activated carbon treatment and use of nanoparticles for adsorption. The cheaper option includes chemical precipitation of metals, alkali chlorination for cyanide and acid reduction for converting hexavalent chromium to trivalent chromium.

Recovery systems used since the 1970s employ evaporation (atmospheric and vacuum), ion exchange, reverse osmosis, electrodialysis and electrowinning (extracting metal in an electrolytic cell). Effluent characteristics vary widely, so the treatment should aim for: (i) cyanide destruction, (ii) removal of metals and (iii) flow equalization and neutralization. Cyanide destruction is usually attained by alkali chlorination. The pH cyanide-bearing water/wastewater is elevated by adding sodium hydroxide or calcium hydroxide (lime) followed by the addition of chlorine as sodium hypochlorite solution or bleaching powder. The use of chlorine gas is possible but is avoided as it necessitates precautionary safety measures that entrepreneurs usually do not invest in. If hexavalent chrome (Cr^{+6}) is present in a wastewater stream it should be converted to a trivalent form using a reducing agent like a sulfide. The process typically known as acid precipitation involves reducing pH by adding sulfuric acid followed by adding lime to achieve a precipitation of chromium.

The usual treatment processes used in electroplating are (i) equalization, (ii) pH adjustment, flocculation and (iii) sedimentation/filtration. pH adjustment is made prior to metal precipitation. The optimum pH for metal precipitation depends on the metals present and it is around 8.5–11. The presence of oil and grease may affect the metal precipitation process so it is preferred to treat degreasing baths separately. Flocculating substances are used to enhance the efficiency of removal of suspended solids.

5.1.2.3 Inorganic Chemical Manufacturing

Inorganic compounds are not of biological origin. Inorganic compounds do not have carbon and hydrogen atoms and they are synthesized in geological systems. Organic chemists traditionally refer to molecules with carbon as organic compounds and therefore inorganic chemistry deals with molecules lacking carbon. Inorganic chemical industry includes the chemical industry that manufactures: (i) inorganic chemicals on a large scale such as sulfuric acid, chlor-alkalis, sulfates, (ii) fertilizers such as nitrogen, phosphorus and potassium products and (iii) fine chemicals used to produce high-purity inorganics on a smaller scale.

Metals are considered to be chemicals. They are manufactured from ores and are used in the manufacture of inorganics. If they are commercialized in a pure form or as alloys they are normally considered as products of the metallurgical industry.

The larger the scale of operations ('heavy' inorganic chemicals) use continuous processes whereas 'specialties' are done in batch processes, from 'intermediates' that have already gone through many steps of synthesis and purification. Basic chemicals like phosphate rock, common salt, limestone, sulfur, potassium compounds and sodium carbonate form the starting point for manufacturing inorganic industrial chemicals. The sulfur, nitrogen, phosphorus and chloralkali industries produce basic inorganic chemicals.

There are numerous sources of raw materials for the manufacture of inorganic chemicals and few of them occur in their elemental form. Air contains molecular nitrogen, oxygen and argon which are separated by liquefaction and fractional distillation. Salt or brine are used as sources of chlorine, bromine, sodium hydroxide and sodium carbonate. Metals, phosphors, calcium, potassium and fluorine are extracted from mineral ores. Recovery and recycling provide some metals/chemicals.

Sulfuric acid and sodium carbonate are industrial chemicals. Sulfuric acid was an essential chemical for bleaching, dyeing activities and the manufacture of other chemicals.

Sustainable water management in the chemical industry should follow these principles: (i) adopting low-waste technology; (ii) substitution of less harmful chemicals wherever possible; (iii) adopting closed-water systems; (iv) segregating waste streams so that different streams can be treated more easily.

5.1.2.4 Boiler Blowdown

As steam formation continues in a boiler, the concentration of dissolved solids increases. Twenty-four per cent of the feed water is therefore discharged as boiler blowdown to control the concentration of dissolved solids. The boiler blowdown contains a maximum of about 200 ppm dissolved solids. The BOD and COD content in boiler blowdown is negligible; hence, boiled blowdown may be added to the circulating cooling water wherever a circulating water cooler is installed. Otherwise it is usually evaporated in a solar evaporator or a forced evaporator.

5.1.2.5 Thermal Power Plant

The usual cooling systems employed in thermal power plants are: (i) once-through cooling systems; (ii) closed-circuit dry cooling systems and (iii) closed-circuit wet cooling systems.

Facilities using once-through cooling systems discharge water at an elevated temperature. This has now been discontinued in many parts of world due to the threat to water resources and the ecology. Closed-cycle, cooling water systems using closed-circuit dry cooling systems or natural forced draft cooling towers using air-cooled condensers are sustainable approaches and have been used in many parts of the world.

The wastewater from coal-fired thermal treatment plants arises from (i) ash handling, (ii) the wet flue gas desulfurization (FGD) system, (iii) metal cleaning wastewater, (iv) material storage runoff, (v) air heater and precipitator wash water, (vi) floor and yard drains and sumps, (vii) boiler blowdown, (viii) laboratory wastes, (ix) boiler chemical cleaning waste (x) regeneration of ion exchangers. Other waste streams can be expected in oil-fired or gas-fired power plants. Contaminants in waste streams include traces from lubricating and auxiliary fuel oils; demineralizers; trace contaminants in fuel and chlorine, biocides and chemicals used in cooling systems. Cooling-tower blowdown will have very high total dissolved solids (TDS), residual chlorine and toxic chemicals due to additives (corrosion-inhibiting chemicals containing chromium and zinc).

Sustainable water management in thermal power plants includes: (i) recycling effluents for use as fgd makeup in coal fired thermal power plants, (ii) collecting fly ash and bottom ash in dry form, (iii) using dry methods to remove fireside wastes, (iv) using infiltration and runoff-control measures from coal piles and (v) spraying coal piles with bacteria inhibitors to avoid bacterial growth and reduce acidity of leachate.

5.1.2.6 Surface Coating

Surface-coating activity involves powder coating and spray painting of metal components to protect them. For example, surfaces are coated for the manufacture of nonstick cookware. Powder coating is a dry process whereas spray painting is carried out by moving components to be painted across the paintspray room where paint is sprayed continuously. Water curtains are generally used to curb air pollution by capturing paint mist. The emulsion of paint and water is treated to settle to obtain paint sludge for final disposal.

Treatment of wastewater from painting booths needs pH adjustment while a detackifying chemical is added. pH is adjusted to around 8 with caustic soda and detackifying agent added to avoid sludge sticking to pipes, clarifier plates and so on. The wastewater is then clarified to obtain clear water. The clarifier waste sludge is periodically sent to sludge-drying beds or dewatered using a filter press. If sludge is dewatered using a filter press, it is mixed with a filter aid to improve porosity and filterability to improve cake dryness and avoid premature blinding of the filter cloths.

5.1.2.7 Phosphating

Phosphating (Figure 5.13) is used as a metal pretreatment process for the surface treatment of metals to impart wear resistance, corrosion resistance, lubrication properties and adhesion. It plays a major role in the automobile, processing and appliance industries.

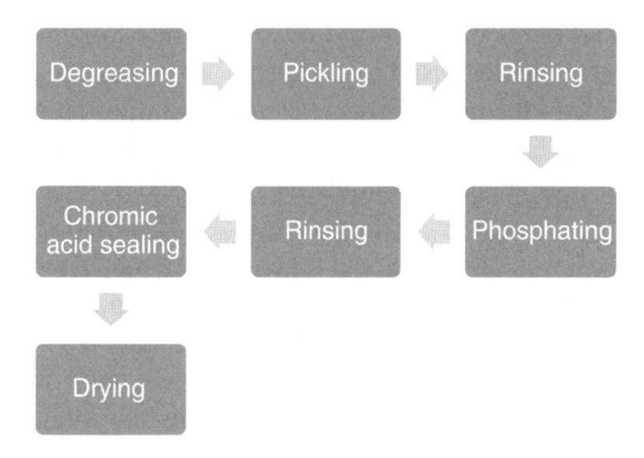

Figure 5.13 *Flow chart showing the operating sequence involved in phosphating process.*

Dilute phosphoric acid-based solutions of heavy metal/alkali metal ions are usually employed, which essentially contain free phosphoric acid and primary phosphates of the metal ions. Oxidizing substances and metals (such as copper and nickel) that are resistant to corrosion and oxidation are used to accelerate the process. Apart from chemical acceleration, mechanical or electrochemical acceleration are also used in some phosphating plants.

Wastewater streams arise from acid pickling, alkaline cleaning, rinsing stages and the chromic acid sealing stage. The quantity of wastewater generated is about 20 800 g/m^2 (Narayanan, 2005). As the wastewater characteristics are similar to those of the wastewater generated during pickling the treatment followed is discussed in section 5.1.2.8.

5.1.2.8 Pickling

Pickling is a metal chemical process used to remove impurities, like stains, rust or scale or inorganic contaminants from ferrous, copper and aluminium metal and alloy products.

Acid has been used in the form of lemon juice, tamarind and vinegar to clean metals for centuries – hence, the name 'pickling' was and continues to be used. Soft drinks with phosphoric acid are also good rust removers. An early application of acid cleaning in industry was the use of hydrochloric acid with zinc added (Hasler and Stone, 1997).

When steel is cooling subsequent to hot rolling, the oxygen in the air reacts with the iron on the steel and forms any/all of following compounds: (i) ferric oxide – rust (brown, powdery), (ii) ferrous oxide – scale (black, scaly), (iii) ferrous-ferric oxide – magnetite (bluish black, magnetic). In order to further process it, scale must be removed. Mechanical scale removal methods like shot blasting, brushing, grinding and scraping damage the steel surface. Hence, pickling is carried out by dipping the object to be pickled in a solution of strong acids, called pickle liquor. The reaction of acid with metal/contaminants produces dissolved metal salts, generating considerable volumes of spent pickle liquor (SPL). This contains the dissolved metal salts and residual free acid, which must be neutralized with alkali before disposal, or can be regenerated.

As extensive metalworking industries grew up in the middle of the nineteenth century, requiring large amounts of clean and descaled metal, most industries cleaned steel by laboriously dipping metal objects in acid contained in wood vats. The metal objects were then rinsed with large amounts of water and oiled or limed to avoid rusting of the clean surface. Most of the early picklers used sulfuric acid, as it was cheap. With time, various improvements were made to the process by mechanical handling, use of inhibitors, heating of the acid and use of acid-brick-lined tanks. Furthermore, the increase in the demand for strip steel led to the development of continuous strip picklers where the uncoiled strips were drawn continuously through hot sulfuric acid. Early lines had speeds of 30 to 100 fpm (foot per minute). Successive coils were held together with mechanical clips. Speeds of up to 400 fpm were achieved with time; welding was done to join each coil to the next.

In the early 1960s, hydrochloric acid pickling was made to increase speeds. Hydrochloric acid is more expensive (approximately 2.5 times the cost of sulfuric acid) but it has been the acid of choice in many cases as it offers a numerous advantages over sulfuric acid: (i) it pickles much faster than sulfuric acid, (ii) there is lower base metal loss, (iii) it generates a uniform surface on high-carbon steel, (iv) there is less chance of overpickling, (v) it is safer to handle than sulfuric acid, (vi) it can be operated at lower operating temperatures. Apart from the cost, hydrochloric acid has its disadvantages: (i) it is much more corrosive than sulfuric acid and (ii) it emits choking fumes when hot.

Sulfuric acid attacks the metal to remove scale. A sulfuric finish is normally more etched than a hydrochloric finish. Metal inhibitors used with sulfuric acid do not completely stop the reaction of metals. In hydrochloric acid, effective inhibitors almost totally stop the reaction of metals. Hydrochloric acid dissolves scale; hence no breakers are needed as in sulfuric-acid pickling. Scale removed in sulfuric-acid pickling settles in the pickle tanks, whereas hydrochloric acid dissolves scale rapidly, hence, accumulation of scale does not occur. At high temperatures ferrous sulfate monohydrate is formed in sulfuric acid pickling, which settles in the tanks due to low solubility at higher temperatures. This does not happen with hydrochloric acid.

Acid regeneration of spent hydrochloric pickle liquor is done by: (i) pyrohydrolysis, (ii) distillation, (iii) adiabatic absorption, (iv) acid retardation, (v) pickling hydrochloric acid regeneration (PHAR).

5.1.2.9 Hardening

Hardening is a process of increasing the hardness of metal. The five hardening processes that are found in the literature are: (i) grain boundary strengthening, (ii) starin hardening or cold working, (iii) solid solution strengthening, (iv) precipitation hardening and (v) quenching (the act of quickly cooling the hot steel to harden it).

Quenching and tempering are used for hardening steel. Quenching can be done by dipping the hot steel in water. The water adjoining to the hot steel vaporizes so water will not have direct contact with the steel, slowing down cooling until the bubbles break and water comes into contact with the hot steel. Bubbles can be prevented on the surface of steel with good agitation, thereby avoiding soft spots. Quick cooling can sometimes cause cracking or distortion in steel. Bubbles are broken easily in hard water and allow for quick cooling of the part but hard water is more corrosive than soft water and therefore must be rinsed off immediately. Oil is used when a slower cooling rate is required. Oil has a very high boiling and reduces the likelihood of cracking but oil quenching results in fumes and may cause a fire hazard. Polymer quenching will produce a cooling rate between water and oil and the cooling rate can be varied by changing the components in the mixture. Polymer

Figure 5.14 *A typical galvanizing rinsing tank.*

quenches are less of a fire hazard than oil and less corrosion than water. Cryogenics or deep freezing is used to achieve certain qualities in metals.

Some industries prefer a potassium cyanide bath for heating cutting tools, dies and so on, to secure ornamental colour effects. Wastewater from quenching contains a cocktail of chemicals, oil and grease depending on the material hardened and the procedure involved. It is of great importance to monitor the TDS as residual contaminants depend on quenching salts and emulsions issued. Chemical precipitation followed by tertiary treatment will be required to treat wastewater.

5.1.2.10 Galvanizing

Galvanization (Figure 5.14) is the process of providing a protective zinc coating to steel/iron/aluminium to prevent rusting. In hot-dip galvanization iron/steel/aluminium is coated with a zinc layer by passing the metal through a zinc bath at about 460 °C. The zinc reacts with oxygen when exposed to the atmosphere to form zinc oxide, which further reacts with CO_2 to form zinc carbonate, which stops corrosion in many circumstances.

5.1.2.11 Textiles

Textile industries convert raw fibre into cloth. The operations carried out in textile industries are (i) spinning of fibre to yarn, (ii) sizing to improve stiffness, (iii) scouring, (iv) kiering, (v) desizing to eliminate excess sizing materials, (vi) bleaching to remove pectin and wax from the yarn and fabric, (vii) weaving and (viii) colouring/printing. All the operations are carried out in single or different units. Sizing and desizing operations generate wastewater, characterized by high BOD processes and treated by anaerobic and aerobic biological methods.

Textile-manufacturing activities involve the use of hazardous chemicals during pretreatment, dyeing and other processes. Highly hazardous surfactants are usually replaced with bioeliminable and biodegradable chemicals. Pollutants in textile effluents will have suspended solids, mineral oils, spinning lubricants, grease, antifoaming agents, low biodegradable or nonbiodegradable surfactants, phenols and halogenated organics.

Characteristics of the effluent streams depend on the operation or activity within the industry. While some units are involved only in dyeing, others specialize in washing. Such specialized activities are common in the developing countries where there is large-scale industrial outsourcing of washing dyeing activities. Dyeing processes use colours and are carried out at high temperature. Effluents will have heavy metals like chromium, copper, zinc, lead or nickel. Textile units processing cotton are likely to contain pesticides as well. Other pollutants from the textile industries include bacteria, fungi and other contaminants like sheep-marking dye and tar.

Finishing operations, namely desizing, mercerizing, bleaching, dyeing, printing and other specific operations, treat fabrics with chemical and liquor baths and generate significant wastewater effluents. Sustainable options for this operation include: (i) the selection of water-soluble and biodegradable lubricants instead of mineral oil, (ii) the use of organic solvent washing, (iii) performing the thermo fixing step before the washing step.

5.1.2.11.1 Desizing

The presence of sizing ingredients can affect operations such as printing, dyeing and finishing. For example, starch can obstruct the penetration of the dye into the fibre. Hence, starch is removed or transformed into simple water-soluble products by hydrolysis with enzymes/mineral acids or by oxidation by sodium chlorite, sodium bromide and so on (Batra, 1985). Typically about 50% of the water pollution in cotton textile occurs due to the desizing operation, which has a high BOD. The problem can be overcome using enzymes that degrade starch into ethanol. The ethanol can be recovered by distillation. Alternatively, a strong oxidizing agent like H_2O_2 can be used to degrade starch completely to CO_2 and H_2O. Electro-oxidation with PbO_2/Ti or RuO_2/Ti electrodes is another effective method for the treating starch effluent. Synthetic sizing chemicals based on polyvinyl acrylic (PVA) are economic considering the cost of effluent treatment. Ultrafiltration and nanofiltration technology allow recovery/reuse of PVA (Babuu *et al.*, 2007; Yu *et al.*, 2001). COD and BOD_5 loads from desizing operations generate 35% to 50% of the total load from textile operation with COD concentrations up to 20 000 mg/L. Pollution prevention and control desizing include: (i) selection of bioeliminable sizing agents, (ii) application of enzymatic or oxidative desizing, (iii) integration of desizing/scouring and bleaching, (iv) recovery and reuse of synthetic sizing agents by ultrafiltration.

5.1.2.11.2 Scouring

Scouring of fibres involves the use of detergents and hot water to remove contaminants from fibres. Scouring wool uses water and alkali, or organic solvents. Scouring with alkali generates effluent that is alkaline. Most BOD and COD loads in textile manufacturing come from scouring processes. Use of readily biodegradable surfactants/detergents is more environmental friendly. Adoption of dirt-removal/grease-recovery loops for water-based wool scouring reduces water consumption by 2–4 l/kg of greasy wool (IFC, 2007).

5.1.2.11.3 Mercerization

In order to improve dye uptake, increase strength, impart lustre, cotton fibre and fabric are mercerized after bleaching. During mercerizing, cotton fibre reacts with caustic soda. Hot water is used to eliminate the caustic solution from the fibre, followed by neutralizing with acid. Several rinses are made to remove the acid from the fibres. Hence, wastewater from mercerizing is highly alkaline. Reuse of alkali from mercerizing effluent will reduce pollution load and save expenditure on chemicals. Mercerization is done by treating cotton material with a strong NaOH and washing off the NaOH after 1 to 3 min. Cotton will undergo a longitudinal shrinkage after impregnation with this solution, hence material is kept under tension. The NaOH in the effluent can be recovered using membranes. Use of $ZnCl_2$ during mercerization allows easy recovery of NaOH and does not require neutralization by formic acid or acetic (Karim *et al.*, 2006).

5.1.2.11.4 Bleaching

Natural colour in the yarn gives a creamy colour to the fabric. It is decolourized by bleaching to obtain white yarn. Conventionally, the textile industry uses hypochlorite as bleaching agent. Chlororganic compounds formed during the bleaching are reduced by adsorbable organically bound halogen (AOX). Hypochlorite is being replaced by other environmental friendly bleaching agents like peracetic acid. A one-step preparatory process combining desizing, scouring, and bleaching will reduce the volume of water. Electrochemical mercerization and bleaching of textiles carried out in an electrochemical cell is environmental friendly but has yet to be adopted over the world. Common bleaching reagents used in the textile industry are sodium chlorite, sodium hypochlorite, hydrogen peroxide and sulfur dioxide. The use of chlorine-based bleaches is likely to produce organic compounds of chlorine, hence the use of a hydrogen peroxide bleaching agent is a sustainable ecofriendly option.

5.1.2.11.5 Neutralization

The process of bringing pH to around 7 after scouring is neutralization. Neutralization with formic acid instead of acetic acid is considered economical and environment friendly as the procedure allows sufficient neutralization in short time with low quantity of water and discharge of low levels of BOD (Bradbury *et al.* 2000).

5.1.2.11.6 Dyeing

Treatment of/fabric/fibre with chemical pigments to provide colour is called dyeing. The wastewater from dyeing contains colour pigments and may contain halogens, metals, amines in spent dyes and other chemicals (dispersing and antifoaming agents, alkalis, salts and reducing/oxidizing agents). Dyeing process effluents will have highboy and COD values, with a COD value normally above 5000 mg/l. Salt concentration may vary between 2000 and 3000 ppm. The sustainable options for water conservation and pollution prevention include (i) automation of dosing and dispensing dyes, (ii) adoption of continuous and semicontinuous dyeing processes, (iii) use of bleaching systems that reduce liquor-to-fabric ratios, (iv) implementation of mechanical liquor extraction, (v) adoption of optimized process cycles, (vi) substitution of conventional dye carriers and finishing agents with lesser toxic chemicals.

5.1.2.11.7 Printing

Printing involves the use of sprint paste components that contain colour concentrates (pigments or dyes), solvents, defoamers and binder resins. Printing cloths that are washed with water prior to drying, generate wastewater contain VOCs. The pollution control measures include: (i) reducing printing paste losses by recovering and recycling printing paste; (ii) reusing left-over rinsing water, (iii) avoiding the use of urea by addition of moisture by two-step printing methods or adding moisture, (4) using printing pastes with no or low VOC emissions (IFC, 2007).

5.1.2.11.8 Finishing

Textiles are subjected to a range of finishing activities to improve properties. They use numerous finishing agents for softening, waterproofing and cross-linking. All of the finishing activity contributes to water pollution. The most used chemicals are formaldehyde-based cross-linking agents that give desired properties like stiffness, softness, smoothness and enhanced dimensional stability. Unreacted formaldehyde necessitates removal of the chemical from fabric. The formaldehyde resin is a known carcinogen and, hence, much effort is under progress for a substitute (Hashem *et al.*, 2005). Easy-care, wrinkle-resistant (durable press), waterproof, mothproof, antibacterial fabric uses a number of chemicals that are highly toxic to aquatic life. Hence, proper care needs to be taken to avoid entry into waste streams.

Process wastewater treatment involves (i) chemical oxidation of high COD streams, (ii) chemical precipitation, coagulation and flocculation for removing heavy metal and (iii) reverse osmosis to remove high TDS wastewater streams.

A typical wastewater treatment system consists of (i) grease traps, oil water separators or skimmers; (ii) filtration for the separation of filterable solids; (iii) equalization; (iv) sedimentation; (v) biological treatment; (vi) biological nutrient removal; (vii) chlorination and (viii) dewatering of sludge.

5.1.2.11.8.1 Manmade Fibres Manmade fibre includes synthetic and semisynthetic materials. Synthetic materials are those made from organic polymers and semisynthetic forms of natural polymers. Manmade fibres include nylon, polyester, acrylic and acetate rayon. Some units produce polymers whereas others procure polymers in the form of chips, which act as raw material for fibre manufacture. Continuous filaments are made by melt-spinning the polymer chips where molten polymers are pumped through holes in a spinnerette. Wastewater generated varies typically from 17 to 364 m^3/t of product in case of nylon and polyester. The quantity of effluent generated varies from 142–175 m^3/t with respect to viscose and 1140–1360 m^3/t for rayon. The wastes from spinning machines and acid baths contain sulfuric acid, zinc sulfate, sodium sulfate, waste fibre and carbon bisulfide and hydrogen sulfide. Some streams are not easily biodegradable and others need supplementary nutrients to degrade. Hence, it is recommended to segregate streams prior to treatment.

5.1.2.11.8.2 Silk Silk is manufactured from silk cocoons. Once the worms start pupating inside cocoons, these are dissolved in boiling water to extract individual long fibres. In other words cocoon cooking is done to unwind the cocoons filament spun by the silkworms. The sericin around the cocoon filament is softened by heat, water and steam.

About 850–1000 m^3 of water is used to manufacture 1 ton of raw silk.

Silk cooking is followed by reeling (unwinding) the filaments from four to eight cocoons at once. Reeling sometimes is done with a slight twist, to get a single strand. To prepare a silk yarn for dyeing it is necessary to remove sericin and natural oils and organic impurities by scouring. At the industrial level, scouring in alkaline conditions is most common. The silk is later dyed to produce the desired colour.

5.1.2.11.8.3 Cotton Textiles The textile industry receives and prepares fibres; converts fibres into yarn, thread, or webbing; converts the yarn into fabric and dye and finishes these materials (Ghosh and Gangopadhyay, 2000). The cotton textile dyeing industry consumes huge quantities of water and generates large quantities of wastewater. Table 5.4 shows sources of wastewater in a textile mill. Interest in environmentally friendly goods has increased in recent years as consumers in the developed world are demanding biodegradable and environmentally friendly textiles (Chavan, 2001).

Table 5.4 *Sources of wastewater in textile mill.*

Sl. no.	Process	Significance with respect to wastewater quantity	Major characteristics
1.	Fibre preparation	Little or none	
2.	Yarn spinning	Little or none	
3.	Slashing/sizing	Significant	BOD, COD, metals
4.	Weaving	Little or none	
5.	Knitting	Little or none	
6.	Tufting	Little or none	
7.	Desizing	Significant	BOD, COD, metals, lubricants, biocides, antistatic compounds
8.	Scouring	Significant	Insecticide disinfectants/residues, NaOH, fats, detergents, pectin, oils, lubricants, wax, spent solvents, spin finishes.
9.	Bleaching	Significant	Sodium silicate or organic stabilizer, hydrogen peroxide,
10.	Singeing	Little or none	
11.	Mercerizing		NaOH
12.	Heat setting	Little or none	
13.	Dyeing	Significant	Metals, salt, cationic materials, colour, BOD, sulfide, acidity/alkalinity, surfactants, toxics, organic processing assistance, spent solvents.
14.	Printing	Significant	Metals, suspended solids, heat, foam, BOD, urea, solvents, colour.
15.	Finishing	Significant	Suspended solids, BOD, COD, spent solvents, toxics
16.	Product fabrication	Little or none	

Table 5.5 Advantages and disadvantages of different treatment process.

Sl. no.	Processes	Advantages	Disadvantages
1	Biological processes	BOD can be reduced economically	Low biodegradability of dyes
2	Chemical precipitation	Elimination of insoluble dyes	Sludge disposal
3	Adsorption on activated carbon	High efficiency	Cost of activated carbon
4	Ozone treatment	Good decolourization	COD unaffected
5	Electrochemical processes	Capacity of adaptation to varying pollution loads	Sludge disposal
6	Reverse osmosis	High efficiency	High capital cost and fouling of membranes
7	Nanofiltration	High efficiency	High capital cost and fouling of membranes
8	Ultrafiltration /microfiltration	High efficiency	High capital cost and fouling of membranes

Cotton requires 70–150 l of water to dye 1 kg of cotton with reactive dyes, 30–60 g of dyestuff, 0.6–0.8 kg of NaCl (Chakraborty *et al.*, 2005). Wastewater from printing/dyeing of cotton fibres is often rich in colour and requires proper treatment before discharging into the environment. Treatment options include biological treatment, electro-oxidation, ion exchange, photochemical processing and membrane techniques (Babuu *et al.*, 2007). Table 5.5 shows the advantages and disadvantages of different treatment processes.

5.1.2.12 Dyeing Industries

Dyeing involves (i) wetting, (ii) bleaching, (iii) neutralizing, (iv) washing, (v) colouring and (vi) washing. The dyeing operation requires 10 m^3 of water for one ton of cotton yarn whereas 1 ton of polyester yarn consumes only 4 m^3 of water in each step (Ranganathan *et al.*, 2007). The dye-bath solution requires alkali sodium salt and dyes in the process. The quantity of sodium chloride used usually depends on the colour shade required. Dyeing generates voluminous quantities of wastewaters. The effluents are characterized by high sodium content, TDS, sulfate, chloride, high BOD, hardness and dye ingredients (Table 5.6).

5.1.2.13 Drugs and Pharmaceuticals

Pharmaceutical manufacturing involves two major stages: (i) the production of the active ingredient – known as bulk drug manufacturing; (ii) the conversion of the active ingredients into products suitable for administration – known as pharmaceutical formulation.

The major manufacturing steps are: (i) preparation of intermediates; (ii) introduction of functional groups; (iii) coupling and etherification; (iv) separation processes like washing and stripping and (v) purification of the final product. Other steps in the pharmaceutical industry are fermentation, crystallization, chemical synthesis, purification and natural

Table 5.6 Characteristics of typical wastewater from dyeing industry.

Sl. no.	Characteristics	Value
1.	BOD (mg/l)	213
2.	Ca-hardness (mg/l) as $CaCO_3$	115
3.	Chloride (mg/l)	5715
4.	COD (mg/l)	702
5.	Percent sodium	97
6.	pH	8.9
7.	Potassium (mg/l)	84
8.	Sodium (mg/l)	3900
9.	Sodium absorption ratio (SAR)	122
10.	Sulfate (mg/l)	1419
11.	Total dissolved solids (mg/l)	13770
12.	Total hardness (mg/l) as $CaCO_3$	192

extraction/biological, drying, granulation, printing, tablet pressing, filling, coating and packaging.

Wastewater arises from equipment cleaning and characteristics of wastewater depend on the materials used and the process. Some wastewaters may contain mercury (in the range of 0.1–4 mg/l) or cadmium (10–600 mg/l). Typical amounts of organic matter released in the wastewater of pharmaceutical industry are 3 kg/t of suspended solids; 25 kg of BOD/t of product or 2000 mg/l; up to 0.8 kg/t of phenol; 50 kg/t of COD or 4000 mg/l (World Bank, 1998).

Effluent treatment usually includes coagulation, filtration, settling, neutralization, flocculation, flotation and ion exchange, Reverse osmosis/ultrafiltration carbon adsorption, detoxification by wet air oxidation/oxidation using ozone/peroxide solutions, biological treatment air/steam stripping to remove organics, chemical precipitation.

Pollution prevention and control measures recommended by the International Finance Corporation (2007c) include: (i) waste reduction by material substitution, (ii) process modifications, (iii) spent solvent recycling and reuse, (iv) salts/organic material recovery and (v) inactivation of potentially pathogenic waste.

5.1.2.14 Industries/Activities using Material with High Organic Contents

Industries using material with high organic content pose a greater challenge for treatment. Traditional aerobic treatments like the activated sludge process, trickling filters and lagoons exhibit difficulties in direct treatment of strong effluents such as distillery, slaughterhouse and pharmaceutical effluents. These difficulties arise due to (i) high aeration costs, (ii) high levels of waste sludge and (iii) oxygen transfer problems. For these reasons, anaerobic pretreatment is considered to be an efficient and cost-effective solution but anaerobic treatment is not the final solution to all wastewater treatment problems, especially those producing wastes with high colour such as distillery wastes. In such cases incineration has been proven effective, especially with respect to effluent with high calorific values where heat recovery makes them more sustainable.

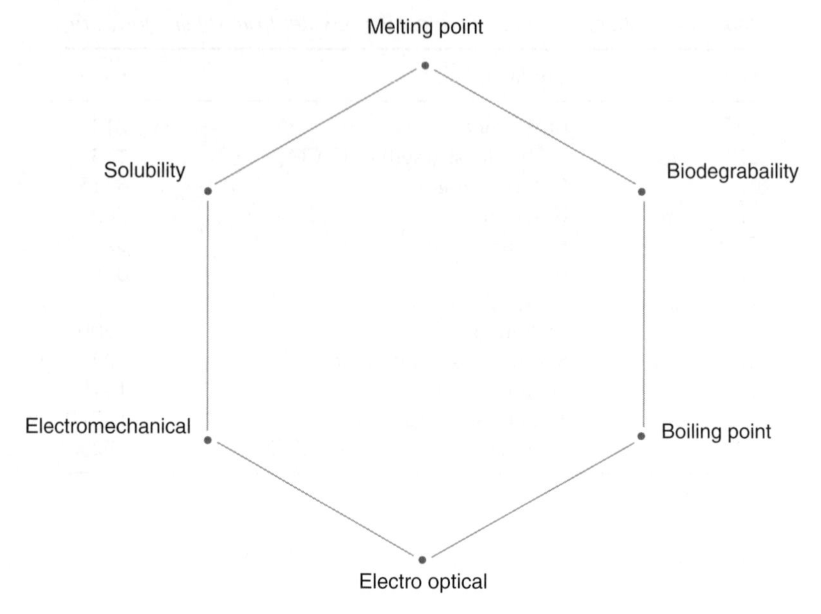

Figure 5.15 Properties to be considered to make decision about treatment of organic chemicals in wastewater.

5.1.2.15 Organic Chemical Manufacturing

An organic compound is a chemical containing carbon. For historical reasons, some types of carbon-containing compounds like simple oxides of carbon, carbides, carbonates, cyanides and forms of carbon like graphite and diamond are considered inorganic. Organic chemicals can be classified into aliphatic, aromatic, biomolecules, fullerenes, polymers and heterocyclic substances. Chemical properties considered for wastewater treatment are given in Figure 5.15 and Figure 5.16 shows the techniques used for removing organic matter in wastewater.

5.1.2.16 Sugar Manufacturing

Sugar is manufactured from sugar cane or sugar beet. Juice is extracted from sugar cane/beet and heated to 70–75 °C followed by treatment with lime solution. The juice is again heated to 100–115 °C. The hot juice is clarified and evaporated in multiple-effect evaporators to obtain a syrup of 60% sugar concentration. The syrup is further concentrated in vacuum pans where sugar grains are formed. This syrup mass with sugar particles is called massecuite. It is dropped in crystallizers for cooling and completion of crystallization. Sugar crystals are separated in a high-speed centrifuge sent to driers. Noncrystallizable matter in the syrup is called molasses and forms raw material for the manufacture of alcohol. Dried sugar is graded by sieving in vibrating sieves and is then bagged.

Sugar manufacturing demands large quantities of high-quality water for cleaning raw material, sugar washing, sugar extraction, steam generation, cooling and for cleaning equipment. Steam is required for evaporation and heating during sugar processing. Beet and cane

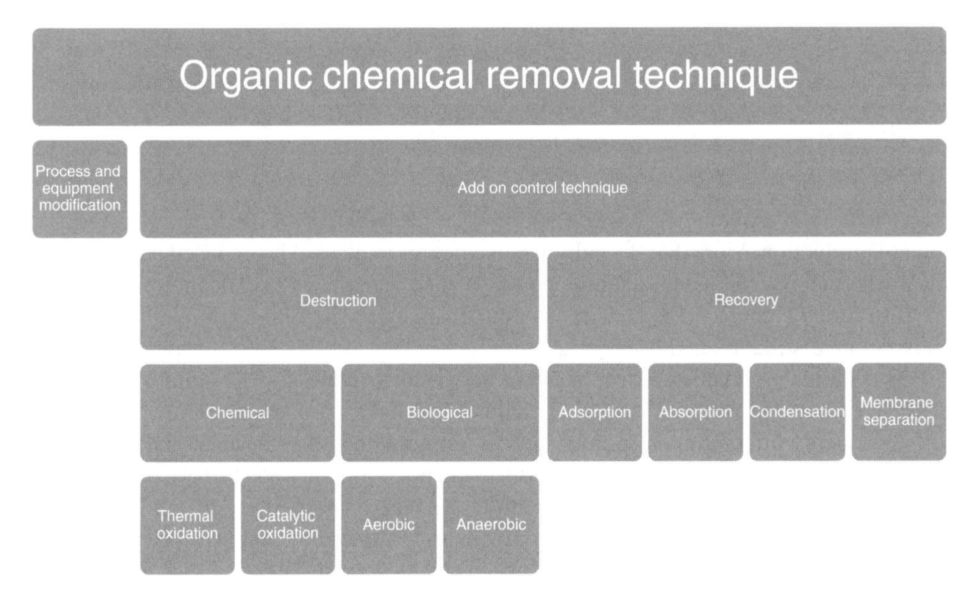

Figure 5.16 *Organic chemical removal techniques.*

contain a high water content that can be recovered for reuse. Industry-specific measures that can be used for sugar production include: (i) recycling process water and using it for washing raw material, (ii) using closed loops for production cane and beet washing and flue gas scrubbers.

Sugar processing wastewater has high BOD due to the presence of sugars and organic material arriving with the cane or beet. Wastewater from washing beet/cane contains pesticide residues, crop pests and pathogens.

Sustainable options for managing wastewater from the sugar industry are: (i) segregation of noncontaminated wastewater streams and contaminated streams, (ii) reducing entry of solid wastes and concentrated liquids into effluent streams, (iii) implementing dry precleaning of raw material, production areas and equipment before wet cleaning, (iv) allowing beet to dry and reducing breakages during collection and transport by use of lined containers and rubber mats, (v) using dry techniques for unloading beet, (vi) using grids/screens/traps to reduce the beet parts entering the wastewater and (vii) preventing direct runoff to watercourses. Industrial process wastewater treatment usually includes filtration for filterable solids, equalization, sedimentation, biological treatment (usually anaerobic followed by aerobic treatment), biological nutrient removal, chlorination, dewatering and disposal of residuals. Composting or land application of sludge is practised with press mud.

5.1.2.17 Distilleries

Alcohol can be produced from an array of feedstocks, which include beet/sugarcane molasses, corn, cane juice, potato, cassava, wheat, fruits (like grapes, cashew fruit), rice, barley, sugarcane bagasse, crop residues, municipal solid wastes, wood and others. Various countries in Asia and South America manufacture sugar from sugar cane, generating

Figure 5.17 Steps in distilleries.

byproducts such as molasses, bagasse and press mud. Molasses is used as raw material in ethanol production. More than 13 million cubic metres of alcohol is produced per year from sugarcane molasses alone (Jiranuntipon, 2009). The manufacture of alcohol in molasses-fed distilleries (Figure 5.17) consists of four main steps as follows: fermentation, feed preparation, distillation and packaging.

Fermentation is done in either batch or continuous mode. Feed is diluted to obtain the desired sucrose level and the pH is adjusted to below 5 using sulfuric acid. This is supplemented with a nitrogen source (such as ammonium sulfate or urea) and phosphate. Yeast culture is added to molasses to produce ethanol. Normally 8% to 10% of ethanol of is formed. The sludge is removed and fractionated (the process in which solid and liquid suspension is divided into a number of components) followed by rectification (the process of repeated distillation). The residue of the fermented mash is called distillery slop or spentwash.

Distillation is done in a series of bubble-cap fractionating columns in two stages: the first stage consists of the distillation column followed by rectification columns. Rectified spirit (95–97% ethanol by volume) is used for the manufacture of acetone, oxalic acid, acetic acid, absolute alcohol and beverages.

In grain-based distilleries the process involves milling (grinding of flour), mashing (formation of slurry), addition of α-amylase (an enzyme), cooking (heating with steam 8 to 10 kg/cm^2 pressure and at a temperature of 115 to 120 °C), liquefaction (hydrolysis into amylose and amylo pectin by adding α-amylase), saccharification (conversion of starch into sugars – viz. dextrins and oligosaccharides), fermentation (conversion of sugars into alcohol) and distillation.

Grain-based distilleries will have COD in the range of 60 000–62 000 (mg/l), BOD in the range of 20 000–22 000 mg/l, pH will be in the range of 3–4, total solids vary in the range of 40 000–42 000 mg/l and TSS will be in the range of 15 000–15 200 mg/l.

5.1.2.17.1 Wastewater Generation and Characteristics

Effluents generated during the manufacture of alcohol from grain/fruit/wood/solid wastes do not pose as much of an environmental challenge as that caused by effluent streams from sugarcane molasses distilleries.

Effluent streams from sugarcane molasses distilleries are about 12–15 times the volume of the produced alcohol. They are characterized by high COD (80 000–140 000 mg/l) and BOD (40 000–50 000 mg/l). Characteristics of the sugarcane molasses wastewater are variable and depend on the raw materials used. They are characterized by comparatively higher inorganic impurities because chlorides/(10 000–11 000 mg/l), sulfates (3000–4000 mg/l), phosphorus (300–400 mg/l), potassium (1400–1500 mg/l) and calcium (6000–7000 mg/l) cause serious environmental pollution. Its refractory nature is due to the presence of ceramal, melanoidins, polyphenols and a range of sugar decomposition products like anthocyanin, tannins and different xenobiotic compounds. The unpleasant odour is due to the presence of indole, skatole and other sulfur compounds, which are not completely decomposed during fermentation and distillation. The high nitrogen and phosphate content can result in eutrophication if it enters natural water bodies without treatment. Melanoidins have antioxidant properties and are toxic to many micro-organisms.

Wastewater from ethanol manufacturing from sugarcane molasses contains dark-brown coloured effluents with high organic content. Even after conventional biological treatment, dark brown colour persists and can even increase due to repolymerization of coloured compounds. The key coloured compounds are melanoidins formed via the complex 'Maillard reaction'. Alcohol-production from molasses generates about ten times as much wastewater as the quantity of alcohol produced.

Wastewaters from wine distilleries will have COD in the range of 7000 to 40 000 g/l and BOD^5 will be in the range of 5500 to 20 000 mg/L. The pH of wine distillery wastewaters varies from 3.5 to 5.0. Wine-distillery effluent contains phenolic compounds (mainly *p*-coumaric acid, gentisic acid and gallic acid), which impart antibacterial activity.

5.1.2.17.2 Treatment

The treatment of distillery effluents has been challenge and is a major source of pollution in the countries that produced alcohol from molasses. Plants that have tried lagoons, on-land application, aerobic treatment and anaerobic treatment have failed to bring down colour. The undiluted spent wash/effluent is applied to the field well before the planting of crops or diluted at a ratio of 1:10 to 1:50 with normal water prior to application (Baskar *et al.*, 2003). Use of effluents for composting is limited to sites used by the sugar industry where a supply of press mud is assured. Other approaches using physico-chemical treatment, such as coagulation, adsorption and flocculation, membrane treatment, chemical oxidation process, electrocoagulation have been tried with limited success. Another approach is one-time land application, where effluent is spread on land and tilled once each year or two. Here entrepreneurs have to look out for farmers who are willing to accept waste. Due to limited success with these conventional approaches entrepreneurs usually discharge partially treated wastewater, damaging the environment. Hence, thermal combustion is currently practised in many countries. Molasses spent wash containing 4% solids is concentrated to about 40% solids in a multiple-effect evaporation system. The concentrated mother liquor is spray dried to obtain a desiccated powder having a calorific value of about 3200 kcal/kg. The powder is usually mixed with 20% agricultural waste and burnt in a boiler.

The concentration of spent wash and its subsequent incineration creates problems due to scaling and shutdown of boilers for maintenance. Hence, co-incineration in cement kilns is considered to be a sustainable approach. Co-processing of spent wash concentrate in a cement kiln has the following benefits: (i) wastes are destroyed at a temperature of around 1200–1400 °C with longer residence time, (ii) inorganic content will be fixed with the clinker leaving no residue, (iii) the acidic gases are neutralized, (iv) it reduces fuel requirements for cement manufacture. The most common feed points for incinerating spent wash in cement manufacturing are: (i) burner at the rotary kiln outlet and (ii) precalciner.

5.1.2.18 Dairy

The dairy industry was established to fulfil urban demand for milk and milk products. A sophisticated dairy collects milk and processes and packs it before distributing to outlets/customers, leading to the generation of waste. Unsophisticated dairies in many parts of the world collect and distribute without any processing. In some places, the owner of the cattle sells milk to the customer without any processing. Hence, environmental problems magnify with the quantity of milk handled.

Dairy processes vary from country to country and case to case. Unlike the developed countries where hundreds of cattle are reared in single location, many developing countries will have one or two cows for a family. Milk from thousands of milk producers who generate as little as 1 l per cow are collected in hundreds of farms/villages and transported to dairies. Sometime chilling centres are used at an intermediate stage before delivering these to the dairy.

5.1.2.18.1 Process

The process in the dairy includes pasteurization, skimming, chilling and packing. Some dairies will have additional facilities to manufacture, butter, yogurt, flavoured yogurt, flavoured milk, buttermilk, spiced butter milk, ice cream, milk powder, cheese, ghee (clarified butter) and sweets. Figures 5.18–5.22 show the activities in the dairy industry.

A dairy consists of a chain of operations that include receiving, storing, processing and packaging. The operations involve heat transfer and cleaning during the processing of materials.

Wastewater arises from spills, leaks, cleaning and sanitizing of equipment during the transfer of milk from bulk carriers into refrigerated tanks for storage. Clarification and separation (of cream) are achieved by large centrifuges/filters of special design.

Pasteurization is accomplished by (i) heating the material for a long period in a vat followed by cooling (vat pasteurization) and (ii) passing the milk through heated and cooled plates or tubes (high-temperature, short-time (HTST) pasteurization).

Depending on the products manufactured, the dairy employs tanks/vats for mixing ingredients and culturing products, evaporators, driers, homogenizers (enclosed high-pressure spray units), churns and freezer.

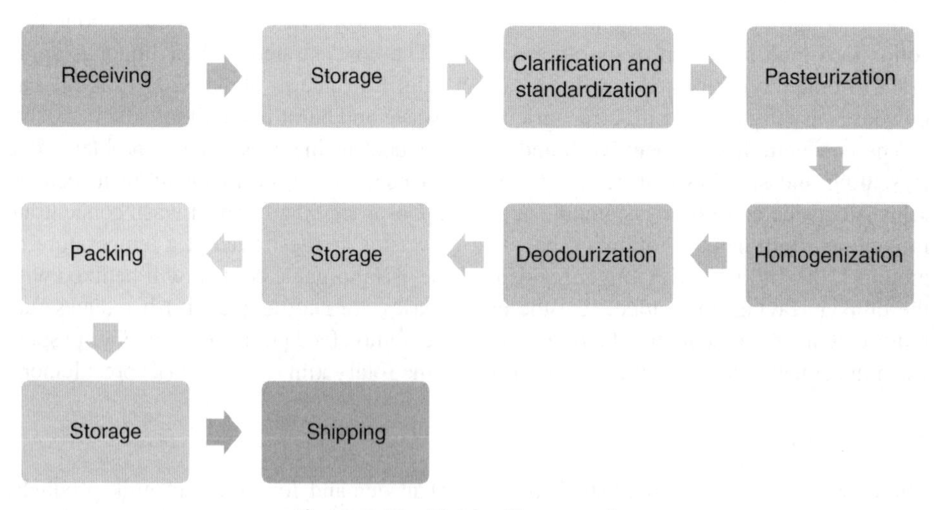

Figure 5.18 *Fluid milk processing.*

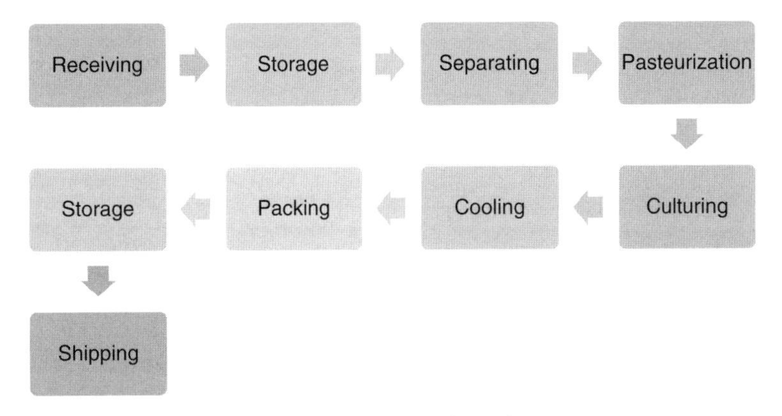

Figure 5.19 Cultured products.

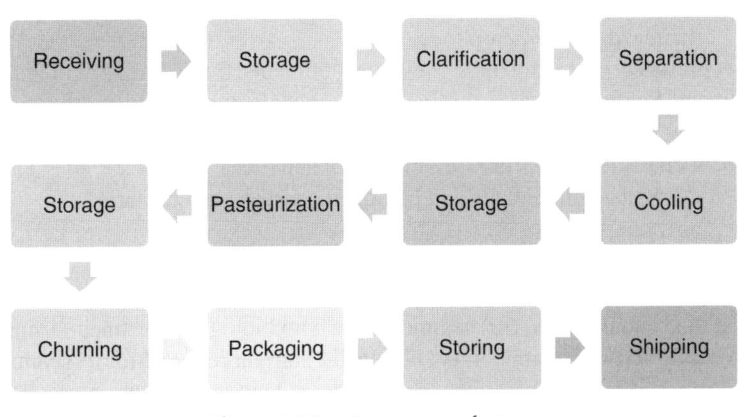

Figure 5.20 Butter manufacture.

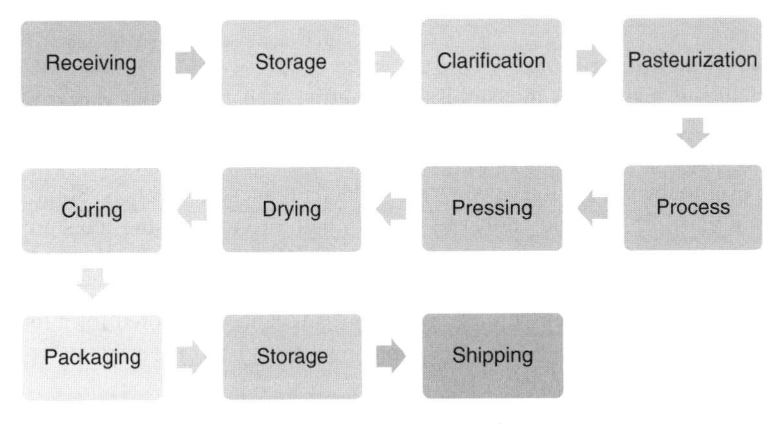

Figure 5.21 Cheese manufacture.

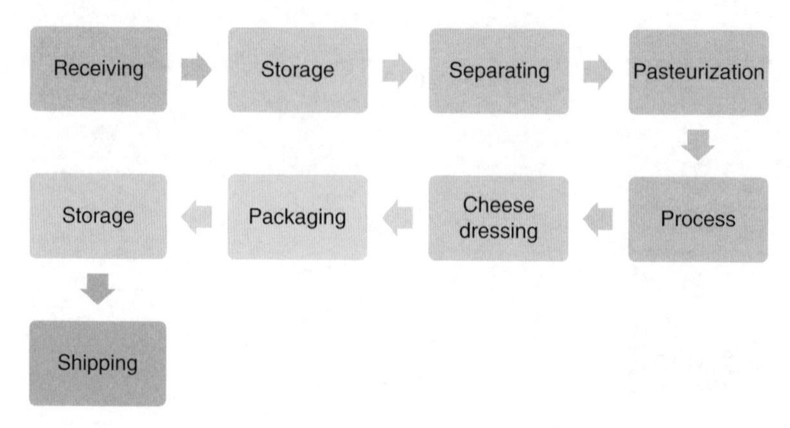

Figure 5.22 *Cottage cheese manufacture.*

5.1.2.18.2 Wastewater Characteristics

Due to the presence of milk solids, untreated effluents from dairies will have a high organic content. Salting activities during cheese manufacturing contributes to salinity in effluent streams. Table 5.7 shows wastewater characteristics of dairy waste and Table 5.8 shows effluent characteristics from different dairy activities. Wastewater may also contain alkali/acids/detergents/disinfectants and pathogenic viruses and bacteria.

5.1.2.18.3 Conventional Approaches to Wastewater Treatment

Techniques for treating dairy wastewater include (i) grease traps or oil water separators, (ii) flow and load equalization, (iii) sedimentation, (iv) biological treatment, usually anaerobic, followed by aerobic treatment, (v) biological nutrient removal for lessening nitrogen and phosphorus and (vi) disinfection. In some cases land application is practised. Additional controls may be needed to reduce odour nuisance. Source-separation treatment based on the characteristics of wastes from different sources is used especially for streams with high salinity that adds to TDS levels in the effluents.

5.1.2.18.4 Practical Problems

A major challenge in the dairy industry is that the effluent characteristics vary from country to country just as the end products vary substantially in different parts of the world. While large quantities of cheese are produced in European countries, dairies in India are switching over to the production of new value-added products like sweets. Some countries manufacture ghee (or clarified butter) by heating butter to remove the moisture in it but clarified butter is not normally used in many countries. Further, the situation varies within a country due to changes in capacity/activity in individual dairies.

Another problem associated with dairy effluent is it cannot be stored for a long period (like nonbiodegradable effluents from electroplating activity). Storing effluents from dairies would lead to an odour nuisance.

Unless treated effluents are reused due to a shortage of water, ETPs are not operated optimally to cut down the overall expenditure of dairies. The aerators are switched off to save energy in the industry. Switching off all aerators will affect the population of aerobic

Table 5.7 *Wastewater characteristics of dairy waste.*

Sl. no.	Characteristics	Range	Inference
1	pH	6–9	Neither acidic nor alkaline. No need to neutralise.
2	BOD_5, mg/l	350–8000	In the range of medium strength sewage. Can be treated in the methods that are used for sewage.
3	COD, mg/l	1500–18000	BOD to COD ratio is around 0.2. This means there is high fraction of nonbiodegradable organic matter. Skimmed milk will have COD value of 1 09 000 mg/l and that of whey is 65 000 mg/l
4	Total suspended solids, mg/l	800–7000	Can be easily settled and, hence, burden on biological process can be reduced
5	Oil and grease, mg/l	250–5000	Most of the nondegradable fraction is due to oil and grease. Separation of oil and grease would make effluent biodegradable.
6	Total TKN, mg/l	300–350	Sufficient to support biomass. No need for adding nitrogen source to treatment plant.
7	PO_4^{3-}, mg/l	80–100	Fulfils phosphate requirement of microbes. No need to add micronutrients.
8	K^+, mg/l	550–650	Fulfils phosphate requirement of microbes. No need to add micronutrients.
9	Na^+, mg/l	200–300	Fulfils phosphate requirement of microbes. No need to add micronutrients.
10	Ca^{2+}, mg/l	700–800	Fulfils phosphate requirement of microbes. No need to add micronutrients.
11	Mg^{2+}, mg/l	100–200	Fulfils phosphate requirement of microbes. No need to add micronutrients.

Source: NCAES (1979); Arbeli *et al.* (2006); Sarkar *et al.* (2006); World Bank (2007); Mohan *et al.* (2007); Vourch *et al.* (2008); Watkins and Nash (2010); Kushwaha *et al.* (2011); Qazi *et al.* (2011); University of Hawai (2008) and observation by authors.

Table 5.8 *Effluent characteristics from different activities in dairies.*

Activity	pH	BOD	SS	Volume m^3
Milking centre	6–9	400–10000		270–330 per 100 cows per day
Silage leachate	6–9	12,000–90,000		390–470 per 100 cows per day
Yogurt and buttermilk manufacturing	5–6	800–1000	1700–14000	
Ice cream	6–9	300–400	35000–58000	4.0–5.0 l/kg ice cream
Whey	4–5	3000–35000		
Milk powder, and (or) liquid products	10–11	2500–3000	500–10000	0.8–1.7 l/l processed milk
Cheese manufacturing	6–7	2500–3000		
Condensate from milk evaporation	8–9			

Source: NCAES (1979); Arbeli *et al.* (2006); Sarkar *et al.* (2006); Mohan *et al.* (2007); World Bank (2007); University of Hawai (2008); Vourch *et al.* (2008); Watkins and Nash (2010); Kushwaha *et al.* (2011); Qazi *et al.* (2011) and observation by authors.

bacteria. Switching off or failure of one/some of the aerators will lead to the formation of dead zones within the aeration tank where the wastewater will remain stagnant, not being treated. Such dead zones will lead to formation of *shortcuts* to part of the effluent, which would eventually get out of the aeration tank without treatment.

Failure of process due to unavoidable circumstances would result in discarding the milk and milk products into drains, which would lead to a shock organic/hydraulic load on the ETP. Sudden entry of such effluent into an ETP without equalization will wash out acclimatized biomass. The entry of disinfectants/chemicals used in cleaning in process area/equipment will lead to a shock toxic load, which will kill acclimatized bacteria.

Nonuniformity in process due to the manufacture of products based on demand would also affect the performance of ETP due to a change in the characteristics of effluent, which would lead to a change in the food-to-micro-organism ratio and hydraulic loading. The formation of lactic acid due to the prolonged storage of wastewater or the discharge of waste yogurt or butter milk may also affect performance. Discarding expired products may the vary load on the ETP.

5.1.2.18.5 Sustainable Approach to Wastewater Treatment

One can view this as the treatment method that does not require any chemicals but the treatment demands aeration and pumping. Effluents from chilling plants in rural and semi-urban locations may be treated with septic tanks and soak pits and, hence, may not require sophisticated treatment plants. Wherever aerobic treatment is adopted, the system requires aerators or pumps. Energy consumption can be brought down by regular auditing and tracking energy consumed per litre of effluent treated. Wherever diffused aerators are used, the aeration can be added with attached media like rings/saddles that help retaining biomass in case of shock loading. Care should be taken to immerse blades of diffused aerators at the optimum level to ensure maximum oxygen transfer per specified amount of electricity consumption. Hydraulic and organic shock loads can be overcome by constructing several small units arranged in parallel so that some of them can be switched off during maintenance and lean production stages. Use of an oxidation pond and ducks for aerating partially treated wastewater may be feasible if sufficient area is available for construction of oxidation pond.

Instead of concentrating on end-of-pipeline solutions it is prudent to consider options available during generation points such as: (i) avoiding raw material, product and byproduct losses due to spills, excessive changeovers, leaks, and shutdowns, (ii) separating and collecting product waste, to facilitate recycling or processing for disposal/sale/use, (iii) installing grids to reduce entry of solid substance into the effluent drainage, (iv) keeping the process and foul drains separate, (v) ensuring that tanks and pipes have provision for self-draining, (vi) using procedures for product discharge before cleaning procedures.

A water hyacinth treatment system can be used in tropical or subtropical countries for dairy wastewater but the water hyacinth cannot tolerate cold climates. It is highly susceptible to damage or death due to frost. The very high BOD loading may prove fatal to water hyacinths as the system needs surface water dissolved oxygen of more than 2.0 mg/l for proper growth. The addition of sodium nitrate may be needed to control odour problems and provide the required nitrogen.

The treatment of dairy waste by surface or subsurface treatment systems would also reduce dependence on energy and is highly sustainable in tropical and subtropical countries. Unless water scarcity demands its reuse, surface or subsurface treatment system are highly cost-effective. If sufficient land is not available then the wastewater can be transported to distant places on a single occasion (once in a year or crop cycle) or on multiple occasions (application of wastewater several times in a year or crop cycle) to avoid clogging of soil particles (as soil clogging may lead to failure of crops).

5.1.2.19 Slaughter Houses

Slaughter houses vary from large-scale, sophisticated, high-tech facilities to crude small-scale facilities with poor lighting and low sanitation conditions. Irrespective of how sophisticated the facility is, they all serve the purpose of killing animals. Some religions want animals to be killed by cutting the throat and spilling blood whereas others practise killing animals with sympathy. Many cities throughout the world have meat shops where animals are slaughtered in front of the customer or at the back of the shop. Some facilities kill poultry by passing nitrogen into a chamber. In some slaughter houses animals are tortured before killing. Slaughter houses use special risk materials (SRMs) that will cause death in humans and animals. Sick and dead animals that reach slaughter houses may not be considered as solid waste in some slaughter houses to avoid losses.

A high COD concentration due to blood is a common phenomenon in slaughter houses. The BOD_5/COD ratio of slaughter house effluent is around 0.44 which is below the BOD_5/COD ratio of domestic wastewater (0.60) (del Pozo *et al.*, 2003). Collection of blood for byproducts for use in feed, food or in the pharmaceutical industry can greatly reduce the load on wastewater treatment plants.

The oil and grease content depends on practices of slaughtering. The suspended solids in wastewater are mainly due to intestine and stomach contents. Nevertheless, the stream contains a high particulate fraction (around 53%) and only about 18% will be settleable solids. The soluble fraction of COD is very high (77%) with the COD/TKN ratio being 10.5 (del Pozo *et al.*, 2003).

The wastewater will be characterized by high chloride content if the slaughterhouse has the facility for a skin-salting process. Wastewater from small butcher houses to large slaughter houses located within the city tends to release wastewater into sewers in many parts of world, thereby releasing pathogens, parasitic eggs or spores.

Techniques for treating wastewater from slaughter houses include skimmers, grease traps, or oil- water separators, sedimentation tanks, biological treatment, usually anaerobic followed by aerobic treatment, biological nutrient removal for removing nitrogen and phosphorus and chlorination. Additional precautionary controls may be required to remove parasitic eggs or spores from effluents, which is not always practised all over the world. Hence it is necessary to have strict legislation to slaughter animals (including birds) outside urban settlements.

5.1.2.20 Fish Processing

The expected increase in the global population will place huge stress on the aquatic ecosystem as the demand for seafood is anticipated to reach 183 million tonnes by 2030. But nature can supply only 80 to 100 million tonnes of fish/year on a sustainable basis. All marine taxa currently fished would collapse by 100% by 2048 in the 'business-as-usual' scenario unless efforts are made to reverse the situation at this point of time (Worm *et al.*, 2006). Aquatic fauna plays an important role in conserving the aquatic environment by consuming waste that enters the water bodies. Aquatic fauna also control water weeds and algal blooms. Consumption of aquatic fauna would not only push them towards to extinction; it would also lead to an ecological imbalance due to pollution and the destruction of the aquatic ecosystem.

Major species that are processed and canned are salmon, clams, shrimps, oysters, mackerel, herring, white fish, octopus and crab. Apart from species that are processed, the industry would use the following raw materials depending on the product manufactured and the way in which it is packed: olive oil, salt, soya bean oil, tomato sauce, pepper, cardamom, ginger, onion, ground mustard seed, spirit vinegar, starch (potato flour), curry powder, milk, monosodium glutamate (msg), wine, beer and sugar.

The fish-canning industry uses a large quantity of water for storage and transport, cleaning, preparation of brine, equipment, sprays, freezing and thawing, cooling water, offal transport, floor cleaning and steam generation. Hence, it generates a large quantity of wastewater. Treatment is difficult due to the high level of organic matter, salts, oil and grease (Table 5.9). Effluents from the fish-processing industry vary to a great extent, depending on the process and the raw material.

In view of these facts, some practices that should be considered are vacuum suction systems during reception of raw materials, flow control in cleaning operations, use of humid air or warm water for fish defrosting, use of supernatant from an ice-removal tank for cooling hot water, reuse of scaling fish water in the initial fish washing after filtration, use of low-flow and high-pressure cleaning systems, use of compressed air instead of water, reuse/recycling of water from and to noncritical operations, adjusting the size of nozzles and the dry transportation of waste (Cristóvão *et al.*, 2012).

Table 5.9 Effluent characteristics in fish-processing units.

Parameter	Range
pH	6.5–7.0
TSS	300–1000
COD	2000–18000
BOD_5	1500–19000
Oil and grease	400–2850
P_{total}	15–70
Cl^-	1800–4200
SO_4^{2-}	5–180

Wastewater treatment units generally include screens, grit chamber, oil and grease remover, equalization, sedimentation and biological processes. The selection of a biological process depends on the quality of the wastewater. Anaerobic treatment is adopted if the BOD values are high enough to sustain the treatment option. Water hyacinth and subsurface treatment are advised for sustainable wastewater treatment in tropical and subtropical countries for the reasons already discussed in section 5.1.2.17.

5.1.2.21 Starch Manufacturing

Starch is commercially produced from corn, wheat, potato, rice, tapioca and sago. In addition to starch, the starch industry usually manufactures animal feed, corn oil, corn sweeteners, ethanol, gluten dextrin, dextrose, glucose and fructose as byproducts.

Corn starch production involves soaking shelled and cleaned kernels in water for 24–48 hours at 50 °F with a small quantity of sulfur dioxide to prevent fermentation. Figure 5.23 shows a schematic diagram of starch manufacturing from corn. This process is called steeping and the water generated in the steeping process is used in animal feed products. The kernals are ground in attrition mills and slurry is formed. The slurry is passed though hydrocyclones and reminder of the kernels are ground again. Starch slurry is separated by

Figure 5.23 Schematic diagram of starch manufacturing from corn.

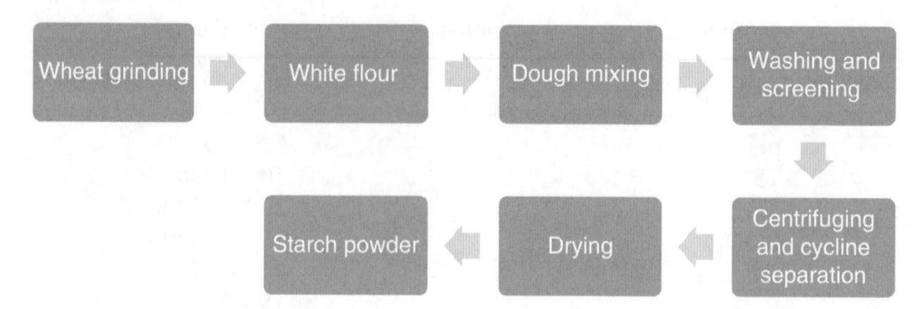

Figure 5.24 Schematic diagram of starch manufacturing from wheat.

screening and centrifugation followed by starch drying to obtain unmodified corn starch. Starch slurry can be given an enzyme/chemical treatment to produce modified corn starch (Murray *et al.*, 1994).

Figure 5.24 shows a schematic diagram of starch manufacturing from wheat. When wheat is used as raw material to manufacture starch it is first taken to a flour mill where white flour is generated. The white flour is made into stiff dough, adding water. The dough is rolled or kneeded and starch is washed off by water sprays. The gluten is separated from starch slurry on screens, ashed and dried (Murray *et al.*, 1994).

A schematic diagram of starch production from potato is shown in Figure 5.25. The potatoes are cleaned by water and crushed to disintegrate potato cells, releasing starch. The crushed potatoes are screened to separate fibre and skin. The starch solution is purified to remove impurities. Purified starch is dewatered, dried and packed.

Tapioca is produced from cassava root. The starch from the production of cassava root involves: (i) washing of roots, (ii) chopping, (iii) grinding, (iv) fibrous residue separation, (v) dewatering and protein separation, (vi) dehydration, (vii) drying and (viii) packaging. According to Tanticharoen and Bhumiratanatries (1995), the generation of wastewater from tapioca starch plants is around 20 m^3 for every ton of starch produced and the wastewater generation is about 12 m^3 per ton of starch produced (Hien *et al.*, 1999). Characteristics of wastewater from tapioca starch plants are: (i) 11 000–13 500 mg COD/l, (ii) 4200–7600 mg SS/l and (iii) a pH of 4.5–5.0 (Hien *et al.*, 1999).

Good housekeeping measures can save water to the maximum extent. The solution adopted in the industries of Thailand, as described by Chavalparit and Ongwandee (2009), are: (i) installation of flow meters and recording water usage, (ii) use of high-pressure

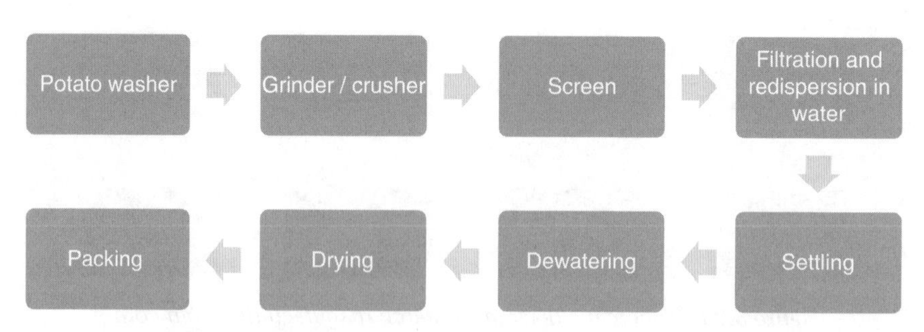

Figure 5.25 Schematic diagram of starch manufacturing from potato.

pumps for filter/mechanical floor cleaning, (iii) collecting product spills before cleanup, (iv) regular checking and repair of pipe leakages and (v) collecting leftover starch from machines. The technology adopted for wastewater treatment is usually anaerobic treatment followed by aerobic treatment. Recovery of biogas from effluent treatment system can be sustainable due to energy recovery.

5.1.2.22 Leather Processing

Leather is processed to convert the raw hide of dead animals to usable leather. Water consumption and effluent discharges vary among tanneries, depending on the processes adopted, raw materials and products. The potential for enhancing efficiency through changes in the process is high and should be identified during the design of the facilities and processes. Water consumption is usually highest in the pretanning areas, even though significant quantities of water are consumed in the post-tanning processes. Effluent from soaking, fleshing, dehairing, liming and from associated rinsing contains hide substances, blood, dirt or dung and, hence, has significant quantities of organic matter and suspended solids. Effluent from deliming, tan yard processes, and bating contains ammonium salts, sulfides and calcium salts and is weakly alkaline. The main wastewater contaminants after pickling and tanning processes depend on the tanning techniques used. Finishing effluents contain solvents, lacquer polymers, colour pigments and coagulants.

Wastewater prevention measures from leather processing include (i) decreasing water consumption by recycling process streams, (ii) using 'batch' instead of 'running water' washes, (iii) segregation of effluent streams, (iv) use of short floats in the tanning cycle, (v) chemical substitution with more biodegradable chemicals

Measures to reduce the organic load include (i) screening wastewater to remove large solids, (ii) using an enzymatic dehairing process, (iii) recycling liming float, (iv) using easily degraded ethoxylated fatty alcohols as surfactants in degreasing, (v) using carbon dioxide (CO_2) deliming.

Measures to reduce TDS loads include (i) use of natural drying warm, dry climates, (ii) use of chilling for short-term preservation of freshly processed hides/skins, (iii) use of antiseptics to enhance storage time, (iv) performing trimming and prefleshing before pretanning operations, (v) use of mechanical/manual removal of salt from skins/hides before soaking, (vi) installing salt-free pickling systems, (vii) use of ammonium-free deliming agents. Measure to reduce sulfide are: (i) use of enzymatic dehairing processes, (ii) using sulfide and lime in a 20–50% overall solution for conventional lime dehairing processes, (iii) maintaining sulfide-containing effluent at pH >10.

Measures to limit use and discharge of chromium are (i) considering using alternative tanning agents, (ii) avoiding the use of chromium (VI), (iii) recycling chrome tanning floats, (iv) using high-exhaustion chromium salts and alkaline products and enhancing the float temperature, (v) avoiding incineration of chrome-tanning sludge as alkaline conditions and presence of excess oxygen can result in conversion of Cr (III) into Cr (VI).

Wastewater treatment for leather processing units includes grease traps, flow and load equalization, skimmers or oil water separators, sedimentation, biological treatment, biological nutrient removal for phosphorus and nitrogen, filtration and chlorination of wastewater when disinfection is required. Additional processes for removal of metals by membrane technology or physical/chemical methods are sustainable as they will enhance/preserve the environment.

5.2 Industries with Low Dissolved Solids

Industries with low dissolved solids will have an advantage in that most of the pollutants can be removed by physical operations but care must be taken to remove toxic chemicals to a safe level.

5.2.1 Industries with Low Amounts of Inorganic Materials

Low amounts of inorganic materials can be treated easily if the inorganic compounds are not toxic to microbes. Sustainable options include acclimatization of micro-organisms to effluents and physico-chemical treatment methods. Good manufacturing practice, which separates different streams of effluent and separate toxic streams, can reduce the burden on the biological treatment system.

5.2.1.1 Dye and Dye Intermediates

A dye is a chemical used to impart colour to objects. An aromatic ring structure with a side chain is generally required for resonance and to impart colour. Dye is synthesized using a chromogen-chromophore with an auxochrome. A chromogen is an aromatic structure comprising naphthalene, benzene or anthracene rings. A chromophore is a colour giver. Auxochromes are used to impart solubility and to cause adherence of the dye. Auxochromes are carboxyl, hydroxyl, amine and sulfonic radicals, or their derivatives. Auxochromes are important in the classification of dyes (Table 5.10).

Table 5.10 *Types of dyes.*

Sl. no.	Types of dyes	Description
1	*Acetate rayon dyes*	Developed for cellulose acetate and some synthetic fibres
2	*Acid dyes*	Usually used for colouring animal fibres through acidified solutions along with amphoteric protein
3	*Azoic dyes*	Contain the azo group normally application to cotton
4	*Basic dyes*	Amino derivatives. Usually used mainly for application on paper
5	*Direct dyes*	Azo dyes and sodium fixing agents, metallic compounds and salts. Normally used on cotton-silk or cotton-wool combinations
6	*Mordant or chrome dyes*	Metallic salt or lake formed directly on the fibre by the use of aluminium, chromium, or iron salts that cause precipitation *in situ*.
7	*Lake or pigment dyes*	Form insoluble compounds with barium, chromium or aluminium, on molybdenum salts; the precipitates are ground to form pigments used in inks and paint.
8	*Sulfur or sulfide dyes*	Contain sulfur or are precipitated from sodium sulfide bath
9	*Vat dyes*	Impregnated into fibre in reducing conditions and reoxidized to an insoluble colour.

Synthetic dyes are used widely in the paper, textile and printing industries. Most of these dyes involve amination, sulfonation, reduction, nitration, diazotization, halogenation and oxidation using benzene, toluene, anthracene, xylene and naphthalene as raw materials (Abrahart, 1977). Due to the abandonment of manufacturing of toxic substances in the developed countries, most of the world's demand is met by the developing countries. The effluents from these industries vary widely in composition and are characterized by a high level of colour and COD. The effluents from the manufacturing of dyes and dye intermediates contain persistent chemicals, many of which are toxic and carcinogenic. The constituents of these effluents can react with each other leading to the formation of new chemicals.

5.2.1.1.1 Pollution Prevention and Control

Adsorption, coagulation–flocculation, oxidation and electrochemical methods to treat effluents from dye and dye intermediate manufacturing are expensive and have operational problems. Colour removal by bioprocessing is difficult due to the presence of nonbiodegradable substances. Reverse osmosis, ultrafiltration and other filtration methods are used to recover/concentrate process intermediates.

Several advanced oxidation processes (AOPs), like photolysis, ozone, hydrogen peroxide, Fenton's reagent (a mixture of H_2O_2 and Fe^{2+}) and photo-Fenton reaction using Fe^{2+}, H_2O_2 and UV light are effective in mineralizing pollutants like PCBs, chlorophenols, chlorinated herbicides, perhalogenated alkanes and dye effluents (Pignatello and Chapa, 1994; Bigda, 1995; Huston and Pignatello, 1996; Huston and Pignatello, 1999). These technologies may be viable but they are only understood and implemented by a few organizations in the world and they may not be available all over the world, especially in the developing countries where dyes and dye intermediates are manufactured by small and medium-sized enterprises (SMEs). These advanced technologies also need skilled manpower, which will not be not be readily available.

Usual sustainable pollution control measures include: (i) avoiding the manufacture of toxic azo dyes, (ii) measurement and control the toxic ingredients to minimize wastage, (iii) reuse of byproducts from the process, (iv) use of automated filling to reduce spillage, and (v) use of equipment wash down water for makeup solutions for subsequent batches.

5.2.1.2 Petroleum Refineries

The largest volume of wastewater in a petroleum refinery arises from 'sour' processes and alkaline processes. Sour water is generated from topping, desalting, vacuum distillation, pretreating, light/middle distillate hydrodesulfurization, coking, catalytic cracking, hydrocracking, visbreaking/thermal cracking. Sour water consists of ammonia, hydrocarbons, hydrogen sulfide, organic acids, organic sulfur compounds and phenol. Wastewater from sour processes is treated in the sour water stripper unit to remove hydrogen sulfide, hydrocarbons, ammonia and other chemicals before recycling or final treatment and disposal.

Nonoily/nonsour wastewater from boiler blowdown and demineralization plant reject, if incorrectly neutralized, will extract phenolics into the water phase from the oil phase and cause emulsions in the wastewater treatment plant. Wastewater from accidental releases or leaks from equipment/machinery/storage tanks can be prevented by (i) control of accidental releases by regular inspections and maintenance, (ii) construction of wastewater/hazardous materials storage containment area and use of basins with impervious surfaces, (iii) segregation of process water from stormwater and hazardous materials containment basins and (iv) implementation of good housekeeping practices.

Specific provisions required in petroleum refinery are (i) direct spent caustic soda from chemical treating and sweetening units routed to the wastewater treatment system after caustic oxidation, (ii) install a closed-process drain system, (iii) neutralize acidic and caustic effluents from the demineralization plant, (iv) cool blowdown from the boilers, (v) hydrocarbon-contaminated water from cleaning activities and process leaks should be treated prior to discharge.

Typical effluent treatment from processes include: (i) skimmers/grease traps/oil water separators/dissolved air floatation, (ii) flow and load equalization, (iii) sedimentation for suspended solids, (iv) biological treatment, (v) chemical or biological nutrient removal, (vi) disinfection, (vii) disposal of designated residues into hazardous-waste landfills.

Additional engineering controls are required for (i) treatment of volatile organics stripped from an effluent treatment plant, (ii) toxicity/metals removal by chemical precipitation or membranes, (iii) removal of refractory organics by chemical oxidation or activated carbon, (iv) neutralization of nuisance odours.

5.2.2 Industries Dealing with Low Dissolved Organic Material

The quality and quantity of wastewater varies from one industry to another. Industries dealing with material with lower inorganic content face the difficulty that it may need treatment other than biological treatment. Material from the garment-washing (Figure 5.26),

Figure 5.26 *View of jeans-washing unit.*

metal-finishing and surface-coating industries usually contains low amounts of dissolved organic content. Low amounts of dissolved organic contents are mixed with sewage to optimize the nutrient supply to the biomass acting on wastewater. The biological systems need to have optimum biomass to form settleable flocs to enable them to be removed by settling. If the industry discharges organic compounds that need treatment by acclimatized micro-organisms the setup should be commissioned with a gradual increase in concentration of such compounds in the treatment units/operations during the commissioning stage until an optimum condition is reached.

5.2.2.1 Paper Industry

Paper is manufactured from wood, bagasse (sugar cane fibres), esparto grass, cereal straw, flax, bamboo, reeds, agricultural residues and recycled paper containing cellulose fibres. The major steps in paper production are chip making; wood debarking; pulping; paper manufacturing; pulp and bleaching. Pulp and paper may exist separately or together. Manufactured pulp is used for manufacturing paper or cardboard. Depithing is done when bagasse is used as the raw material. Chemi-mechanical pulp-manufacturing processes involve mechanical abrasion and the use of chemicals. Mechanical pulping is done by disk abrasion and billeting. Thermo-mechanical pulps are manufactured by the application of heat and mechanical operations. Chemi-mechanical pulping and chemi-thermo-mechanical pulping (CTMP) use sodium sulfite, carbonate or hydroxide.

Chemical pulps are manufactured by digesting (cooking) the raw materials with the kraft (sulfate). Kraft processes generate pulps used for high-strength papers and board. In chemical pulping material with cellulose is cooked with caustic soda. The product of this process is called brownstock. The brownstock is washed with water to recover chemicals and remove the black colour.

Mechanical pulp without bleaching can be used for newsprint. However the pulp has to be bleached to manufacture paper required for many applications. Bleaching is done with oxygen, ozone, hydrogen peroxide, sodium hypochlorite, peracetic acid, chlorine dioxide, chlorine and other chemicals.

Oxygen is used in modern mills to avoid chlorine compounds. In total chlorine-free (TCF) processes, bleaching effluent is fed to the recovery boiler to generate steam; the steam is used to generate electricity. Chlorine dioxide is used for bleaching in elemental chlorine-free (ECF) processes.

Chemical additives are added to impart specific properties to paper. Pigments may be added for colour.

Chlorinated organic substances generated by use of chlorine in the paper industry are toxic. Hence recovery of chlorinated organics is needed to safeguard the environment and the health of citizens and animals.

Paper and cardboard are manufactured from pulp by deposition onto a moving forming device (Figure 5.27). The residual water in the wet web is eliminated on a series of hollow-heated cylinders

5.2.2.1.1 Waste Characteristics

Wastewater generation varies between 20 and 250 m^3/t of air dried pulp (ADP). Total suspended solids vary between 10 and 50 kg/t of ADP; BOD varies between 10 and 40 kg/t

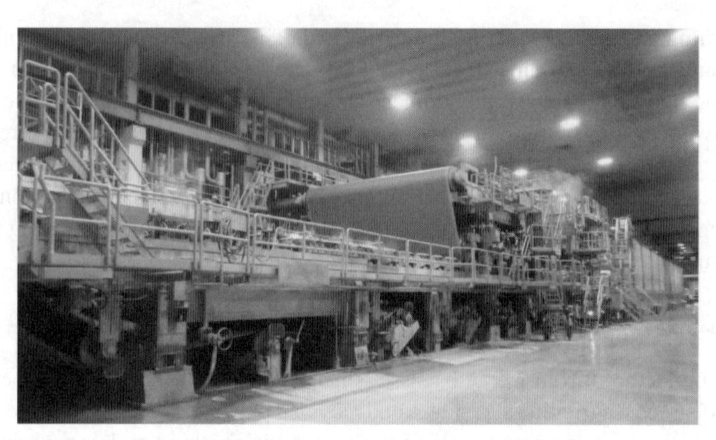

Figure 5.27 *View of paper mill.*

of ADP; COD varies between 20 and 200 kg/t of ADP (Table 5.11); adsorbable organic halides (AOX) varies between 0 and 4 kg/t of ADP (World Bank, 1998).

5.2.2.1.2 Pollution Prevention and Control

Doing away with paper is one of the options to control pollution from the paper industry. Companies are already shifting from conventional paper-based communication to paperless communication not only within the office but also among stakeholders such as shareholders. Electronic communications have helped the world to cut down dependency on paper but paper is still an essential raw material for newspapers, books and magazines. Chlorine-based organic compounds and other toxic organics cause significant environmental problems. Hence the use of paper manufactured by thermo-mechanical processes, and of recycled fibre, should be encouraged.

Significant methods of controlling water pollution in a sustainable way are: (i) energy-efficient pulping, (ii) minimizing the generation of effluents by process modifications and recycling wastewaters, (iii) recovering pulping chemicals, (iv) adopting dry debarking instead of wet debarking; (v) recovering cooking chemicals, (vi) using high-efficiency washing and bleaching equipment, (vii) reducing bleaching requirements through proper design and operation, (viii) minimizing unplanned and nonroutine discharges of wastewater and black liquor, (ix) extended cooking and oxygen delignification, (x) optimizing pulp

Table 5.11 *Wastewater characteristics from pulp and paper mills.*

Sl. no.	Parameter	Value
1	pH	6–9
2	COD	300 mg/l for kraft and CTMP pulp mills; 700 mg/l for sulfite pulp mills; 10 mg/l for mechanical and recycled fibre pulp; 250 mg/l for paper mills

washing prior to bleaching, (xi) use of ECF bleaching systems, (xii) use of TCF process, (xiii) use of oxygen, hydrogen peroxide, ozone, enzymes or peracetic acid for bleaching, (xiv) reducing the chlorine charge by splitting the addition of chlorine and controlling pH.

Box 5.1 Sustainable Paper Manufacture – The Swedish Experience

Sweden is a major consumer and producer of paper. Practices followed by the industry to reduce water pollution are (i) dry debarking, (ii) increased delignification by extended cooking and adding oxygen stages, (iii) ECF or TCF bleaching (recycling), (iv) process water recycling, (v) use of black liquor as fuel (vi) efficient and closed brown stock washing, (vii) substitution of potentially harmful substances by less harmful alternatives and (viii) high-efficiency biological treatment of wastewater.

Practices followed to reduce air pollution are: (i) collection and incineration of malodourous gases with SO_2 control, (ii) combustion control, (iii) SO_2 emissions from recovery and auxiliary boilers reduced (dry black liquid), (iv) use of bark or other low-sulfur fuel or scrubber, (v) reduction of NOx emissions from recovery boiler (controlling fire conditions, use of proper design), (vi) reduction of dust emissions from boilers (electrostatic precipitators).

Practices followed to attain energy efficiency and clean production are: (i) installing recovery boilers and auxiliary boilers, (ii) phasing out fossil fuels, (iii) drying bark in order to increase energy efficiency, (iv) increasing internal electricity production, (v) delivery of waste heat to district heat system.

5.2.2.2 *Soft Drinks and Packed Juices*

Soft drinks, packed juices and fruit nectars are popular in the developed world. Fruit juice contains 100% fresh fruit juice whereas fruit nectar is a drink that contains other ingredients such as preservatives and sugar. European Union fruit juice and nectar consumption was 10.7 billion litres in 2011 (European Fruit Juice Association, 2012).

The process of soft drink manufacturing consists of water treatment, diluting ready-made concentrate, dissolving carbon dioxide under pressure, bottling and despatching. The practice of bottle washing is decreasing as glass bottles are replaced by plastic bottles.

Manufacturing fruit juice involves washing, cleaning, grading, peeling, juice extraction, filtration, sterilization and packaging. Sugar and preservative may be added in some products.

Effluent from food and beverages will have high BOD and COD. The effluent will also have chemicals or detergents that are used for cleaning. Effluent may contain pesticides, suspended solids, dissolved solids, nutrients and microbes. Prevention of entry of raw materials, intermediates, product and byproduct will reduce the strength of wastewater.

Techniques for treating wastewater in this sector include screening, flow and load equalization, sedimentation, pH adjustment, biological treatment, usually anaerobic followed by aerobic treatment and disinfection. The flows are frequently seasonal and pond systems are successfully used but soil deterioration, odour nuisance and groundwater pollution need to be avoided.

Food and beverage processing (internal transport of raw materials with water, cooling, washing, equipment cleansing) consume large amounts of water. Dry peeling methods decrease the effluent volume by up to 35% and the organic load on wastewater treatment plants can be reduced by up to 25%. Decrease in effluent quantities of up to 95% have been reported by implementing good practice (IFC, 1998). Water consumption can be reduced by: (i) optimizing product-conveying systems, (ii) using dry-conveying systems instead of wet conveying, (iii) optimizing operations to avoid spilling of raw materials and water, (iv) procuring clean raw vegetables and fruit, (v) using dry methods like vibration or air jets to clean fruit and vegetables, (vi) separating and recirculating process wastewaters, (vii) using steam instead of hot water, (viii) using countercurrent systems, (ix) remove solid wastes without using water, (x) reuse concentrated wastewaters and solid wastes for producing byproducts (IFC, 2007).

5.2.2.3 *Healthcare Establishments*

Healthcare establishment (HCEs) normally are not considered to be sources of pollution. But activities in health care units are highly complex and the environmental literature/legislation in recent years has given them lot of attention. The healthcare of humans and animals in recent years has taken on new dimensions all over the world. The size of healthcare establishments varies from a consultancy clinic (where doctors such as skin specialists do not even touch the patients and provide only prescriptions) to multispecialty hospitals (with thousands of beds and hundreds of specialists providing health services to patients). Healthcare facilities also vary from war camps to specialized hospitals that treat only pregnancy and infertility. Due to cheap labour some countries are emerging as health tourism destination where people come only for treatment.

Wastewater in hospitals arises due to: (i) disinfection of surgical instruments and laboratory glassware, (ii) discharge of chemicals used for development of X-ray films, (iii) flour and equipment washing, (iv) boiler/autoclave blowdown, (v) discarding blood and medicines in drains, (vi) wastewater from kitchens of canteen, (vii) toilets and bath rooms, (viii) wastewater from operating theatres, (ix) radiation rooms, (x) vehicle washing and (xi) massaging activities.

Wastewater is characterized by (i) pathogens, (ii) mercury due to breaking of equipment using mercury, (iii) disinfectants, (iv) pharmaceuticals, (v) chemicals used to develop x-ray film, (vi) radiation activity, (vii) presence of incineration facilities to dispose of solid biomedical waste, (viii) vehicle washing and (ix) chemicals used to preserve body parts (usually formalin). The BOD, COD and suspended solids will be almost same as domestic sewage. The effluents from hospital are bound to have higher pathogen content than domestic effluent and, if not treated, this will multiply in sewers or discharge points, contaminating surface and groundwater bodies.

The quantity of wastewater generated is usually around 350 to 450 l/bed/day for HCEs. The quantity of effluents from laboratories due to washing glassware varies from 0.01 to 1.01 l/patient per day but, the quantity of wastewater depends on (i) the type of patient (cancer patients and patients who have undergone surgery use less water than patients suffering dehydration), (ii) the number of beds, (iii) the number of operating theatres, (iv) the number and type of laboratories, (v) the presence of a blood bank, (vi) the number of patients (outpatients and in patients) associated with the hospital, (vii) the number of staff in the hospital, (viii) canteen facilities, (ix) the presence of an incineration facility to dispose of solid biomedical waste and (x) infection control policy and practice in the hospital.

Containing and treating wastewater for veterinary hospitals poses a great challenge as many veterinary institutes perform operations in an open area and will not have any way to contain the stream arising from washing the areas where animals are treated.

Treatment options include biological treatment for sewage and chemical disinfection for the waste streams that generate from labs followed by discharge into drains. Healthcare establishments use chlorine-based disinfectants like sodium hypochlorite and bleaching powder for waste from laboratories to attain safety from pathogens. This practice leads to the formation of chlor-organic chemicals that are known carcinogens. Poor operation of incinerators in healthcare establishments can lead to the formation of dioxins and furans that may gain entry into wastewater streams from scrubbers.

5.2.2.4 Packed Food Processing

A variety of food grains are used for the manufacture of food. A typical food-grain milling process primarily involves cleaning, conditioning and milling. The cleaning process involves (i) screening or blowing with axial fans, (ii) destining, (iii) scouring dust by passing through rubber rollers, (iv) cleaning with water.

Some cleaned food grains are conditioned (conditioning is the treatment of food grain with water and heat in order to change its structural and biochemical properties). Some food grains need soaking for followed by drying. The soaking time differs depending on the food grain.

Milling involves application of abrasive force to remove the outer shell of food grains followed by pulverizing to obtain flour.

Washing of food grains generates 350–450 mg/l of BOD and 1200 to 1400 COD. The TSS in the cleaning process will be in the range of 200–400 mg/l. pH will vary between 6.5 and 8.5. Hence washing waste will be easily biodegradable. Unlike sewage, where micro-organisms constantly degrade during collection and storage, waste from food grain processing will be devoid of any significant micro-organisms and therefore will require acclimatized micro-organisms in treatment plants. Any shock loads will have implications in terms of reduced efficiency. Subsurface irrigation, on-land treatment, oxidation ponds and treatment with hyacinth and wetlands are best in terms of sustainability in tropical and subtropical countries. Use of wastewater for farming without any treatment will block the pores of the soil, which hence needs proper tilling after each cropping. The frequent loosening of soil, even before harvesting, will help to retain/improve soil health. The use of water for horticulture or forestry is also one of the best options and is practised without adverse impacts on health or environment.

In the case of cold countries, conventional aerobic treatment with suspended or attached growth will provide the desired result but, since both have inherent advantages and disadvantages, a combination of attached and suspended methods like using attaching media (like plastic rings) in aeration tank of activated sludge process is gaining popularity.

5.2.2.4.1 Canning

Salt is used extensively in vegetable canning to increase flavour, to preserve, or for conditioning. Hence this industry produces wastewater with high salt content. The canning process for molluscs also generates huge quantities of wastewater with salt content more than 2% as the molluscs are washed after shelling with 3% to 6% salt solution. They are then drained and cooked or steamed for 10 to 15 minutes at 100 °C followed by packing in cans with 1% to 2% brine. Effluent from processing molluscs will have very high nitrogen, organic and salt content (18.5 g/l of COD, 4.0 g/l of N and above 2% salt) (Mendez *et al.*, 1992). Canning of shrimp involves receiving, peeling and washing. During precooking, shrimp is boiled in a brine solution for 3 to 5 minutes, or it is steamed. The salt content of wastewater from precooking ranges from 2% to 3% (UNEP, 1999). The approach for managing water and wastewater from vegetable, fruit or seafood and meat is similar (even though wastewater from meat and seafood has more pollutants).

5.2.2.4.2 Confectionaries

The confectionery market is divided into: chocolate, sugar confectionery and gum (chewing gum and bubble gum). Chocolate is manufactured from the dried and fermented cacao beans.

The beans from cacao tree are cleaned, shelled, roasted, ground and refined to produce cocoa liquor, which is used to make cocoa butter, powder and chocolate.

Sugar confectionery is manufactured by dissolving sugar in water or milk, which is boiled until it reaches the desired concentration or starts to caramelize.

Chewing gum and bubble gum are manufactured from natural or synthetic gums, which are ground, cooked, flavoured, rolled, cut and packaged.

Large volumes of water are used for cleaning process equipment and work place. Wastewater produced during the confectionery manufacturing will have a high organic content, in the form of sugars and vegetable fats.

5.2.2.4.3 Chips and Other Fried Food

Chips are a favourite food for all ages, manufactured all over the world. Even though small-scale production at restaurants and snack shops does not cause much harm to the environment, the production of chips needs precautions at large manufacturing facilities. In large-scale potato-chip factories, potatoes are stored at a temperature between 4.4 and 7.2 °C, before they are to be used. They are then transferred to a reconditioning room whose tempering is maintained at 21.1–23.9 °C. Potatoes smaller than a baseball and larger than a golf ball are best suited for chips and it takes 100 kg of raw potatoes to make 25 kg of chips. The potatoes are fried in vegetable oil. An antioxidizing agent is added to the oil to avoid rancidity (disagreeable odour/taste of decomposing oils or fats). Some producers treat the potatoes with citric acid, phosphoric acid, calcium chloride or hydrochloric acid to decrease the sugar level and improve the colour of chips.

Potatoes that arrive at the manufacturing unit are examined and tested for quality. The consignment is rejected if it does not meet the required standard. The potatoes are conveyed through a helical screw conveyer, which allows stones to fall to the bottom. Potatoes are peeled in an automatic peeling machine and washed with water. The potatoes are then passed through an impeller, which slices potatoes into paper-thin slices, between 1.7 and 1.85 mm in thickness. The slices fall and are washed in a second water wash to remove the starch released during the slicing of the potatoes. Some manufacturers may not wash the starch from slices. Potatoes are chemically treated if the manufacturers choose to increase colour at this stage by immersing slices in a solution adjusted for hardness, mineral content and pH.

The potato slices are placed under air jets to eliminate excess water and are fired at 176.6–190.5 °C. As the slices tumble, salt is sprinkled. Potato chips are flavoured as per product specification in a drum with powdered seasonings. Excess oil is drained off through a mesh conveyor belt as the chips begin to cool and are passed through an optical sorter to pick burnt slices. The final product is packed using a packing machine and stored in a warehouse before dispatch.

Rejected potatoes and peelings are used as animal feed or can be composted. The starch removed during the rinsing process is sold to a starch processor.

5.2.2.4.4 Breweries

A brewery is a dedicated building for the making of beer. Beer is a fermented beverage made with barley, maize, wheat and other grains. The beer-manufacturing steps include: (i) grain cleaning, (ii) steeping of the grain in water to start germination, (iii) growth of rootlets, (iv) kilning and polishing of the malt to remove rootlets, (v) storage of the cleaned malt, (vi) grinding the malt to grist, (vii) mixing grist with water to generate a mash, (viii) heating of the mash to activate enzymes, (ix) separation of grist residues, (x) boiling of the wort with hops, (xi) separation of the wort from the precipitated residues, (xii) cooling of the wort, (xiii) addition of yeast to cooled wort, (xiv) fermentation, (xv) separation of spent yeast, (xvi) bottling or kegging.

Water consumption for breweries usually ranges between 4 and 8 m^3/m^3 of beer produced with wastewater discharge of 3–5 m^3/m^3 of beer produced (exclusive of cooling waters). Vinasse is wine distillery effluent left after alcohol has been distilled from ferment fruit/food grain, molasses. Suspended solids of vinasse vary in the range of 10–60 mg/l, nitrogen in the range of 30–100 mg/l, BOD in the range of 1000–1500 mg/l, COD in the range of 1800–3000 mg/l and phosphorus in the range of 10–30 mg/l.

Sustainable practices with respect to water in breweries include: (i) clean-in-place (cip) methods to decontaminate equipment, (ii) low-volume, high-pressure hoses for equipment cleaning and (iii) closed-loop cooling-water circuits.

Primary treatment of wastewater includes pH adjustment, screening, grit removal through grit-settling chambers, sedimentation and anaerobic treatment followed by aerobic treatment.

5.2.2.4.5 Pickling

Pickling is one of the ancient methods of preserving food. Pickles are made up of meat, fish, vegetable, unripe fruits, leaves, roots and stems. The main ingredient of pickles is common salt or sodium chloride.

Cucumber pickles have been widely used in Europe and the United States. There are two main *types* of cucumber pickles – cured and fresh. *Cured* pickles are made by naturally fermenting in common salt (sodium chloride) brine. Flavourings such as dill may be added to the pickle after curing. During curing, the salt concentration in the cucumber may reach levels too high for tastiness, in which case the pickle is partially desalted and packed in vinegar solution. *Fresh-pack* pickles are made from uncured, unfermented cucumbers and packed in vinegar of various types and processed by heat for preservation. Sources of wastewaters from cucumber pickle manufacturing are: (i) brining, (ii) 'processing' or 'freshening' and (iii) finishing (Little *et al.*, 1976). The commercial production of vegetable pickles in Thailand has increased due to rising demand and the brine wastewater from the pickle industry is released without treatment, which is a serious potential source of pollution (Duangsri and Satirapipathkul, 2011). In general, effluents from cucumber pickling will have high chloride, high oxygen demand, high total solids, low pH, high suspended solids and high Kjeld-N. The effluent characteristics, as reported in Little *et al.* (1976), are 3400 mg/l of organic carbon, 730 mg/l of total Kjeld-N, 90 mg/l of phosphorus, 330 mg/l of suspended solids and 2300 mg/l of acidity. The spent tank yard brines will have 1.0–1.6 lb NaCl/gal (120–192 kg/m^3).

Pickles made up of vegetables, unripe fruits, fruits, leaves, roots and stems in India do not waste brine solution. The plant parts to be pickled are washed and cut into pieces or placed wholly in a curing jar where salt is added with spices. Vegetable oil is sometimes added, depending on the desired taste. Even though many pickle makers do not use preservatives, some manufactures use them. Wastewater from these does not use brine solution as in cucumber processing. The usual source of wastewater is washing plant parts, jars and packing bottles.

Salinity decreases the efficiency of BOD removal and increases turbidity. Anaerobic digesters are more sensitive to chlorides than aerobic processes. Some industries dilute the saline waste stream with fresh water to overcome operational difficulties. Hence salt-tolerant micro-organisms can be acclimatized in the process by slowly increasing salinity so that the population of tolerant microbes will increase gradually and sustain the treatment. Alternatively, sludge from the existing pickle industry or microbes from the seashore can be extracted by the selective extraction technique.

5.2.2.4.6 Ready-to-Eat Foods

Ready-to-eat foods include sweets, bread, bakery items, curries, rice, spiced snacks and so on. The type of effluent generated depends on the end products and raw materials. While some of the products require washing, others may start with a semifinished product procured from other units. Sustainability with respect to water management in these units depends on the size and combination of products. The wastewater can be mixed with sewage and treated with biological treatment systems.

References

Abrahart, E.N. (1977) *Dyes and their Intermediates*, Edward Arnold Ltd., London.

Arbeli, Z., Brenner, A. and Abeliovich, A. (2006) Treatment of high-strength dairy wastewater in an anaerobic deep reservoir: Analysis of the methanogenic fermentation pathway and the rate-limiting step. *Water Research* **40**, 3653–3659.

Babuu, B.R., Parande, A.K., Raghu, S. and Kumar, T.P. (2007) Cotton textile processing: waste generation and effluent treatment. *The Journal of Cotton Science* **11**, 141–153.

Baskar, M., Kayalvizhi, C. and Bose, MS.C. (2003) Eco-friendly utilisation of distillery effluent in agriculture – a review. *Agricultural Review* **24**(1), 16–30.

Batra, S.H. (1985) Other long vegetable fibres: abaca, banana, sisal, henequen, flax, ramie, hemp, sunn and coir, in *Handbook of Fibre Science and Technology*, vol. IV (eds M. Lewin and E.M. Pearce), Marcel Dekker, New York, pp. 15–22.

Bigda, R.J. (1995) Consider Fenton's chemistry for wastewater treatment. *Chemical Engineering Progress* **91**, 62–66.
Bradbury, M.J., Collishaw, P.S. and Moorhouse, S. (2000) Controlled rinsing: A step change in reactive dye application technology. *Colourage Annual* **5**, 73–80.
Chakraborty, S., Purkait, M.K., DasGupta, S. *et al.* (2003) Nanofiltration of textile plant effluent for color removal and reduction in COD. *Separation and Purification Technology* **31** (2), 141–151.
Chavalparit, O. and Ongwandee, M. (2009) Clean technology for the tapioca starch industry in Thailand. *Journal of Cleaner Production*, **17**, 105–110.
Chavan, R.B. (2001) Environment-friendly dyeing processes for cotton. *Indian Journal of Fibre and Textile Research* **4**, 239–242.
Cristóvão, R., Botelho, C., Martins, R. and Boaventura, R. (2012) Pollution prevention and wastewater treatment in fish canning industries of northern. Portugal, International Conference on Environment Science and Engineering, 7 to 8 April 2012, Bangkok, Thailand, *International Proceedings of Chemical, Biological and Environmental Engineering*, Vol. 3, IACSIT Press, Singapore.
del Pozo, R., Tas, D.O., Dulkadiroğlu, H. *et al.* (2003) Biodegradability of slaughterhouse wastewater with high blood content under anaerobic and aerobic conditions. *Journal of Chemical Technology and Biotechnology* **78**, 384–391 (online: 2003). doi: 10.1002/jctb.753
DoE (Department of Environment, UK) (1995) Metal Manufacturing, Refining and Finishing Works, Electroplating and Other Metal Finishing Works, DoE, London.
Duangsri, P. and Satirapipathkul, C. (2011) Spirulina sp. production in brine wastewater from pickle factory, 2011 International Conference on Bioscience, Biochemistry and Bioinformatics, IPCBEE vol. 5.
European Fruit Juice Association (2012) Liquid Fruit, market report.
Ghosh, P. and Gangopadhyay, R. (2000) Photofunctionalization of cellulose and lignocellulose fibres using photoactive organic acids. *European Polymer Journal* **3**, 625–634.
Hashem, M., Refaie, R. and Hebeish, A. (2005) Cross-linking of partially carboxymethylated cotton fabric via cationization. *Journal of Cleaner Production* **13**, 947–954.
Hasler, F. and Stone, N. (1997) *The Whys and Hows of Hydrochloric Acid Pickling*, Eso Engineering.
Hien, P.G., Oanh, L.T.K., Viet, N.T. and Lettinga, G. (1999) Closed wastewater system in the tapioca industry in Vietnam. *Water Science and Technology* **39**, 89–96.
Huston, P.L. and Pignatello, J.J. (1996) Reduction of Perchloroalkanes by ferrioxalate generated carboxylate radical preceding mineralization by the photo Fenton reaction. *Environmental Science & Technology* **30**, 3457–3463.
Huston, P.L. and Pignatello, J.J. (1999) Degradation of selected pesticide active ingredients and commercial formulations in water by the photo assisted Fenton reaction. *Water Research* **33**, 1238–1246.
IFC (1998) Fruit and Vegetable Processing, Pollution Prevention and Abatement Handbook, World Bank Group, Washington, D.C.
IFC (2007a) Environmental, Health and Safety Guidelines – Pharmaceuticals and Biotechnology Manufacturing, World Bank Group, Washington, D.C.

IFC (2007b) Environmental, Health and Safety Guidelines – Textile Manufacturing, World Bank Group, Washington, D.C.

Jiranuntipon, S. (2009) Decolourization of Molasses Wastewater from Distilleries Using Bacterial Consortium. Doctoral thesis. L'Institut National Polytechnique De Toulouse.

Karim, M.M., Dasa, A.K. and Lee, S.H. (2006) Treatment of coloured effluent of the textile industry in Bangladesh using zinc chloride treated indigenous activated carbons. *Analytica Chimica Acta* **576**, 37–42.

Kushwaha, J.P., Srivastava, V.C. and Mall, I.D. (2011) *Food Science and Nutrition* **51**, 442–452.

Little, L.W., Lamb, J.C. III and Horney, L.F. (1976) *Characterization and Treatment of Brine Wastewaters from the Cucumber Pickle Industry*, UNC Wastewater Research Center Department of Environmental Sciences and Engineering School of Public Health University of North Carolina, Chapel Hill.

Mendez, R., Omil, F., Soto, M. and Lema, J.M. (1992) Pilot plant studies on the anaerobic treatment of different wastewaters from a fish-canning factory. *Water Science and Technology* **25**(1), 37–44.

Mohan, S.V., Babu, V.L. and Sarma, P.N. (2007) Anaerobic biohydrogen production from dairy wastewater treatment in sequencing batch reactor (AnSBR): Effect of organic loading rate. *Enzyme and Microbial Technology* **41**, 506–515.

Murray, B.C., Gross, D.H. and Fox, T.J. (1994) Starch Manufacturing: A Profile, Research Triangle Institute, Research Triangle Park, NC.

Narayanan, T.S.N.S. (2005) *Reviews on Advanced Materials Science* **9**, 130–177.

NCAES (North Carolina Agricultural Extension Service) (1979) Water and Wastewater Management in Food Production, NCAES, Raleigh, NC.

Pignatello, J.J. and Chapa, G. (1994) Degradation of PCBs by ferric iron, hydrogen peroxide and UV light. *Environmental Toxicology and Chemistry* **13**, 423–427.

Qazi, J.I., Nadeem, Md., Bai, S.S. *et al.* (2011) Anaerobic fixed film biotreatment of dairy wastewater. *Middle-East Journal of Scientific Research* **8**(3), 590–593.

Ranganathan, K., Karunagaran, K. and Sharma, D.C. (2007) Recycling of wastewaters of textile dyeing industries using advanced treatment technology and cost analysis—Case studies. *Resources, Conservation and Recycling* **50**(3), 306–318. doi: 10.1016/j.resconrec.2006.06.004

Sarkar, B., Chakrabarti, P.P., Vijaykumar, A. and Kale, V. (2006) Wastewater treatment in dairy industries – possibility of reuse. *Desalination* **195**, 141–152.

Tanticharoen, M. and Bhumiratanatries, S. (1995) Wastewater treatment in agro-industry: a case study in Thailand, in *Waste Treatment Plants* (eds C.A. Sastry, M.A. Hashim and P. Agamuthu), John Wiley & Sons (Asia) Pte. Ltd., Singapore.

UNEP (1999) *Industrial Sector Guide. Cleaner Production Assessment in Fish Processing Industry*, UNEP, Paris.

University of Hawaii (2008) *Waialee Livestock Farm Polluted Runoff Control,* http://www.ctahr.hawaii.edu/wwm/waste.asp (accessed 30 December 2013).

Vourch, M., Balannec, B., Chaufer, B. and Dorange, G. (2008) Treatment of dairy industry wastewater by reverse osmosis for water reuse. *Desalination* **219**, 190–202.

Watkins, M. and Nash, D. (2010) Dairy factory wastewaters, their use on land and possible environmental impacts – a mini review. *The Open Agriculture Journal* **4**, 1–9

World Bank (1998) *Pollution Prevention and Abatement Handbook*, World Bank Group, Washington, DC.
Worm, B., Barbier, E.B., Beaumont, N. (2006) *et al.* Impacts of biodiversity loss on ocean ecosystem services. *Science* **314**, 787–790.
Worm, B., Sandow, M., Oschlies, A. *et al.* (2005) Global patterns of predator diversity in the open oceans. *Science* **309**, 1365. doi: 10.1126/science.1113399
Yu, S., Gao, C., Su, H. and Liu, M. (2001) Nanofiltration used for desalination and concentration in dye production. *Desalination* **140**, 97–100.

6

Sustainable Effluent Disposal

Pollution can occur due to natural contamination, like poison springs and oil seepage, or due to anthropogenic contamination. The sources could be point (ditches, drain pipes, or sewer outfalls) or nonpoint pollution sources (runoff from agricultural fields/feedlots, lawns, gardens, golf courses and atmospheric deposition of air pollutants carried by air currents). One of the most serious concerns about water contaminants in terms of animal/human health is pathogenic organisms that originate from untreated or improperly treated wastes. Water bodies with clear water and less biological productivity (oligotrophic conditions) are often converted to eutrophic conditions, which are rich in organisms and organic materials. Human activities have greatly accelerated eutrophication, leading to imbalances in nature. High biological productivity is seen in 'blooms' of algae and dense growth of aquatic plants. Blooms of deadly microbes called dinoflagellates have become increasingly common in many surface-water bodies, resulting in red tide. *Pfiesteria piscicida*, a poisonous dinoflagellate, which has wiped out millions of fish in polluted surface water bodies. Some toxic chemicals released from rock weathering are transferred by runoffs and anthropogenic activity can increase the rate of release of such inorganic chemicals by mining, use, processing and discarding of minerals. Rivers carry sediment to the oceans and human activities have greatly increased pollution by suspended solids. Sources of suspended solids include grazing lands, forests, and urban construction areas. Sediment fills surface-water bodies, clogs hydroelectric turbines, obstructs shipping channels, causes purification of drinking water to become costly, affects fragile ecosystem in estuaries and affects coral reefs and shoals near shores. Raising/lowering water temperatures due to the discharge of hot water can adversely affect water quality resulting in adverse impact on aquatic life.

The concept of sustainable development should ensure economy; the environment and wellbeing are not separate. One in five of the global population does not have access to safe potable water. About 50% of the global population does not have access to hygienic sanitation leading to the death of 3 to 4 million people due to water related diseases per year. Half of the wetlands in the world were destroyed in the twentieth century, leading to major losses of biodiversity; many rivers and streams are dead or dying. The Yellow River

Sustainable Water Engineering: Theory and Practice, First Edition. Ramesha Chandrappa and Diganta B. Das.

Table 6.1 Sustainable options for wastewater disposal.

Sl. no.	Type of settlement	Population	Sustainable wastewater discharge mode
1	Individual house	1–10	Septic tank and soak pit/trench
2	Village	Up to 1000	Individual septic tank and soak pit/trench; community toilet and DEWATS treatment and reuse
3	Temporary camp	Up to 1000	Community toilet and DEWATS treatment and reuse
4	Township	1000–20 000	Individual toilet/toilets for each residence with treatment in centralized treatment plant with reuse and energy recovery from sludge
5	Town	1000–50 000	Individual toilet/toilets for each residence with treatment in centralized treatment plant with reuse and energy recovery from sludge
6	City	50 000–3 000 000	Individual toilet/toilets for each residence with treatment in centralized treatment plant with reuse and energy recovery from sludge
7	Metropolitan city	3 000 000 and above	Individual toilet/toilets for each residence with treatment in centralized treatment plant with reuse and energy recovery from sludge

in China and the Colorado in North America are drying up and the Ganga in India contains polluted water. While rich people continue to flush toilets with water that could be drunk, poor people do not have enough safe drinking water. Governments cannot be blamed for the current situation as the lack of knowledge within the community has been a cause of the situation.

After treatment, wastewater must either be reused or disposed of into the environment. The most common means of treated wastewater disposal is by discharge and dilution into streams, rivers, lakes, estuaries, or the ocean. Table 6.1 gives sustainable options for wastewater disposal. If adverse environmental impacts are to be avoided, the quality of the treated and dispersed effluent must be consistent with local quality objectives.

For many years, effluent disposal to receiving waters was achieved by open pipes. The efficiency of mixing varied depended on the flow regime and the quality of the receiving water. An important aspect of effluent disposal was that of the assimilative capacity of the receiving waters, often representing the amount of organic matter that could be released without excessively depleting the dissolved oxygen. Greater attention is now being paid to the environmental effects of other constituents, such as suspended nutrients, solids and toxic compounds and how they can be safely assimilated into the aquatic environment.

The Kumbh Mela, the Hindu festival held every 12 years at Allahabad, India, attracts millions of people who take a dip in the holy place where Ganga and Yamuna Rivers meet. The Maha Kumbh Mela happens only once in 144 years. In 2013 the event was visited by more than 100 million bathers between 27 January and 25 February 2013. The visitors occupied a temporary camp that covered 20 km^2 with 35 000 temporary toilets and more than 14 000 policemen/paramilitaries/commandos. During the Kumbh Mela, thousands of workers were hired to clean waste but their services were not continued until the temporary camp was fully cleaned. The government released 71 m^3/s of water from January 1 to February 28 to ensure an adequate depth of water and dilution of pollution loads at the bathing site. The government also released 11.3 m^3/s additional water two days before and one day after six *Shahi snan* (royal bathing) days. A similar event at Haridwar, in India in 2010 resulted in a pH between 7.6 and 8.2; turbidity ranged between 30 NTU and 125 NTU; TDS ranged between 81.9 and 153.5 mg/l; TSS ranged between 1.97 and 9.4 mg/l; DO ranged between 7.0 and 9.8 ppm; BOD ranged between 1.21 and 3.6 mg/l; alkalinity ranged between 122.1 and 159.1 mg/l; hardness ranged between 110 and 140.1 mg/l; chloride ranged between 17.36 and 38.90 mg/l (Sharma *et al.*, 2012).

Effluent disposal focuses on the transportation of contaminants in the environment and the transformation processes that occur. To ensure that effluent disposal is accomplished in conformance with environmental requirements, a rigorous analysis must be performed in many cases. Mathematical modelling involves the application of mass balance transport analysis and kinetic expressions to explain the response of the system. By modelling the river and estuarine systems, it is possible to assess the assimilative capacity of these systems and thus to predict the impacts of the proposed discharge (Nassehi and Das, 2007). Some of the important transformations that occur include oxidation, bacterial conversions, natural decay, photosynthesis and respiration.

6.1 Dissolved Oxygen Sag Curves, Mass Balance Calculations and Basic River Models

Dissolved oxygen (DO) concentration is a measure of the health of the surface water body. Apart from BOD load, other factors that affect DO concentration in water body are (i) turbulence, (ii) quantity of water, (iii) temperature of water, (iv) velocity of water, (v) presence of surfactants/oil/grease, (vi) toxins in the water, (vii) colour of water, (viii) dissolved chemicals and (ix) the ecological health of the stream. Figure 6.1 shows the typical trends in oxygen depletion and changes in water bodies due to the entry of biodegradable wastewater.

Turbulence decides the distribution of BOD while the quantity of water decides the amount of water available for dilution. The temperature of the water is a function of the solubility of DO and the speed at which the BOD is degraded. The velocity of water determines the rate at which pollutants are transported and, hence, the length of the polluted stretch. Surfactants/oil/grease inhibit the transfer of oxygen to the water body. Toxins kill bacteria that are responsible for degradation and plants that are responsible for producing oxygen in water. The penetration of light into water depends on the colour of water. Dissolved chemicals either inhibit or accelerate the oxidation of BOD. The ecological health of the system decides the quantity of BOD it can accept without detrimental effects on its physical/biological components.

Water bodies lose and gain heat slower than air and land. The temperature of water bodies changes gradually with seasons. In temperate zones, heat transfer due to turbulence is negligible due to a phenomenon called *thermal stratification*.

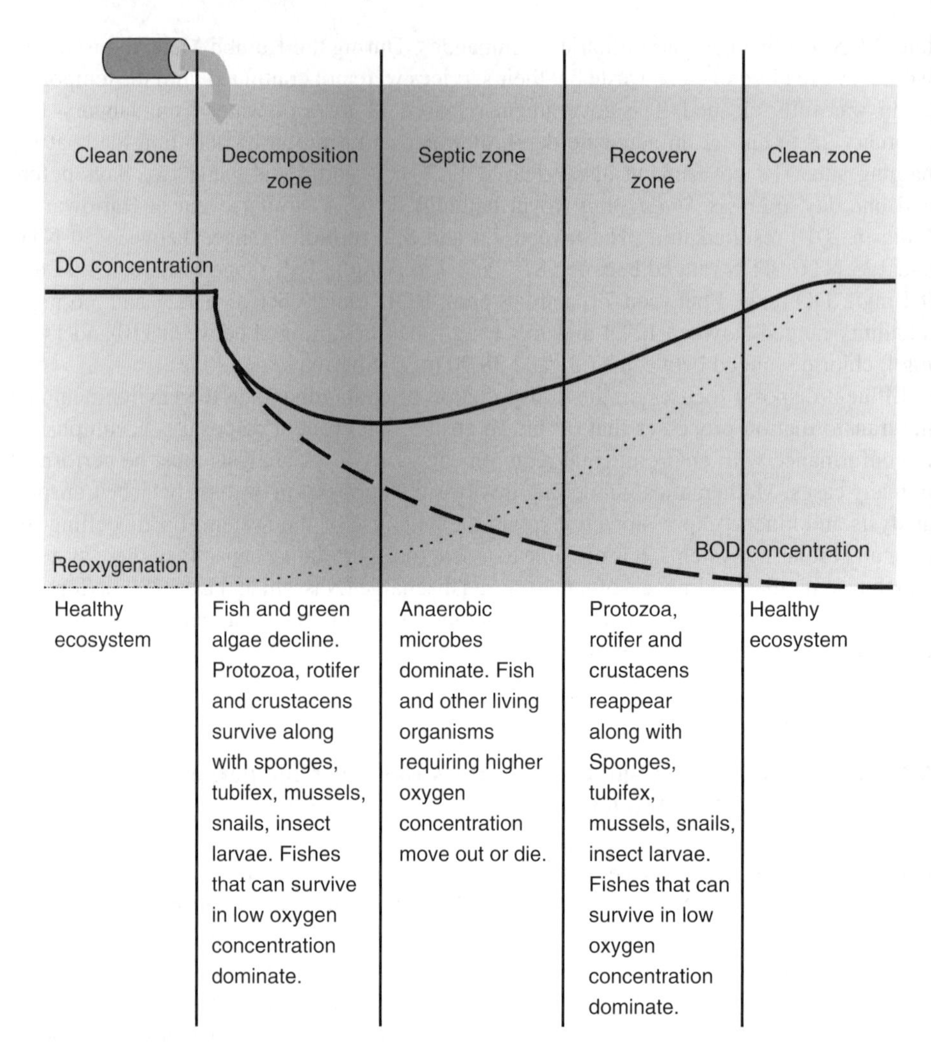

Figure 6.1 *Typical trends in oxygen depletion and changes in water bodies due to entry of biodegradable wastewater.*

Fresh water reaches its maximum density at 4 °C. In a warm season, water divides into an upper warm layer called the *epilimnion* and a lower cool layer called the *hypolimnion.* These two regions are separated by a *thermocline* or *metalimnion.* As the surface of water cools down due to changes in the weather, the warm surface layer gains density and sinks. In locations where the surface of water freezes, the water below ice, which is denser than ice, remains at around 2 °C until the season changes. When a warmer season arrives the ice melts and turnover occurs (Peavy *et al.*, 1986).

Many water bodies that suffered during the Industrial Revolution in Europe have regained their health. Currently the drop in oxygen is not a problem even when the water is flowing within the city, whereas water bodies in the developing countries have deteriorated due to the discharge of industrial effluents, sewage, religious offerings, partially/fully burnt funeral proceedings.

The transfer of oxygen into water is an important part of natural purification and can be calculated based on principles of mass balance. Oxygen lost through degradation is made up by oxygen transfer.

Solubility of gas in equilibrium with liquid can be explained by Henry's law, $x = P/H$, where x is the equilibrium mole fraction of the dissolved gas at 1 atm, H is the coefficient of absorption and P is the pressure of gas above the liquid.

$$x = \text{moles of gas}\,(n_g)/[\text{moles of gas}\,(n_g) + \text{moles of liquid}\,(n_1)]$$

The rate of transfer can be mathematically expressed as $dC/dt = (C_s - C)k_a$ where dC/dt is the instantaneous rate of change of the gas concentration in the liquid, C_s is the saturated concentration of gas and C is the actual concentration of gas.

Many water bodies have suffered from DO deficits that affected aquatic life critically. Scientists have studied the DO uptake characteristics of water streams. Water quality modelling in a stream has developed from the revolutionary work of Streeter and Phelps (1925). If dispersion is considered, the principal equation needs to be changed to a partial differential equation. The dispersion effect on BOD and DO in small rivers is not significant (Li, 1972; McCutcheon, 1989). Many investigators (Thomas, 1957; Clark and Viessman, 1965; Tebbutt and Berkun, 1976) have shown that second-order reactions often describe the stabilization of wastewaters, rather than first-order reactions.

Although other models have been used, the first-order kinetics equation given below has been used widely:

$$Y = L_0[1 - \exp(-k_1 t)]$$

where y = BOD, k_1 = BOD reaction rate constant, L_0 = ultimate BOD and t = time.

The common form of this Streeter and Phelps (1925) equation can be given as follows for stream pollution:

$$D_t = [k_1 L_0/(k_2 - k_1)][\exp(-k_1 t) - \exp(-k_2 t)] + D_0 \cdot \exp(-k_2 t)$$

Toxic chemicals cause serious variations in the self-purification mechanism of water bodies. In the studies conducted by Berkun (2005), inhibition of BOD was observed for 5 days in the presence of $HgCl_2$ at concentration more than 0.3 mg/l, $HgSO_4$ at concentration beyond 1.05 mg/l, $K_2Cr_2O_7$ at concentration 1.25 mg/l, Ni-$(NO_3)_2$ at concentration more than 3.5 mg/l, $CrCl_3$ at concentration more than 4 mg/l. It was also observed that, at low concentrations, reactions started following an acclimatization period.

In addition to toxic substances, light plays an important role in the reoxidation process. During daytime, algae and other plants generate oxygen depending on the time of the day. In the daylight, excess oxygen generated by algae is lost to the atmosphere but at night even algae need oxygen for respiration. Sludge deposition is another phenomenon that needs to be considered in reality. The maximum depth of sludge deposition occurs near the fall, which will degrade over time. Thus there will be diversity of flora and fauna in the decomposition/septic zone whereas a diverse population of plants and animals can be observed after the recovery zone. After entering the stream, bacteria in wastewater will multiply and become abundant. As protozoa prey on bacteria, their number will come

Figure 6.2 *A canal connecting water bodies in Stockholm.*

down. Pollutant-tolerant species are especially well adapted to life in thick sludge deposits and low DO (Ramalho, 1977).

The BOD assimilation is very fast for the Ganga and Yamuna Rivers in the Indo-Gangetic plain as these streams provide noteworthy bioflocculation of the colloidal organic solids discharged into the these water bodies, as a result of which about 60% of the BOD is removed in 30–60 minutes, 6 to 20 times faster than the normal BOD assimilation rate (Bhargava, 1983, 1986).

Figure 6.2 shows the canal connecting water bodies in Stockholm. In consideration of the importance of freshwater bodies, most of the countries have stopped discharging wastewater into streams. Polluted waters are mostly reused or dispersed into the ocean or onto land after treating with the best available technology.

6.2 Disposal Options and Impact on Environment

The decrease in the unregulated release of wastewater and the increase in safe water are important for achieving sustainable development. Reuse of wastewater conserves water and protects water bodies. Unsustainable wastewater treatment and management can limit sustainable development. Worldwide, two million tons of domestic, industrial and agricultural waste is released into the water bodies (Thomas and Matiasi, 2012). About 90% of all wastewater in the developing nations is released untreated into water bodies (Corcoran *et al.*, 2010).

Food production uses 70% to 90% of global fresh water, returning it with additional nutrients and contaminants along with industrial waste, contaminating freshwater and coastal ecosystems. Unsustainable disposal affects food security, safe drinking and bathing water. Around 90% of wastewater in the developing nations flows into the coastal zone resulting in growing marine dead zones, which currently cover an area of 245 000 km^2 (nearly the same as the area of world's coral reefs), affecting marine ecosystems, fisheries, livelihood and the food chain. Worldwide, two million tons of wastewater from domestic, industrial and agricultural activities is discharged into the water bodies. Discharge of untreated wastewater is the reason for climate change due to emissions of methane and nitrous oxide.

Mercury, which is one of the toxic elements, has raised concerns since the Minimata episode in Japan. Mercury finds its way into water by more than one route. The largest anthropogenic sources of mercury are artisanal/small-scale gold mining and coal burning, which account for around 62% of the total annual anthropogenic emissions to the atmosphere. Other sectors include ferrous/nonferrous metal production and cement production. Mercury emissions to the atmosphere peaked in the 1950s to the 1970s and subsequently declined due to reductions in Russia, Europe and North America. Natural processes convert less toxic elemental mercury into more toxic methyl mercury in aquatic systems, which results in high concentrations in some species of seafood. Mercury that is deposited on land is retained in soils and vegetation and re-emission from soils is a major addition to the environment. Concentrations of mercury are currently around 10–12 times higher than in preindustrial times. Mercury levels in the upper 100 m of the oceans have doubled compared to a century ago and intermediate/deeper waters have 10 to 25% more mercury (UNEP, 2013a).

Large urban settlements with low sanitation infrastructure can be affected very easily by human waste. More than a million septic tanks in Jakarta have contaminated the groundwater with faecal coliform. The contents of septic tanks are dumped untreated into waterways when septic tanks are cleaned. Since the 1980s, the economic boom in China has resulted in significant environmental pollution, which has resulted in deterioration of surface and groundwater bodies resulting in chronic mercurialism, arsenism, cancers related to microcystins and other diseases and pollution accidents (Lu *et al.*, 2008).

Changes in the sediment loading of water bodies can affect downstream habitats. Changes in the sediment supplies can affect sea grasses and coral reefs. Improper disposal of nutrient-rich wastewater into water bodies can cause eutrophication. Some chemicals, like pesticides, were developed to kill insects, rodents, weeds or other organisms. As the environment is an open system these chemicals also affect nontarget organisms including bees/insect eaters (Gil *et al.* 2012; Tu *et al.* 2013) and ultimately humans.

For radioactive wastes, various barriers are provided to hold the radionuclides and to restrict release of radiation into the environment. Geological repositories are provided to (i) protect the environment, (ii) isolate waste from human activities, (iii) limit release from the degrading engineered barrier system and (iv) dilute and disperse the flux of radionuclides.

Disposal of radioactive waste involves placing radioactive waste with safety guarantee in a disposal facility. Safety is ensured by placing natural or engineered barriers around the waste to avoid the release of radiation into environment. A system of multiple barriers is adopted to ensure isolation and to minimize the release of radionuclides into environment. High level waste (HLW) management and disposal is the most difficult problem in the nuclear power industry. According to Amaral *et al.* (2007), around 190 000 metric tons of HLW in temporary storage was awaiting disposal. Low-level waste streams are treated to reduce their level to statutory limits by filtration, ion-exchange, chemical treatment, solar evaporation, steam evaporation and membrane processes (IAEA 1983).

Pollution-induced impacts will not be uniform all over the world. The least developed countries (LDCs) and small island developing states (SIDS) are more vulnerable to pollution, along with impacts due to climate change. Pollution with severe floods/droughts will affect people in LDCs and SIDS. More than 70% of the population of the LDCs reside in rural areas.

The LDCs are a group of 49 nations: Angola, Bangladesh, Afghanistan, Bhutan, Burkina Faso, Benin, Burundi, Comoros, Cambodia, Chad, Democratic Republic of Congo, Central African Republic, Djibouti, Ethiopia, Eritrea, Equatorial Guinea, Guinea, Gambia, Guinea-Bissau, Kiribati, Haiti, Lesotho, Lao People's Democratic Republic, Liberia, Madagascar, Maldives, Malawi, Mauritania, Mozambique, Mali, Myanmar, Republic of Nepal, Niger, Rwanda, Sao Tome and Principe, Sierra Leone, Samoa, Senegal, Somalia, Sudan, Solomon Islands, Timor-Leste, Togo, Tuvalu, United Republic of Tanzania, Vanuatu, Uganda, Yemen, Zambia. Around 860 million people in these countries will be affected by climate change and pollution will increase the magnitude of the impact so that many of them will become environmental refugees. There are 51 SIDS: Antigua and Barbuda, Anguilla, American Samoa, Bahamas, Aruba, Barbados, Belize, British Virgin Islands, Cape Verde, Commonwealth of Northern Marianas, Comoros, Cook Islands, Dominican Republic, Cuba, Dominica, French Polynesia, Federated States of Micronesia, Fiji, Guam, Grenada, Guyana, Haiti, Guinea-Bissau, Jamaica, Nauru, Kiribati, Maldives, Marshall Islands, St Kitts and Nevis, Montserrat, Netherlands Antilles, Mauritius, Niue, Palau, New Caledonia, Puerto Rico, Samoa, Papua New Guinea, Sao Tome and Principe, Seychelles, Singapore, Solomon Islands, St Vincent and the Grenadines, St Lucia, Suriname, Tuvalu, Tonga, Trinidad and Tobago, US Virgin Islands, Timor-Leste and Vanuatu. Eleven countries amongst the SIDS are also LDCs – Comoros, Tuvalu, Kiribati, Sao Tome and Principe, Maldives, Guinea-Bissau, Vanuatu, Solomon Islands, Haiti, Samoa and Timor-Leste. Although SIDS, they have some common challenges: (i) limited physical size; (ii) relatively thin water lenses; (iii) high susceptibility to natural hazards; (iv) limited natural resources; (v) small economies and high sensitivity to market shocks and (vi) high population densities (Chandrappa *et al.*, 2011).

6.2.1 Ocean Disposal

Oceans cover around 70% of the earth's surface. Until the year 2000, four oceans were recognized: Atlantic, Pacific, Indian and Arctic and, in the year 2000, the International Hydrographic Organization delimited the Southern Ocean which surrounds Antarctica that extends to 60° latitude. Marine ecosystems are the largest of the earth's aquatic ecosystems, which include salt marshes, oceans, estuaries, intertidal ecosystems, mangroves, lagoons, coral reefs, the seafloor and the deep sea.

The oceans have been subject to anthropogenic activities to varying extents, which have negative impacts on the marine ecosystem. Negative environmental impacts depend on the type of human interference, which consists of two types: pollution and physical destruction. Sources of pollution are classified into coastal sources, including shipbuilding (Figure 6.3), ship breaking, river influxes, offshore inputs and atmospheric deposition. Coastal sources can be either diffuse sources or point sources. Point sources are discharged from sources like an industry whereas the diffuse sources include runoffs from forest/agricultural/urban areas.

Figure 6.3 *Shipbuilding units without wastewater treatment.*

Over the years ocean diffusers have been used as a solution for disposing of wastewater. But changing climate can prove that these practices are not sustainable and need to be stopped and replaced by water reuse. Marine outfall is a pipeline or tunnel that discharges wastewater/stormwater/sewage/cooling water/brine effluents to the sea. The outfalls vary from 5 cm to 8 m in diameter and a few km in length. The depth of the deepest point varies from 3 m to 60 m. Outfall materials include stainless steel, polyethylene, carbon steel, glass-reinforced plastic, cast iron, reinforced concrete, or tunnels through rock.

The impact of ocean disposal cannot be seen in isolation. Climate change is melting ice masses around the world, resulting in two major impacts: (i) variation in surface runoff, and (ii) a rising sea level. The rising sea level leads to (i) deep ocean overturning, (ii) damage to property along the coastal area, (iii) seawater intrusion to ground/surface water bodies, (iv) damage to flora/fauna in the ocean, and (v) land subsidence (Chandrappa *et al.* 2011).

Nitrogen and phosphorous from domestic and industrial activity are major issues for coastal waters and freshwater. It is evident that coastal ecosystems are endangered and this is very significant for human wellbeing. The recycling of nutrients is extremely important where nutrient use is low due to economic constraints. Separation of urine and faeces at source is an efficient option for recycling of nutrients. Coastal areas without existing infrastructure hold large nutrient potential; urine can be collected separately from them and disposed of on agricultural land.

The stress on mangroves, sea grasses and coral reefs has increased remarkably over the years due to decreasing water quality, sewage discharge, overexploitation of land and change in land use, decreasing resources, deforestation, destructive fishing and acidification of the oceans. The south-east Asian region, where there is shipping traffic and there are many ports, has led to oil spills, ballast/bilge discharge, pollution from ports, anchor damage and groundings and garbage disposal, resulting in the destruction of coral reefs. About 21% of the coral reefs in south-east Asia are in danger due to runoff; 35%% of coral reefs are in danger due to sediments and nutrients (Burke *et al.* 2002).

The ocean has become 30% more acidic in the preceding two centuries because of anthropogenic CO_2 emission, heat budgets, freshwater balance and land-ocean exchange (UNEP, 2013a).

Desalination of sea water has been used to provide drinking water in coastal regions, arid regions and small islands since the 1950s. Around 2006, about 24.5 million m^3 of water were

Figure 6.4 *Boulder placed on seashore to protect from tides.*

being produced every day for drinking water, tourism, industry as well agriculture (UNEP, 2008; Lattemann and Hopner, 2008) and production is likely to rise to 98 million m^3/d by 2015 (UNEP 2008). The process discharges concentrated brine into sea that can lead to local ecological changes. The process uses descaling and antifouling products that contain heavy metals and toxic chemicals and require treatment before disposal. Ecologically, the Asian mega deltas, which are diverse ecosystems of unique plants/animals, have been threatened due to anthropogenic activity, which includes marine pollution. The impact on mangrove forests that at present occupy about 14 650 000 ha of coastline globally would threaten human sustainability and livelihoods due to marine pollution as they are used traditionally for timber, fuel, medicine and food. Mangroves are also breeding sites for many mammals, birds, fish, shellfish, reptiles and crustaceans. Mangroves provide protection from tidal bores, waves and tsunamis and can reduce the erosion of the shoreline. Several locations around the world have experienced rising sea levels and tidal floods bringing pollution from elsewhere. Hence many coasts are protected by artificial barriers (Figure 6.4) to protect them from pollution and coastal erosion.

6.2.2 Disposal into Fresh Water Bodies

Surface water bodies have been used for centuries and many civilizations have been established on the banks of rivers as they provide an easily accessible source of quality water. Rivers have been used for drawing water for drinking, irrigation and fishing but they have also been used for waste disposal. In 1388 the English Parliament banned waste dumping in ditches and public waterways and around 1407 England passed a law requiring waste to be stored inside until rakers came to remove it. London city authorities prohibited throwing rubbish/gravel/earth/dung into the Thames in 1357 but butcher's waste was deposited at the centre of the Thames from 1392 to avoid pollution of the river bank. Waste management in Europe was given utmost importance but, despite colonization by Europeans, Africa and

Figure 6.5 *Solid waste dumped into river.*

Asia stood still in this respect. Different forms of land use affect rivers through loading of nutrients, suspended solids, acidifying substances and metals. Environmental impacts can be reduced with proper land use planning in the river basin. Buffer zones, sedimentation basins and wetlands are some effective ways to control pollution.

Some plants, like mosses and liverworts, attach to solid objects whereas other plants, like duckweed or hyacinths, are free floating and form dense mats. Still other plants are rooted and are submerged or emergent. Taxa of surface water bodies include molluscs, crustaceans, insects, fishes, mammals, birds, amphibians and reptiles.

Solid waste (Figure 6.5) and wastewater from communities, farms, homes, urban areas, villages and industry contain harmful suspended and dissolved substances. Indiscriminate discharge of wastewater affects natural flexibility, biological diversity and the ability of the earth to provide ecosystem services affecting both rural and urban populations. In all cases, the poorest are affected severely (Corcoran *et al.*, 2010).

The ecosystem of a river includes living organisms and nonliving things that interact with each other physically, chemically and biologically. Lotic ecosystems are ecosystems associated with flowing water and they range from springs that are a few centimetres wide to rivers that are kilometres in width. On the other hand lentic ecosystems are ecosystems associated with relatively still surface-water bodies.

The following characteristics make running water unique compared to other aquatic habitats: (i) continuous physical change, (ii) flow is unidirectional, (iii) there is a high degree of temporal and spatial heterogeneity, (iv) the flow is affected by climate change and so are living organisms associated with it and (v) the biota are acclimatized to live in flowing water.

Water flow is the major factor in lotic ecosystem that influences their ecology. The flow can vary from slow backwaters to torrential rapids. The speed and turbulence can vary within a system due to obstructions, sinuosity and the gradient. The quantity of water into the system depends on snowmelt, groundwater and/or precipitation characteristics.

Flowing waters alter the shape of the stream bed and the path of flow through erosion and deposition, creating a range of habitats that include riffles, pools and glides. Light also plays an important role in aquatic ecosystem because it supplies the energy required to drive primary production by photosynthesis. Dark places provide hiding places for prey species.

The internal temperatures of most lotic species vary with their environment. Spring-fed systems will have little variation as they are usually from groundwater sources that are close to ambient temperature. Many systems show seasonal/diurnal variations and are most extreme in desert, arctic and temperate systems.

Water chemistry in surface-water bodies varies tremendously and is determined by inputs from geology and anthropogenic activity. Oxygen is an important chemical constituent of surface-water ecosystems, which enters the water by diffusion at the air-water interface. The solubility of oxygen in water reduces as temperature rises and as turbulence in water exposes more water to the air, thereby absorbing more oxygen compared to slow backwaters. Oxygen is also a byproduct of photosynthesis and, hence, the presence of plants results in high DO concentration in the day, which decreases during the night.

Bacteria in surface water are associated with the decomposition of organic matter. Periphyton and phytoplankton are important sources of primary production in surface-water bodies. Phytoplankton that float freely in surface-water bodies fail to maintain their populations in fast-moving streams as against slow-moving water bodies and stagnant water bodies.

Floating-treatment wetland (FTW) is a new treatment concept that use rooted plants growing on the floating mat of the water (Tanner and Headley, 2011; Fonder and Headley, 2010). The plant roots below the floating mat give a wide surface area for the growth of biofilm and for entrapping suspended particulate matter. Plants are forced to fulfil nutrition requirement directly from the water, thereby reducing pollution. The buoyancy of FTWs enables them to bear fluctuations in water depth, thereby providing the potential to improve performance (Tanner and Headley, 2011; Weragoda *et al.*, 2011).

Floating-treatment wetland is still not popularly accepted across the world (Weragoda *et al.*, 2011). Artificially created floating wetlands are mostly used for water-habitat enhancement, quality improvement and aesthetic improvement. Sewage, pond water, piggery effluent, dairy manure effluent, urban lake water and water supply reservoirs.

6.2.3 On-Land Disposal

On-land disposal (Figure 6.6) has been widely accepted due to its potential for treatment by soil bacterial and nutrient recycling by agriculture. Urban agriculture, comprising the production, processing and distribution of a food and ornamental plants within intraurban and periurban areas, has been a highly sustainable option catching on speedily across the globe. Reuse of wastewater for agriculture, which is discussed in more detail in section 6.3.3, is reported all over the globe (Mills *et al.*, 1992; Salgot and Pascual, 1996; Angelakis *et al.*, 1999; Bonomo *et al.*, 1999; Bahri, 1999; Faby *et al.*, 1999; Friedler, 2001). Reuse is important in locations with limited water resources and high urban water demand is fulfilled by reducing water.

Where groundwater and soil conditions are favourable for anthropogenic groundwater recharge by infiltration, partially treated effluent can be allowed to infiltrate to reach groundwater. The unsaturated zone, called the 'vadose' zone, acts as a natural filter and

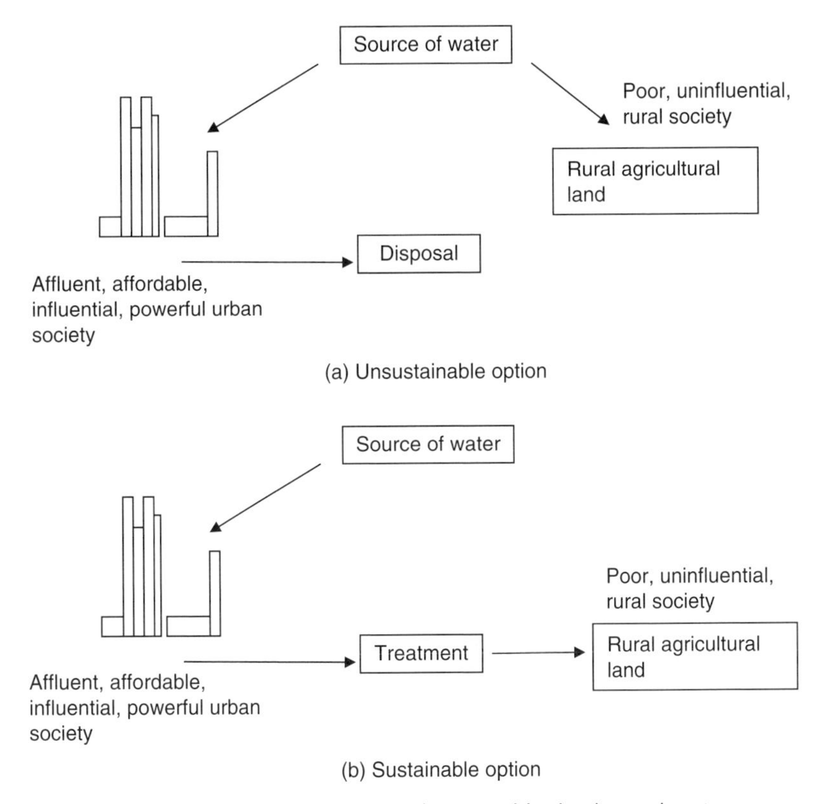

Figure 6.6 *Unsustainable (a) and sustainable (b) disposal options.*

removes suspended solids. Major reductions in phosphorus, nitrogen and heavy metals can be achieved through this disposal method.

The recharged water is allowed to flow in the aquifer before it is collected to allow purification by filtration, precipitation and adsorption. Such systems are called soil-aquifer treatment systems (SAT systems).

The simplest type of SAT system is carried out by infiltration basins on high ground so that it drains into a lower area, like a seepage area, a natural depression, a stream, a lake, or a surface drain. The SAT can also be achieved by allowing water through drains. Where the groundwater is too deep, pumped wells must be used.

Soil-aquifer treatment systems should be installed in soils with permeability that allows high infiltration rates. Groundwater tables should be at least 1 m below the infiltration basins. Vadose zones should not contain impermeable soils/rocks that could hinder the movement of water.

Infiltration basins/drains (Figures 6.7 and 6.8) in SAT systems are flooded intermittently from 8 hour to 2 week flooding/drying cycles. Annual infiltration typically varies from 15 to 100 m per annum depending on climate, soil and quality of water infiltrated and frequency of basin cleaning.

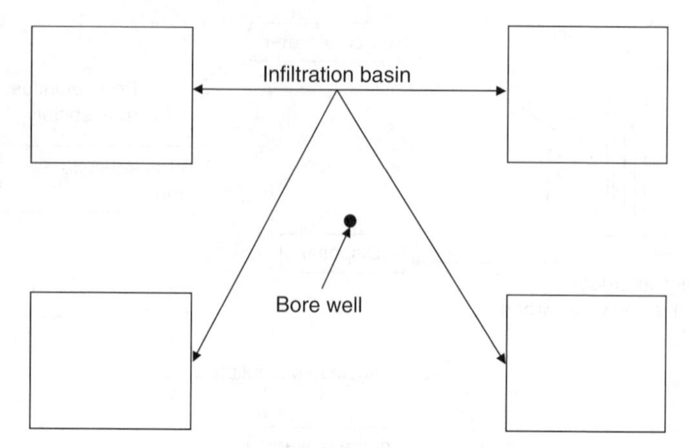

Figure 6.7 *Infiltration through basins.*

Low-water depths of around 20 cm are preferable to large water depths thus giving less time for algae. During flooding, suspended solids in water accumulate and clog the bottom of the basins causing a decline in infiltration rates. Drying of the basins leads to cracked, curled-up flakes and the decomposition of organic matter, restoring the hydraulic capacity close to the original levels. Removing the clogged substances by raking is a better option than mixing with soil to avoid accumulation of clogging materials in the soil.

6.3 Sustainable Reuse Options and Practice

As discussed in Chapter 1, the majority of water resources in Earth are saline, with only 2.5% being fresh water. As most of the fresh water is frozen in icecaps of Poles, green land and mountains only 0.001% is accessible, of which nearly 50% is withdrawn. Only about

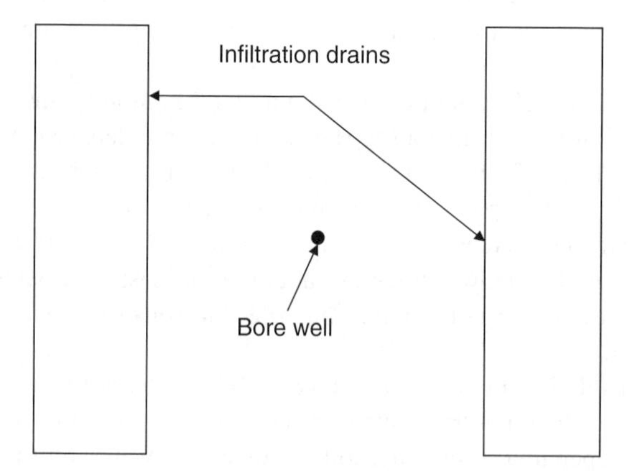

Figure 6.8 *Infiltration through drains.*

5% of global wastewater was collected and treated worldwide in 2009 (Lazarova *et al.*, 2013).

Nations of the southern Mediterranean and Middle East face fresh water-shortage problems. The desalination of brackish water and seawater is being implemented or planned in some of the countries despite high costs. Exploitation ratios of 50% or more in many Mediterranean countries like Gaza, Egypt, Libya, Israel, Malta and Tunisia show that the consumption of water exceeds the renewable capacity.

Reuse has become an attractive option in the Mediterranean region as the total population of the region is expected to rise from 420 million in 1995 to 508–579 million in 2025. Per capita renewable water varies from more than 100 m^3 per annum to more than 1000 m^3 per annum (Margeta and Vallée, 2000).

The megacities face global climate-change impacts that affect hydrology. Industrial water use in Shanghai increased from 20% to 69.3% of the total water requirement since the 1960s, requiring large capital investment (Geiger, 2006). Another problem in megacities is that the data is not collected in pace with the growth and changes, making forecasts difficult. The increasing water problems of large cities cannot be solved with conventional supply and disposal systems cannot solve the increasing water problem in the megacities of the Third World (Corcoran *et al.*, 2010). Arab countries, which cover around 10% of the area of the world, receive about 2% of the world's annual precipitation and contain about 0.3% of the world's annual renewable water resources. By 2015, almost all Arab countries are expected to be below the level of severe water scarcity of 500 m^3 per capita per annum; with six below 100 m^3 and nine countries below 200 m^3, (World Bank, 2011).

Figure 6.9 shows a correlation between natural water quantity/quality, demand and time. Water stress in many areas has resulted in direct/indirect recycling (Figure 6.10). Water problems of large cities are water scarcity, inefficient distribution, wastewater collection, flooding and pollution.

Figure 6.11 shows the relation between available global freshwater and water that is reused. Major constraints for reuse are:

- environmental;
- mechanical;
  - corrosion;
  - scaling;
  - biological control;
- cost constraints;
  - capital cost;
  - operational cost.

With time, many interesting reuses are being made apart from agriculture, pisciculture, flushing and so on (Figure 6.12). Reclaimed wastewater is being used in Japan for sprinkling roads to reduce urban heat islands in cities. Sprinkling of reclaimed wastewater can decrease the temperature on road surfaces by 3 °C at night and by 8 °C during the day. Several municipalities in Japan are also creating artificial streams using reclaimed water to improve the environment. Quality is always a limiting factor in deciding reuse options.

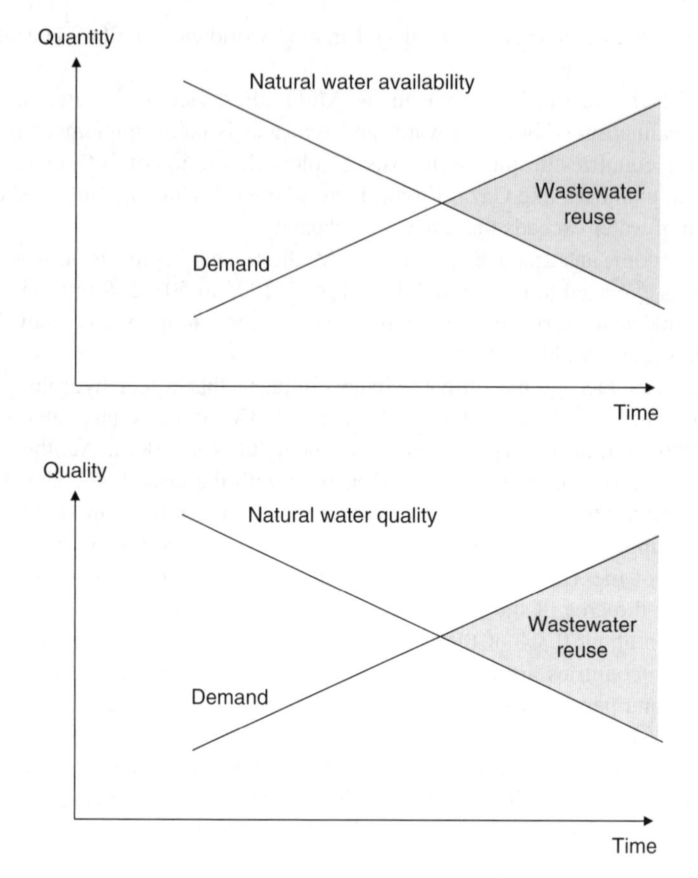

Figure 6.9 *Correlation between natural water quantity/quality, demand and time.*

While drinking water and processing of food/pharmaceutical industries require water of potable quality, other uses like recreation/construction/fire protection need lower quality water (Figure 6.13). Figure 6.14 shows a correlation between quality and reuse options. Several countries still practise the use of wastewater without treatment for agriculture.

The most common operation in farms houses – diversion to crops – has also been used in urban settlements. A 60 m^2 engineered wetland in a courtyard in Berlin, Germany, is operating successfully (Allen *et al.*, 2010). Wherever sufficient land is not available, grey water can be treated physically and chemically using a sand media filtration tank followed by disinfection. Biological grey-water treatment with membrane bioreactors (MBR), which has been employed since the early 1990s, uses membranes directly immersed into the bioreactor for separation of solids.

One of the characteristics of water is its ability to preserve heat energy. Water can warm or cool objects when it comes into contact with them. The temperature in Japan has a large variation from below zero to nearly 40 °C, but the temperature of sewage varies from 12 °C to 30 °C. The temperature of wastewater is hence lower in the summer and higher

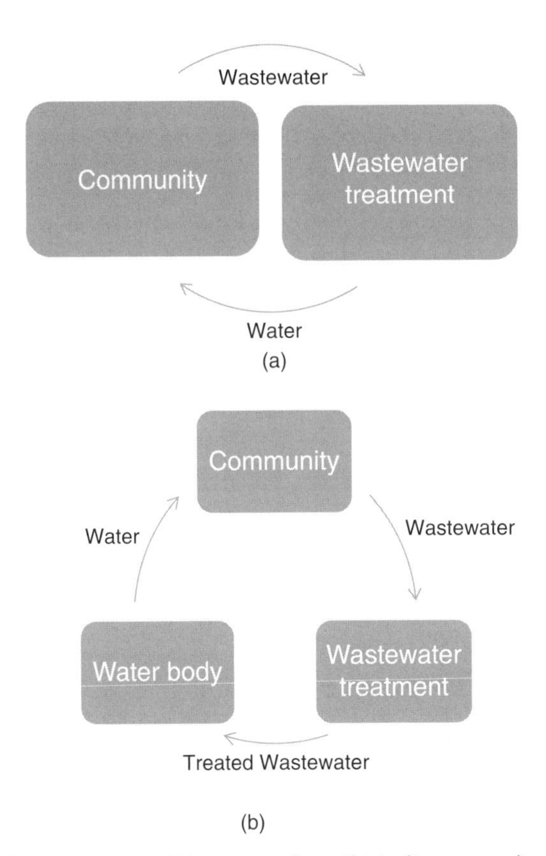

Figure 6.10 *(a) Direct recycling; (b) indirect recycling.*

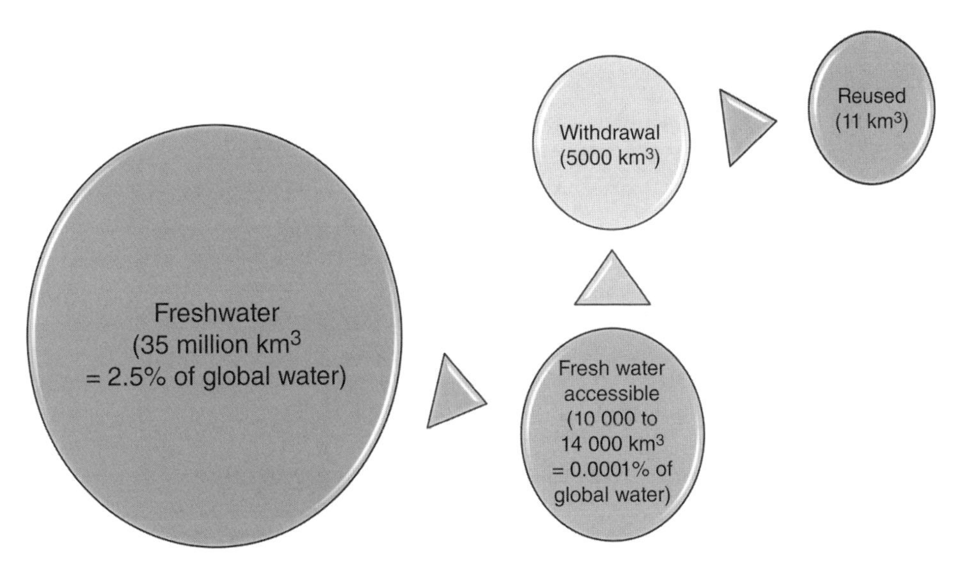

Figure 6.11 *Relation between global freshwater available and reused.*

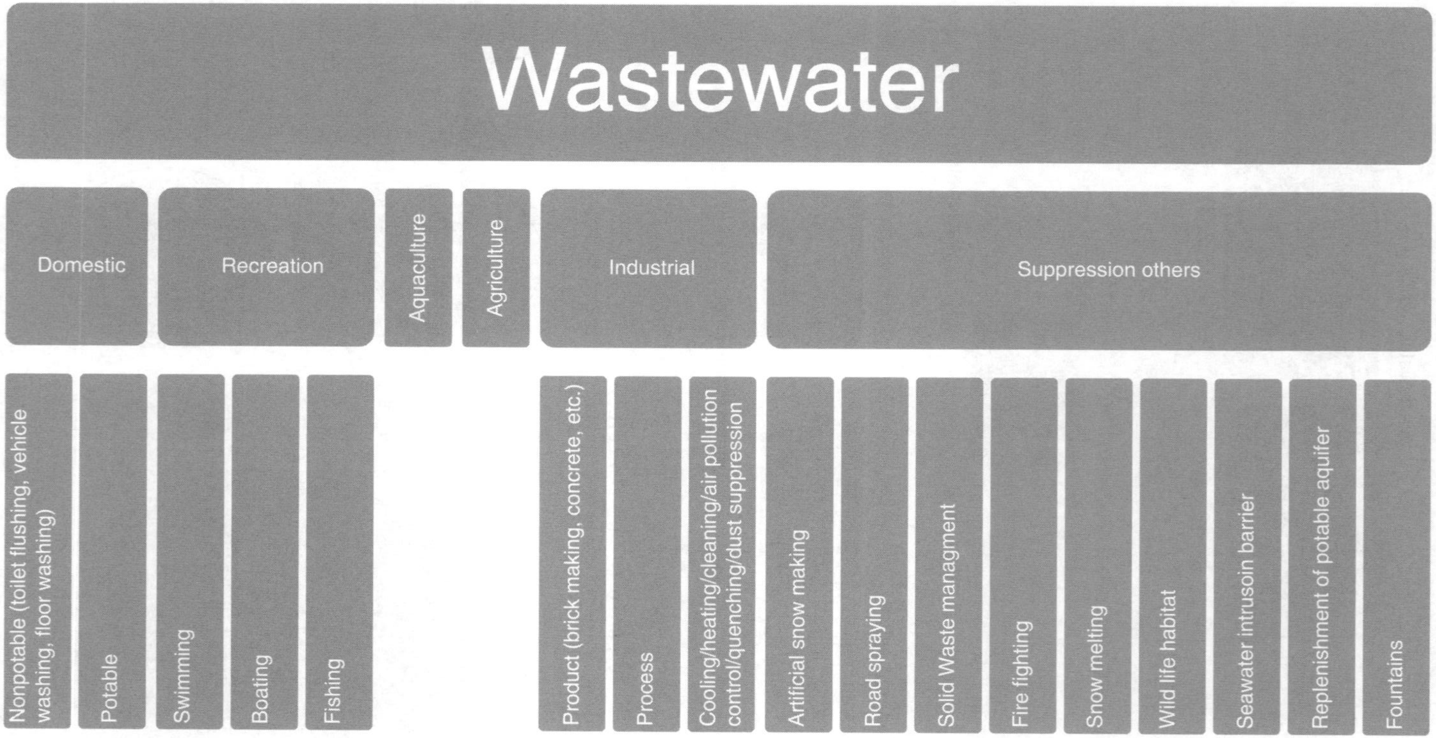

Figure 6.12 *Wastewater reuse possibilities.*

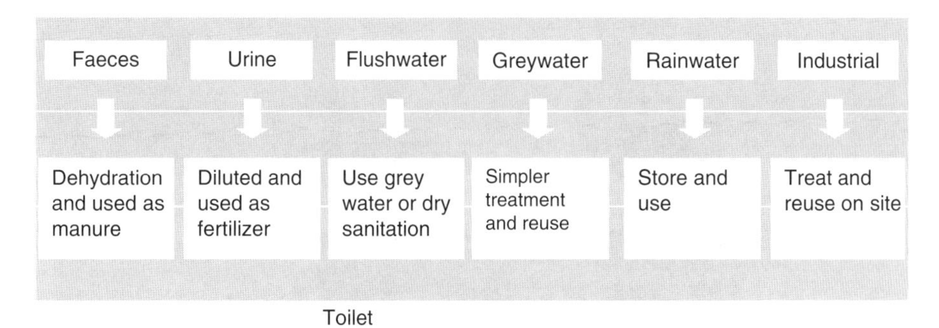

Figure 6.13 *Wastewater reuse, alternative possibilities.*

in the winter compared to the ambient temperatures. Based on this phenomenon, a heat exchange system has been developed in Osaka, Japan. In the City of Sapporo, where there is high snowfall during winter, treated wastewater is used for melting snow by providing snow-damping ditches/tanks, to which treated wastewater is supplied.

Sustainable reuse of water systems varies from low-cost options, which divert grey water for direct reuse, to complex treatment processes. One of the most simple reuse options could be to use water from washbasins or sinks for toilet flushing; alternatively wastewater from washing machines can be used for flushing in households. Uses like fountains (Figure 6.15) for ornamental purposes or artificial snow (Figure 6.16) do not need high-quality water. Chlorine tablets or other disinfectants can be put in the grey-water storage tank to disinfect it. There is a range of commercial systems that recycle grey water from washbasins, kitchen sinks and showers into toilet water tanks (Allen *et al.*, 2010).

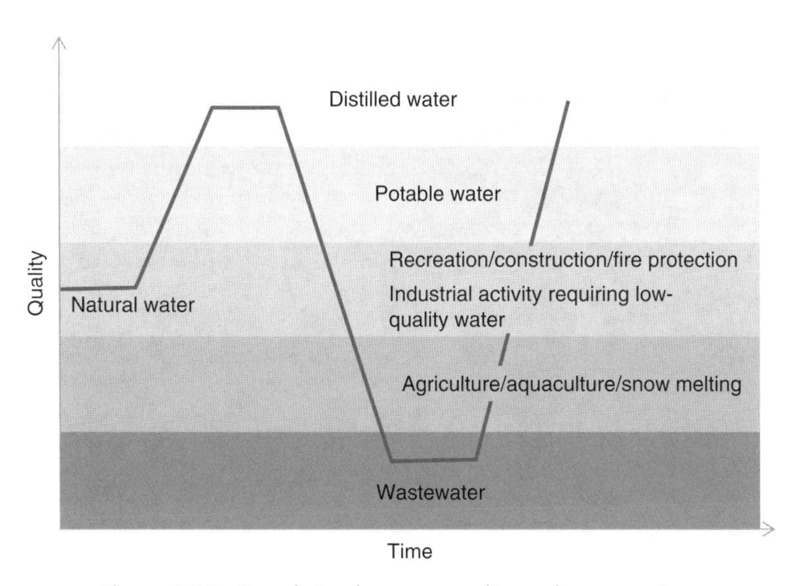

Figure 6.14 *Correlation between quality and reuse options.*

Figure 6.15 *Fountains for ornamental purposes do not need potable water quality.*

Figure 6.16 *Artificial snow does not need potable water quality.*

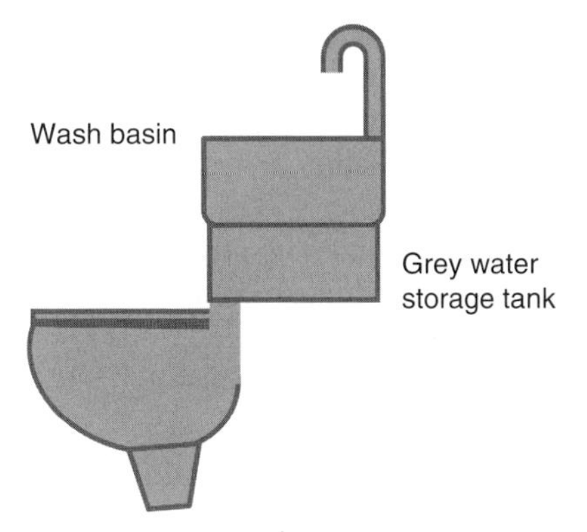

Figure 6.17 *Simple straightforward reuse option for grey water.*

6.3.1 Toilet Flushing

Reuse of wastewater for toilet flushing is the most viable option as it does not need potable water. The simplest way grey water can be used for wash basins is given in Figure 6.17. Another option is to use untreated wastewater from washing clothes. The only requirement for water for flushing is that it should not leave stains on the water closet and should not cause odour problems. If the grey wastewater cannot be used immediately the other option would be to store and reuse after filtration. Black water (sewage) can be reused after secondary treatment followed by sand filtration and activated carbon filtration.

6.3.2 Floor Washing

Floor washing does not need potable-quality water. The required quality depends on the location of the floor. The floor washing can be done either by mopping or with water jets or with water and a broom. The quantity of water required would depend on the magnitude of cleaning. A relatively clean area needs less water while dirty floors need more water. Places like international airports and luxury hotels are cleaned more frequently than solid waste-disposal sites. On average, 0.1 l/m^2 is required for mopping a five-star hotel as against 1.0 l/m^2 for mopping industrial floors.

6.3.3 Sustainable Wastewater Irrigation

Effluent irrigation has been practised throughout the world for centuries. There is a shift from 'disposal ' mentality to an approach of reusing water in a sustainable manner, especially in countries where water is scarce. Agriculture uses about 66% of all recycled water, which represents in approximately 2% of total water consumption in Australia (ABS 2006). Forage crops are grown using sewage irrigation in Canada. Even though the trend is the same in many water-stressed countries, it is usually not recorded properly. The effluent from major cities in India has been used to grow vegetables and fodder whereas in Australia it is spread systematically across about 270 recycled water schemes catering for the irrigation of pastures and fodder crops. Recycling water for agriculture provides a sustainable water source capable of assuring greater certainty and offers a valuable nutrient source for healthy crop production. Reuse of water for irrigation often requires practical management solutions to avoid risk to consumers and the environment.

The design approach to irrigation with treated wastewaters depends on whether emphasis is placed on water supply for irrigation or wastewater treatment via land treatment (Tchobanoglous *et al.*, 2003).

Health risks from reuse of effluent include the spreading of infectious diseases by bacteria, virus infection and worm infection. Effluent used on food crops must be coagulated, clarified, filtered, disinfected and oxidized to achieve total coliform counts within a value of 25/100 ml in a single sample or a median value of 2.2/100 ml (coliforms must be monitored daily) (Ongerth and Jopling, 1977).

The key principles of wastewater irrigation are: (i) lower costs to the community, (ii) plants perform best when adequate nutrients are supplied in aerated soil, (iii) evapotranspiration from plants demand should drive the supply of effluent, (iv) the application of wastewater should not exceed the capacity of soil to provide the plant's growing conditions, (v) the skills of the operator (Standen, 2000).

The five main components of sustainable wastewater irrigation are: (i) treated water quality, (ii) site, (iii) irrigation type, (iv) plant type, (v) management. It is essential to consider wastewater volume, nutrients, viability, toxicants, salt and sodicity to protect soil, public amenities, surface water, native vegetation, human/animal health, groundwater and cultural heritage (Standen, 2000). To attain sustainable wastewater irrigation it is essential that: (i) each element has to be considered, (ii) the desired outcome should be recognized, (iii) the associated risk level should be determined, (iv) suitable action should be selected to achieve the result.

Gardening and agriculture consume a substantial amount of water. Adjusting sprinklers so that the water falls only on the lawn and not the sidewalk, or street or any other place where water is not required can also save water. Many parks and gardens in water-scarce areas can be planted with shrubs and trees instead of planting with lawns and turf. Spreading a layer of organic mulch or mulching sheets can save water as they retain moisture. If water easily runs off of a lawn, splitting irrigation time into shorter periods can provide for better absorption. Many gardeners apply water when plants do not need water. Checking moisture in garden soil prior to gardening can save water substantially. A taller lawn shades roots and helps save water. Same people install automatic irrigation controllers. Installing a rain sensor to avoid watering while it is raining will save water. Watering your plants deeply and less frequently helps deep root growth and drought tolerance. Grouping plants with similar watering needs avoids overwatering or underwatering of plants.

Both fresh and treated water have been used negligently over the years in most parts of the world. There is an attempt in many parts of the world towards sustainable irrigation so that water is used prudently without compromising future generations. Increasing water use efficiency has been important for government, commercial and agricultural endeavours for several decades.

Sustainable irrigation should consider soil, crops and water simultaneously so that the quality of any of these three is not compromised. Some of the techniques adopted in recent years are shown in Table 6.2.

Reuse of wastewater is the most used option so far. Investments in reclaiming water in water-scarce regions are generally paid back and profitable in the long term (Heinz *et al.*, 2011; Norton-Brandão *et al.*, 2013). The agricultural community saves costs in

Table 6.2 Some of the techniques used in sustainable irrigation.

Sl. no.	Technique	Description
1	Insurance irrigation	Plants grow optimally at a certain water-supply rate. Irrigation at optimum feed will save water, maintain health of soil and increase yield of plant output
2	Oxygenation	Soils with low infiltration rates become waterlogged and roots do not receive enough oxygen. Hence air is injected into the irrigation supply. Roots close down their function if oxygen is insufficient in root zone and they do not take up water and therefore cannot take up nutrients
3	Fertigation	Fertilizer in required dose is added to water to improve yield of crop
4	Canopy management	Canopy of trees is pruned to save water

the reclamation area, in water pumping and fertilization, while an increase in incomes is generated. The Lobregat river basin of Spain proves that water reclamation ensures economic and environmental benefits (Heinz *et al.*, 2011) and at the same time fresh water resources are preserved (Norton-Brandão *et al.*, 2013). Reclamation projects are more common in southern Europe, (Marecos do Monte, 2007) including Spain, Greece, Italy, Cyprus, Portugal and France, which use wastewater for agriculture. Of the 964 million m^3/year of treated wastewater used in Europe, 347 million m^3/year are used by Spain and 233 mm^3/year by Italy (MED WWR WG, 2007).

Water shortages occur every 5 to 7 years in Greece, due to droughts. The available natural water of Greece is about 14 340 hm^3 and the country has water demands of 8243 hm^3/year with the water demand for cultivation being 6833 hm^3 (Pedrero *et al.*, 2010). Analysis of water use wastewater treatment plant areas revealed that more than 80% of the treated effluent is generated in regions with a scarce water balance (Angelakis *et al.*, 2003). Cyprus uses all of the treated wastewater (MED WWR WG, 2007) mainly on irrigation for agriculture. The treated wastewater from the WWTP of Clermont-Ferrand, France, is partially used for irrigation. Constructed wetlands in Portugal have been tested for irrigation and have been found suitable for the purpose (Norton-Brandão *et al.*, 2013).

More than 130 agricultural areas in Korea use effluent for irrigation (Rhee *et al.*, 2011) and reclamation technologies for agriculture will be included in agricultural water development plans to overcome severe drought (Jang *et al.*, 2010). Sydney, Australia, has 15 existing schemes and 25 new schemes for irrigation are being planned. Many projects have been developed, like the Gippsland Water Factory, where wastewater from eight Victorian towns and a paper mill will treat 14 000 m^3 per annum for many uses, including agriculture (Radcliffe, 2010). California uses 656 000 hm^3 of treated municipal wastewater per year (Norton-Brandão *et al.*, 2013). Jordan, Tunisia and Israel use an array of conventional and nonconventional systems (Qadir *et al.*, 2010). In Tunisia, treated effluent reuse is increasingly included in planning and development (Ayni *et al.*, 2011). In Jordan, treated wastewater has been used for irrigation for several decades and the government of Jordan has standards for the use of sewage. A separate government agency was established for regulating wastewater use there (Norton-Brandão *et al.*, 2013).

In Israel, integrated wastewater use enables wastewater to utilize 20% of the water used for agriculture (Pedrero *et al.*, 2010). In India, untreated/treated wastewater have been in use since the 1960s (Singh *et al.*, 2010) and 26% of the vegetables in Pakistan are cultivated with wastewater (Ensink *et al.*, 2004). Wastewater is used for irrigation in periurban regions of Hanoi, in Vietnam.

6.3.4 Nonpotable Industrial Use

Fresh water resources are under increasing pressure. Hence industries are planning to reuse water within their facilities to minimize the quantity of water discharged. There are numerous reasons for this pressure, such as: (i) the cost of fresh water, (ii) the cost of treatment to reach discharge limits, (iii) water availability, (iv) environmental awareness and (v) community relations. Nonpotable industrial uses include (i) air conditioning, (ii) fire protection, (iii) construction and (iv) manufacturing processes.

Any discussion of wastewater reuse is incomplete without a consideration of wastewater reuse in the mining sector. Mining produces significant environmental pollution and, hence, requires efforts to protect the environment. The mining sector uses recycle/reuse principles to achieve cleaner production goals. The main sources of wastewater from mining are: (i) minewater, (ii) process wastewater, (iii) sewage and (iv) surface runoff. To preserve the environment, a concept called 'low-' or 'no-waste technology' (LNWT) or 'cleaner production (CP)' is being adopted in many parts of the globe to prevent waste emission at the source.

Wastewater is generated in mines through seepage from excavated areas. The mine water is collected in sumps with an opportunity for recycling for firefighting, dust suppression within the mining area (Figure 6.18) in crushers (Figures 6.19 and 6.20). The quantity and quality of the mine water depends on the level of the groundwater table and the ground conditions. The major pollutants in the mining sector are dissolved solids. In the case of acid mine drainage, the pH can be in the range 2 to 3 due to the oxidation of sulfides and it needs neutralization. Radiological pollutants are specific to mining of radioactive materials.

Air-pollution control is another important are where wastewater can be of much use without affecting efficiency of air-pollution control equipment. Particulate matter can be scrubbed from processes by partially treated water whose function is to trap particles and settle. On the other hand. alkaline water is preferred to scrub acidic fumes. The evaporation loss from the scrubber can be made up by partially treated water.

6.3.4.1 Boiler Feed

Boiler feed water at the refineries need a much higher level purity. It is therefore treated by using reverse osmosis preceded by microfiltration. This process reduces mineral content. Total dissolved solids (TDS) can be reduced from 800–900 to 50 ppm and second-stage reverse osmosis can be installed if greater purity is required. It is essential that corrosion-free piping and storage arrangements be made to avoid building up of TDS due to the corrosion or solubility of constructed material (Lazarova *et al.*, 2013).

Figure 6.18 *Water spraying for suppressing dust.*

6.3.4.2 Cooling Tower

Cooling towers use lot of water. Recycled water has been used in many industries successfully since early 1970s. It is now used extensively in chemicals/power/steel/refineries/auto and commercial/institutional systems. A 1000 ton air conditioner machine evaporates about

Figure 6.19 *A crusher hopper in mining area.*

Figure 6.20 *Water spraying to suppress dust during crushing.*

115 m^3/day discharges about 35 m^3/day and, hence, requires makeup of 150 m^3 per day. A refinery circulating 500 m^3 per vapour minute evaporates about 16 000 m^3/day, discharges about 4000 m^3 /day, requiring makeup of 20 000 m^3/day. A large power plant circulating 1500 m^3 per vapour minute evaporates about 43 500 m^3/day, discharges about 9500m^3/day requiring makeup of 53 000 m^3/day.

Reusing wastewater saves fresh water, reduces wastewater in the environment and lowers the consumption of fresh water. At the same time critical parameters should be considered during operation of cooling towers (Table 6.3).

Table 6.3 *Maximum concentration allowed in circulating water.*

Constituent	Concern	Maximum permissible limit (except pH)
Silica	Scaling	150 mg/l, SiO_2
Calcium	Scaling	400 mg/l as Ca, 1000 mg/l as $CaCO_3$
Phosphates	Scaling	12 mg/l
Chloride	Corrosion	900 mg/l Cl
Suspended Solids	Deposition	10 mg/l
pH	Scaling, corrosion	6.5 to 8.5
Alkalinity	Scaling	200
Total hardness	Scaling	500 mg/l as $CaCO_3$
Carbonate hardness with addition of stabilizers	Scaling	1000 mg/l as $CaCO_3$
Iron	Staining	0.3 mg/l
TDS	Scaling	2000 mg/l

Cooling towers can use treated municipal effluent, in-plant 'used' water like softener rinse, boiler blowdown, condensate, RO reject, grey water and cooling tower blowdown with pretreatment.

The thermal efficiency and longevity of the cooling tower depends on the proper management of recirculated water. Water from cooling towers leaves in four ways:

1. **Evaporation**: conversion of water to water vapour that transfers heat from water in the cooling tower to the environment.
2. **Drift**: water carried away as mist or small droplets from a cooling tower.
3. **Blowdown or bleed-off**: water let out from a cooling tower to control dissolved solids.
4. **Basin leaks or overflows**: loss of water due to cracks/holes and overflow.

The following steps can help increase water-use efficiency:

- Install a side-stream filtration system with a high-efficiency cartridge filter or rapid sand filter to clean the circulated water.
- Install a side stream softening system or make-up water when dissolved solids are the limiting factor.
- Install covers to block sunlight.
- Consider water treatment options, like ionization or ozonation, to reduce usage of chemicals and water.
- Install automated chemical feeding in the case of large cooling towers.

6.3.4.3 Quenching

'Quench' in English means 'extinguish'. In material science, quenching means rapid cooling of a work piece to get certain properties. Quenching increases the toughness of alloys and plastics. Quenching ash and slags do not need high quality water. The ash handling water can be of poor quality and, hence, wastewater from other streams can be used for the purpose. The water requirement varies from operation to operation. Coke quenching needs 2.5 cum/tonne coke whereas steel making needs 2.0 kilolitres per tonne.

6.3.4.4 Vehicle Washing

Washing of car/passenger vehicles and food-carrying vehicles need water without toxins/pathogens and should not leave stains. Waste-transportation vehicles need lesser quality water for washing, making it possible to use low-quality water. Some countries have progressed in reuse of wastewater by setting up laws, whereas other nations lack adequate planning and laws. Switzerland, Germany and the Netherlands do not allow car washing at home (Janik and Kupiec, 2007).

6.3.4.5 Dust Suppression

Dust suppression is required when crushing stones, dusty roads, in mining areas (Figure 6.21), when handling ash or waste and some raw materials like coal. Water sprinkling

Figure 6.21 *Water sprinkling in an open cast mining area.*

in such circumstances does not need high-quality water and, hence, wastewater can be used easily with minimum or no treatment depending on the quality of wastewater.

6.3.4.6 Solid-Waste Management

Solid waste includes municipal solid waste, hazardous waste, biomedical waste, end-of-life vehicles and waste from electrical equipment. Water is required for composting, dust suppression, fire extinguishers and air pollution control. The water is required at the point of generation, transfer, transportation, waste processing and disposal. At the point of generation/transfer/transport it may be required for dust suppression but waste with organic content including wastewater with high BOD (like wastewater from distilleries) can be used for composting thereby using wastewater and solid waste to treat each other.

Figure 6.22 shows an ash stabilization and solidification plant where ash is mixed with appropriate solidifying agents/cement before being sent to landfill sites so as to avoid toxic leachate formation. The unit does not need high-quality water and wastewater can be easily used after preliminary treatment.

Figure 6.23 shows a landfill used for the final disposal of solid waste. An engineered (or sanitary) landfill facility is a waste-management disposal system providing long-term confinement of solid waste. A typical landfill will undergo planning, site selection, site preparation, bed construction, leachate and gas-collection system construction, land filling, monitoring, closure of landfill operations and postclosure monitoring. Each stage of landfill construction (except planning and site selection) needs water, which can be sourced from wastewater treatment. Similarly, water is required for suppression during handling of waste in a waste-to-energy plant (Figure 6.24). Partially treated wastewater can be used for dust suppression.

Figure 6.22 *Ash stabilization and solidification unit.*

Figure 6.23 *Landfill.*

Figure 6.24 *A waste-feeding place for converting waste to energy.*

References

ABS (2006) Water Account Australia 2004–05. Australian Bureau of Statistics, Sydney.

Allen, L., Christian-Smith, J. and Palaniappan, M. (2010) Overview of Greywater Reuse: The Potential of Greywater Systems to Aid Sustainable Water Management, Pacific Institute, Oakland.

Amaral, E., Brockman, K. and Forsström, H.G. (2007) International perspectives on spent fuel management, in Management of Spent Fuel from Nuclear Power Reactors, Proceedings of the International Conference on Management of Spent Fuel from Nuclear Power Reactors, organized by the International Atomic Energy Agency in cooperation with the OECD Nuclear Energy Agency, Vienna, 19–22 June 2006.

Angelakis, A.N., Bontoux, L. and Lazarova, V. (2003) Challenges and prospectives for water recycling and reuse in EU countries. *Water Science and Technology. Water Supply* **3** (4), 59–68.

Angelakis, A., Marecos, M., Bontoux, L. and Asano, T. (1999) The status of wastewater reuse practice in the Mediterranean basin: need for guidelines. *Water Research* **33** (10), 2201–2218.

Ayni, F.E., Cherif, S., Jrad, A. and Trabelsi-Ayadi, M. (2011) Impact of treated wastewater reuse on agriculture and aquifer recharge in a coastal area: Korba case study. *Water Resources Management* **25** (9), 2251–2265.

Bahri A (1999) Agricultural reuse of wastewater and global water management. *Water Science and Technology* **40** (4–5), 339–346.

Berkun, M. (2005) Effects of Ni, Cr, Hg, Cu, Zn, Al on the dissolved oxygen balance of streams. *Chemosphere* **59**, 207–215.

Bhargava, D.S. (1983) Most rapid BOD assimilation in Ganga and Yamuna rivers. *Journal of Environmental Engineering* ASCE, **109** (1), Feb., 1983, Paper No. 17674, pp. 174–188.

Bhargava, D.S. (1986) DO Sag model for extremely fast river purification. *Journal of Environmental Engineering* **112**, 572–585.

Bonomo, L., Nurizzo, C. and Rolle, E. (1999) Advanced wastewater treatment and reuse: Related problems and perspectives in Italy. *Water Science and Technology* **40** (4–5), 21–28.

Burke, L., Selig, L. and Spalding, M. (2002) Reefs at Risk in Southeast Asia, World Resources Institute, Washington DC.

Chandrappa, R., Gupta, S. and Kilshrestha, U.C. (2011) *Coping With Climate Change*, Springer-Verlag, New York.

Clark J W and Viessman W (1965) *Water Supply and Pollution Control*, International Textbook Company, Seranton, pp. 387–390.

Corcoran, E., Nellemann, C., Baker, E. *et al.* (eds) (2010) Sick Water ? The Central Role of Wastewater Management in Sustainable Development. A Rapid Response Assessment. United Nations Environment Programme, UN-HABITAT, Arendal.

Ensink, H.H., Mehmood, T., Van Der Hoeck, W. *et al.* (2004) A nation-wide assessment of wastewater use in Pakistan: an obscure activity or a vitally important one? *Water Policy* **6**, 197–206.

Faby, J.A., Brissaud, F. and Bontoux, J. (1999) Waste water reuse in France: water quality standards and wastewater treatment technologies. *Water Science and Technology* **40** (4–5), 37–42.

Fonder, N. and Headley, T. (2010) Systematic nomenclature and reporting for treatment wetlands, in *Water and Nutrient Management in Natural and Constructed Wetlands* (ed. J. Vymazal), Springer, Dordrecht, pp 191–220.

Friedler, E. (2001) Water reuse – an integral part of water resources management: Israel as a case study. *Water Policy* **3**, 29–39.

Geiger, W.F. (2006) Sustainable water management – chance for mega cities? Sustainable Water Management, Problems and Solutions under Water Scarcity, International Conference, Beijing, People's Republic of China, 6–8 November 2006.

Gil, R.J., Ramos-Rodriguez, O. and Raine, N.E. (2012) Combined pesticide exposure severely affects individual and colony-level traits in bees. *Nature* **491**, 7422, 105-108.

Heinz I, Salgon M and Koo-Oshima S (2011) Water reclamation and intersectoral water transfer between agriculture and cities – a FAO economic wastewater study. *Water Science and Technology* **63** (5), 1067–1073.

IAEA (1983) Treatment of low and intermediate level liquid radioactive waste, technical report series no. 236, IAEA, Vienna, p. 62.

Jang, T., Lee, S.B., Sung, C.H. *et al.* (2010) Safe application of reclaimed water reuse for agriculture in Korea. *Paddy and Water Environment* **8** (3), 227–233.

Janik, H. and Kupiec, A. (2007) Trends in modern car washing. *Polish Journal of Environmental Studies* **16** (6), 927–931.

Lattemann, S. and Hopner, T. (2008) Environmental impact and impact assessment of seawater desalination. *Desalination* **220** (1–3), 1–15.

Lazarova, V., Asano, T., Bahri, A. and Anderson, J. (2013) *Milestones in Water Reuse The Best Success Stories*, IWA Publishing, London.

Li, W.H. (1972) Effect of dispersion on DO-Sag in uniform flow. *Journal of Sanitary Engineering Division, American Society of Civil Engineers* **98** (SA1), 169–182.

Lu, W.-Q., Xie, S.-H., Zhou, W.-S. *et al.* (2008) Water pollution and health impact in China: a mini review. *Open Environmental Sciences* **2** (1), 1–5.

Marecos do Monte, M.H. (2007) Water reuse in Europe, *E-Water* 1–18, http://www.ewa-online.eu/tl_files/_media/content/documents_pdf/Publications/E-WAter/documents/21_2007_07.pdf (accessed 6 January 2014).

Margeta, J and Vallée, D. (2000) Mediterranean Vision on Water, Population and the Environment for the 21st Century. Blue Plan. Mediterranean Action Plan, UNEP, GWP/MEDTAC, Montpellier.

McCutcheon, S.C. (1989) *Water Quality Modeling, Vol. I Transport and Surface Exchange in Rivers*, CRC Press, Boca Raton, FL, p. 141.

MED WWR WG (Mediterranean Wastewater Reuse Working Group) (2007) *Mediterranean Wastewater Reuse Report*, http://ec.europa.eu/environment/water/water-urbanwaste/info/pdf/final_report.pdf (accessed 13 January 2014).

Mills, S.W., Alabaster, G.P., Mara, D.D. *et al.* (1992) Efficiency of faecal bacterial removal from waste stabilization ponds in Kenya. *Water Science and Technology* **26** (7–8), 1739–1748.

Nassehi, V. and Das, D.B. (2007) *Computational Methods in the Management of Hydro-Environmental Systems*, IWA Publishing, London, ISBN: 1-84339-045-0, ISBN_13: 978-1-84339-045-9.

Norton-Brandão, D., Scherrenberg, S.M. and van Lier, J.B. (2013) Reclamation of used urban waters for irrigation purposes - A review of treatment technologies. *Journal of Environmental Management* **122**, 85–98.

Ongerth, H.J. and Jopling, W.F. (1977) Water reuse in California, in *Water Renovation and Reuse* (ed. H.I. Shuval), Academic Press, New York.

Peavy, H.S., Rowe, D.R. and Tchobanoglous, G. (1986) *Environmental Engineering*, McGraw-Hill International, Singapore.

Pedrero, F., Kalavrouziotis, I., Alarcon, J.J. *et al.* (2010) Use of treated municipal wastewater in irrigated agriculture e review of some practices in Spain and Greece. *Agricultural Water Management* **97** (9), 1233–1241.

Qadir, M., Bahri, A., Sato, T. and Al-Karadsheh, E. (2010) Waste water production, treatment and irrigation in Middle East and North Africa. *Irrigation and Drainage Systems*, **24** (1–2), 37–51.

Radcliffe, J.C. (2010) Evolution of water recycling in Australian cities since 2003. *Water Science and Technology* **62** (4), 792–802.

Ramalho, R.S. (1977) *Introduction to Wastewater Treatment Process*, Academic Press, New York.

Rhee, H.P., Yoon, C.G., Son, Y.K. and Jang, J.H. (2011) Quantitative risk assessment form reclaimed wastewater irrigation on paddy rice field in Korea. *Paddy and Water Environment* **9** (2), 183–191.

Salgot, M. and Pascual, A. (1996) Existing guidelines and regulations in Spain on wastewater reclamation and reuse. *Water Science and Technology* **34** (11), 261–267.

Sharma, V., Bhadula, S. and Joshi, B.D. (2012) Impact of Mass Bathing on water quality of Ganga River during Maha Kumbh-2010. *Nature and Science* **10** (6), 1–5.

Singh, A., Sharma, R.K., Agrawal, M. and Marshall, F.M. (2010) Health risk assessment of heavy metals via dietary intake of foodstuffs from the wastewater irrigated site of a dry tropical area of India. *Food and Chemical Toxicology* **48** (2), 611–619.

Standen, R. (2000) Developing irrigation guidelines for wastewater irrigation, 63rd Annual Water Industry Engineers and Operators' Conference Civic Centre – Warrnambool, 6–7 September 2000.

Streeter, H.W. and Phelps, E.B. (1925) A Study of the Pollution and Natural Purification of the Ohio River, III. Factors Concerned in the Phenomena of Oxidation and Reaeration. Bulletin 146, US Public Health Service, Washington, DC.

Tanner, C.C. and Headley, T.R. (2011) Components of floating emergent macrophyte treatment wetlands influencing removal of stormwater pollutants. *Ecological Engineering* **37**, 474–486.

Tchobanoglous, G., Burton, F.L. and Stensel, H. (2003) *Wastewater Engineering, Treatment and Reuse*, Tata McGraw-Hill Ltd, New Delhi.

Tebbutt, T.H.Y. and Berkun, M. (1976) Respirometric determination of BOD. *Water Research* **10**, 613–617.

Thomas, H.A. (1957) Hydrology and oxygen economy in stream purification. Seminar on Waste Water Treatment and Disposal, Boston Society of Civil Engineers, Boston, MA.

Thomas, M. and Matiasi (2012) Wastewater management and reuse for sustainable development. *Scholarly Journal of Agricultural Science* **2** (11), 269–276.

Tu, W., Niu, L., Liu, W. and Xu, C. (2013) Embryonic exposure to butachlor in zebrafish (*Danio rerio*): Endocrine disruption, developmental toxicity and immunotoxicity. *Ecotoxicology and Environmental Safety* **89**, 189–195.

UNEP (2008) Desalination Resource and Guidance Manual for Environmental Impact Assessments, United Nations Environment Programme, Regional Office for West Asia, Manama and World Health Organization, Regional Office for the Eastern Mediterranean, Cairo.

UNEP (2013a) Global Mercury Assessment 2013 Sources, Emissions, Releases and Environmental Transport, UNEP Chemicals Branch, Geneva.

UNEP (2013b) *Water and Wastewater Reuse, An Environmentally Sound Approach for Sustainable Urban Water Management,* http://www.unep.or.jp/Ietc/Publications/Water_Sanitation/wastewater_reuse/Booklet-Wastewater_Reuse.pdf (accessed on 16 May 2013).

Weragoda, S.K., Jinadasa, K.B.S.N., Zhang, D.Q. *et al.* (2011) Tropical application of floating treatment wetlands. *Wetlands* **3** (4), 955–961.

World Bank (2011) Water Reuse in the Arab World from Principle to Practice. A Summary of Proceedings Expert Consultation Wastewater Management in the Arab World 22–24 May 2011 Dubai-UAE.

7

Sustainable Construction of Water Structures

Water is not just required for day-to-day activities. It provides livelihoods to more than 50% of the world's population. Occupations depending on water include plumbers, researchers, water-treatment plant operators, ships' crews, water vendors, farmers, fishermen, crews in water amusement parks, construction labourers, vehicle washers, sanitary fixture manufacturers/traders directly depend on water. Further, fish/flower/vegetable/fruit and bottled-water traders depend on farmers and manufacturers for survival. As discussed earlier, metal manufacturing and other activities including software development need water in some form.

A wide range of human/ecological crises affect freshwater resources. As human settlements grow, these problems become more serious and frequent. Construction activities have considerable impacts on the environment including the water resources. Construction has been accused of causing environmental degradation through excessive consumption of resources, carbon emission and ecological destruction. However, little or no concern has been given to the selection of more environmentally friendly designs at the project appraisal stage when environmental issues are best incorporated (Ding, 2008).

There have been noteworthy trends worldwide towards urbanization but in some of the developed countries some people are abandoning the cities and migrating to smaller communities for better living, which can become basis of sustainable living. In most African and Asian nations, people migrate from rural to urban areas for many reasons, for example, conflict with growing families, more dignity/equality in the workplace in urban areas, a larger number of employers to choose from, more opportunities, higher salary for comparatively less work, dignity associated with urban living and so on. With a greater number of people migrating to urban areas, new problems arise for the original inhabitants and the new migrants. Instead of designing policy to encourage people to stay in rural areas or small towns, governments are often mesmerized by the incomes generated in mega cities.

Sustainable Water Engineering: Theory and Practice, First Edition. Ramesha Chandrappa and Diganta B. Das.

Construction poses economic challenges at several levels. In both developing and developed countries, sustainable construction is more difficult to achieve as the sector depends on the goals of the national economy, as infrastructure is a key necessity for development. Supply chains feeding construction activity are long and consume natural resources, stressing ecological settings. Buildings are created in shorter time periods in response to demand and are dominated by mechanical, electrical and electronic equipment (Bon and Hutchinson, 2000). As human settlements are the key polluters of water resources, water and wastewater management is necessary to limit pollution and reduce health risks.

The biophysical environment, including the atmosphere, underground resources, land, flora, fauna and the built environment, should not be sacrificed to fulfil water requirements. Industrial pollution and waste are endangering water resources and ecosystems worldwide. Even though wastewater treatment plants aim to control water pollution, construction activity itself can cause water pollution apart from air and noise pollution. Conventional materials like asphalt are being mixed with old rubber from tires and ground glass to achieve sustainable construction. Drainage systems with pieces of polystyrene, held in place with plastic netting, make constructions more ecofriendly. The use of plywood and dimensional lumber for formwork often results in the cutting of forests. Hence, permanent formworks with rigid plastic foam are used to make construction eco-friendly.

Discussion of sustainable water engineering is not complete without considering hydroelectricity, which depends on water for the generation of electricity. Hydropower, which contributes 19% of global electric power generation, is known to cause heavy destruction of the environment, which can be avoided with run-of-river hydropower stations, which are affordable and sustainable. Energy-intensive water delivery systems have many different impacts. They not only have energy demands but also encourage unsustainable practices. Reductions in energy use in the water/wastewater sector need to be encouraged for sustainable water management.

Climate change is not considered seriously while designing and choosing sites, material and technology. Unsustainable practices during construction, like freshwater use for buildings and washing fertile soil to separate sand for construction, will deplete water resources and reduce the fertility of the soil. There is also pressure on the mountains and hills due to mining and quarrying activity necessary for the supply of construction material. Development planning usually neglects climate change and, as a consequence, disasters due to climate change cause economic and ecological disruption. Hence, planners should consider probable hazards due to climate change; this can reduce risk from extreme events. Although humans cannot change the magnitude/frequency of natural disasters, they can reduce the severity of hazards. Humans can save mangroves and forests, which act as barriers against disasters. Devastation of endangered species can be avoided by restricting development/tourism in the areas where they live. Risk assessment helps to (i) identify and understand risks in the area, (ii) earmark funds needed, (iii) identify alternative solutions, (iv) identify knowledge gaps, (v) make decisions to deal with risks, (vi) identify risks to new projects and (vii) plan mitigation.

Important information needed in hazard and risk assessment is (i) the location and size of the area of interest, (ii) the severity of hazards and (iii) the probability and frequency of hazard occurrence. Planners should be aware of secondary hazards (like landslides and soil erosion) and problems that could occur due to hazards outside the project location (like power disruption).

7.1 Sustainable Construction – Principles

The concept of technology assessment (TA) (Figure 7.1) was introduced in the 1960s due to the enhanced importance of technology in society and its potential consequences. Technological developments are usually driven by market forces and government support. The process of TA requires that it foresee future scenarios and evaluates the impact on society and the environment. Technologies that can be detrimental can be eliminated or altered so that they no longer present unacceptable risks (Khan, 2005).

There is no specific model for carrying out TA and the methodology varies from case to case. One such example is given in Table 7.1. Theoretically TA may look straightforward but it is essential that one should have sufficient knowledge of the subject to analyse the impact. Structures are often designed by international consultants or architects who usually will not know the local situation well. The construction boom and the demand for sand in Bangalore resulted in the alteration of the natural shape of many rivers and banks located within about 100 km from Bangalore. Some of the sand suppliers even used fertile silt in farmlands to supply sand for construction, thereby reducing the fertility of the land and wasting water in the locality. The demand for coarse aggregates resulted in the establishment of stone crushers surrounding Bangalore. Many hillocks that were part of the natural terrain were destroyed to fulfil the demand for coarse aggregates.

Table 7.2 gives a number of conservative estimates of the weights of selected Indian cities, which have been built by bringing building material (cement, steel, sand, stones and so on). The impact of these cities on environment is still not fully understood, nor are the possible effects of placing so much weight on geological structures without assessing the ability of the structures to sustain the load placed on them. The withdrawal of groundwater and weakening geological structures due to the drilling of innumerable bore wells is yet to be understood.

Green or sustainable buildings have become prevalent not only in the United States but all over the world. Buildings that are designed, constructed, maintained and demolished without harming the environment are known as 'green buildings'. The US Green Building Council recognizes such buildings and rates buildings using the Leadership in Energy and Environmental Design (LEED) rating, which was released in 2000. LEED-certified

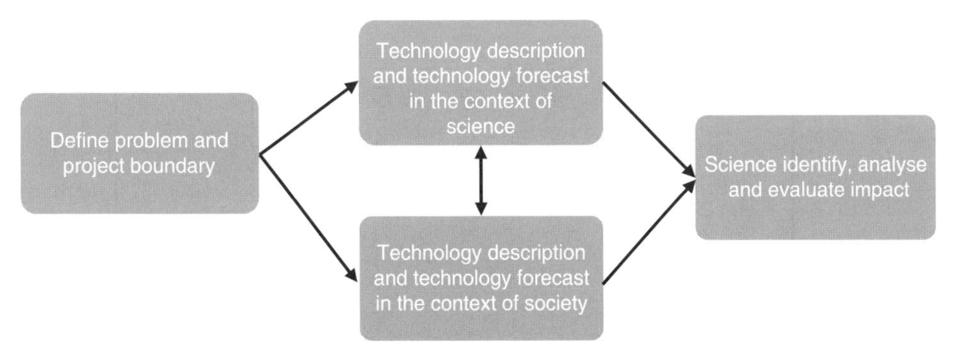

Figure 7.1 *Technology assessment.*

Table 7.1 *Example of technical evaluation.*

Process	Factor	Impact	Rating			Risk value	Risk condition
			Severity	*Probability*	*Period*		
Anaerobic treatment plant	Effluent	Human/animal health,	3	2	2	12	Medium
		Environment	3	2	4	24	High
	Methane and carbon dioxide	Human/animal health,	3	2	4	24	High
		Environment	4	4	4	64	High
	Hydrogen sulfide	Human/animal health	4	3	4	48	High
		Environment	4	4	4	16	Medium
	Construction	Human/animal health	4	3	4	48	High
		Environment	4	4	4	16	Medium

Severity of hazard	Probability of exposure to hazard	Period of exposure to hazard
1: No impact	1: Impossible	1: Very brief
2: Short term	2: Possible but unlikely	2: Short period
3: Long term	3: Possible and likely	3: Long period
4: Permanent	4: Definite	4: Continuous

Table 7.2 *Conservative estimation of weights of selected Indian cities.*

Sl. no.	City	Approximate area of city (km^2)	Approximate average height of city (m)	Specific gravity of buildings (t/m^3)	Weight of the city million (t)
1	Delhi	1484	10	2	29 680
2	Mumbai	603	20	2	24 120
3	Chennai	1189	10	2	23 780
4	Bangalore	741	10	2	14 820
5	Pune	710	10	2	14 200
6	Hyderabad	650	10	2	13 000
7	Jaipur	112	6	2	1344

buildings are on the rise but are usually restricted to residential/commercial and industrial buildings but not water and wastewater infrastructures.

The main principles of sustainable construction consist of:

- assessment of proposed activities;
- involving people affected in decision making;
- using lifecycle frame work;
- sustainable site management;
- extracting fossil fuels/minerals at rates not faster than their redeposit;
- selection of ecofriendly construction material;
- sustainable transportation;
- reducing the use of energy, materials, water and land;
- use of non-ozone-depleting substances;
- low carbon footprint;
- reuse of construction and demolition debris;
- avoiding unnecessary construction;
- minimizing soil/air/water/noise pollution;
- minimal disturbance to native flora and fauna by maintaining and restoring ecological diversity;
- maximizing resource reuse/recycling, as this leads to a reduction in waste;
- preferential use of renewable resources instead of nonrenewable resources and
- elimination/management of the use of hazardous or toxic products.

It is often thought that environmental impact assessment (EIA) can achieve sustainable projects and minimum degradation of environment due to construction. It has been incorporated into legislation and enforced globally in several countries but, due to poor understanding of the concepts and lack of professionalism of people who prepare and appraise EIA reports, it has resulted in huge damage to the environment. Environmental impact assessment legislation may have become another way to raise funds for political parties in power and to increase the wealth of corrupt bureaucrats. The EIA procedure needs a thorough understanding of baseline data, the proposed project and the impact due to the project but, as the EIA studies are conducted by consultants who are paid by clients, the reports do not reject upcoming projects on the proposed site as investor usually buys land

Figure 7.2 Entrance to wastewater treatment plant located underground at Stockholm. Such a location can avoid freezing of water during a severe winter and save construction costs.

and appoints consultants to prepare the EIA. It is also common practice to split the projects into many smaller projects to avoid statutory obligations for EIA studies. Even the projects funded by governments may have pressures to utilize funds and complete the project. Due to nonavailability of quality baseline data in the developing countries, false baseline data are often generated to project 'all-is-well' scenarios to appraisal committees/authorities.

Water and wastewater treatment plants are often built on land. The plant in Stockholm shown in Figures 7.2 and 7.3 is unique in the way that it is carved out in the tunnel underneath the ground, thereby saving construction, space, material and energy to keep the treatment plant warm during winter.

Figure 7.3 Inside view of a wastewater treatment piping located underground in Stockholm.

Table 7.3 Causes of construction waste.

Sl. no.	Construction material	Cause	Unsustainable practice
1	Stone tablets	Cutting, shaping, storage and handling, order too much	Waste-causing choices in design, imperfections in the product, choices, method of transportation, lack of knowledge during design activities, nonavailability of quality supply, order too much, improper storage and handling, unpacked supply
2	Concrete	Ordering/mixing too much, scraping off, loss during transportation	Quantity of products not known due to improper planning
3	Mortar	Scraping out, messing	Negligent practice.
4	Steel	Overordering/oversize	Lack of design/quantification knowledge, unethical practice
5	Brick	Overordering, poor quality	Negligent practice
6	Cement	Improper handling/storage	Negligent practice
7	Sand	Improper handling/storage	Negligent practice

Source: based on Bossink and Brouwersz, 1996, Chandrappa and Das, 2012.

Drinking-water treatment/distribution and wastewater collection/treatment/disposal/reuse projects usually do not come under the purview of EIA legislation in almost all the developing world. As a result, bore wells are drilled and dams are built without any consideration of the future, which is a cause of conflict in many parts of the world.

Construction and demolition (C and D) activities generate waste for numerous reasons listed in Table 7.3. The quantity of C and D waste depends on size of the activity. Most of this waste is used within the site for filling lower areas. When the waste exceeds the capacity at the site it needs to be transported to other another site.

7.1.1 Green Building

Green building involves restoring damaged and/or polluted environments and shifting to cyclical processes (Miyatake, 1996). To achieve sustainable construction, construction activity must change the processes of constructing built environments to a cyclical (reusing construction and demolition waste) process. Energy and materials from the earth or the natural environment is converted into built environments, generating debris and requiring demolition after the structures have finished their lives.

The principles of sustainable construction are (i) use of fuel-efficient vehicle spaces, (ii) sustainable stormwater management, (iii) adoption of lower of exterior lighting, (iv) water-efficient landscaping, (v) reduction in water use, (vi) optimizing energy performance, (vii) enhanced refrigerant management by using non-ozone-depleting substances, (viii) use of

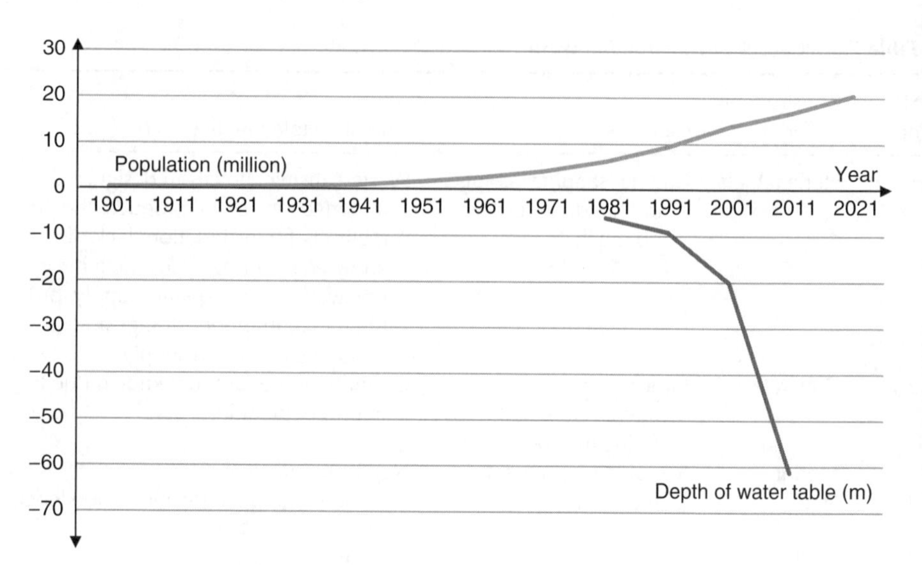

Figure 7.4 *Population growth versus groundwater depletion in Delhi (Source: based on India Today, 2012; India Online pages, 2013). (For a colour version of this figure, see the colour plate section.)*

low air-pollutant-emitting materials (including adhesives, sealants, paints and coatings), (ix) reduce wastage of water consumption.

The Yamuna River catchments up to Delhi consist of two parts – the upper catchment from the Himalayas and lower catchment. The flow of the Yamuna depends on discharge from Tajewala Headwork located 240 km upstream. During heavy rain in the catchment area, surplus water is released from Tajewala, creating flooding in Delhi. A rise in the water level causes backflow into the city's drains. Delhi also experiences floods due to an increase in the permeability of soil and the overflow of drains in the city. Protection from the river by dams leads to a false sense of safety. Development takes place in the shadow of embankments and in the event of failure of these protective works, people on downstream side suffer.

Delhi occupies an area of about 1500 km^2, which is supplied with water from the Yamuna River and groundwater. The quantity of water from the ground is roughly 1/8th of the total water drawn. The depth of the water in the city varies from 1.2 m to more than 64 m. The groundwater is declining at the rate of 1–2 m/year in some areas of the city on account of overexploitation. Figure 7.4 shows a graph of population growth versus groundwater depletion at most groundwater-dependent areas in Delhi (source based on India Today, 2012; India online pages, 2013).

Bank filtration is used along with wastewater reclamation and artificial aquifer recharging, enhancing Berlin's groundwater resources. About 65 750 m^2 of green roofs were installed through a greening programme to reduce the following concerns: (i) hydraulic stress on stormwater drains, (ii) a lack of humidity and excess warming, (iii) a decrease in flora and fauna, (iv) a high degree of soil sealing, (v) inadequate replenishment of groundwater. Berlin's residents received subsidies for investing in green roofs. The public water companies established three groundwater recharge plants to increase groundwater quantities. Water authorities in West Berlin introduced a water demand-management strategy to reduce per capita consumption by introducing higher tariffs for water, effective publicity campaigns, temporary subsidies for the installation of water-saving equipment, and measures to reduce leakages and losses due to pipe breakages (Salian and Anton, 2013).

7.1.2 Cementless Construction

Since its invention, cement has been important in binding various construction materials. The advantage of concrete resulting from its ability to be shaped and moulded has made cement the main construction material. About 260 million tons of cement are required every year all over the world, which will be increased by 25% by 2022 (Abdul Aleem and Arumairaj, 2012). Acute shortage of limestone, which is the main raw material of ordinary Portland cement, may occur between the years 2035 to 2065. Cement production is one of the major carbon emitters and energy consumers.

There are several ways in which one can achieve cementless construction in water/wastewater treatment plants (Figure 7.5). Some of the these are: (i) construction of earthen tanks lined with geomembranes (Figure 7.6), (ii) using metal sheets/fibreglass sheets for small water /wastewater treatment plants, (iii) using geopolymer concrete, (iv) constructing water/wastewater plants in tunnels or on rocks (Figure 7.7), (v) using alternate materials to cement.

Many ecofriendly construction materials have been invented over the year to conserve resources and the environment. Aluminosilicate gel, produced with fly ash and alkaline solution, uses binding material to produce geopolymer concrete, which is an alternative construction material that is gaining popularity worldwide. Advantages of geopolymer concrete are: (i) it sets at room temperature, (ii) it has a long working life before stiffening, (iii) it is nontoxic and bleed free, (iv) it is impermeable, (v) it has higher compressive strength, (vi) it has higher resistance to heat and inorganic solvents. Limitations of geopolymer concrete are: (i) the need for transportation of fly ash to the required location, (ii) safety risks, (iii) the high cost of the alkaline solution, (iv) practical difficulties in applying the high-temperature, steam-curing process.

Apart from construction material, construction technology has also witnessed changes. During construction, manmade elements are being integrated in the soil to improve its behaviour and this is known as inclusion. Examples of inclusions are: (i) steel strips, (ii) geotextile sheets, (iii) steel or polymeric grids, (iv) steel nails and (v) steel tendons between anchorage elements. Multiple layers of inclusions act as reinforcement in soils; they are placed as fill and this technology is referred as a 'mechanically stabilized earth wall'.

Mortar prepared by Hwangtoh (clay in South Korea created by the weathering of rocks) can act as a highly effective binder (Yang *et al.*, 2007). An alkali-activated, ground-granulated blast-furnace slag binder with a fly ash-based geopolymer binder concrete is an environmental-friendly structural concrete (Yang *et al.*, 2011). Other cementless construction material includes oil shale fly ash binder (Freidin, 2002), pressed blocks from the ashes of a coal-fired power station (Freidin, 2005) and mortars activated by sodium silicate (Yang *et al.*, 2008; Metakaolin, Rashad, 2013).

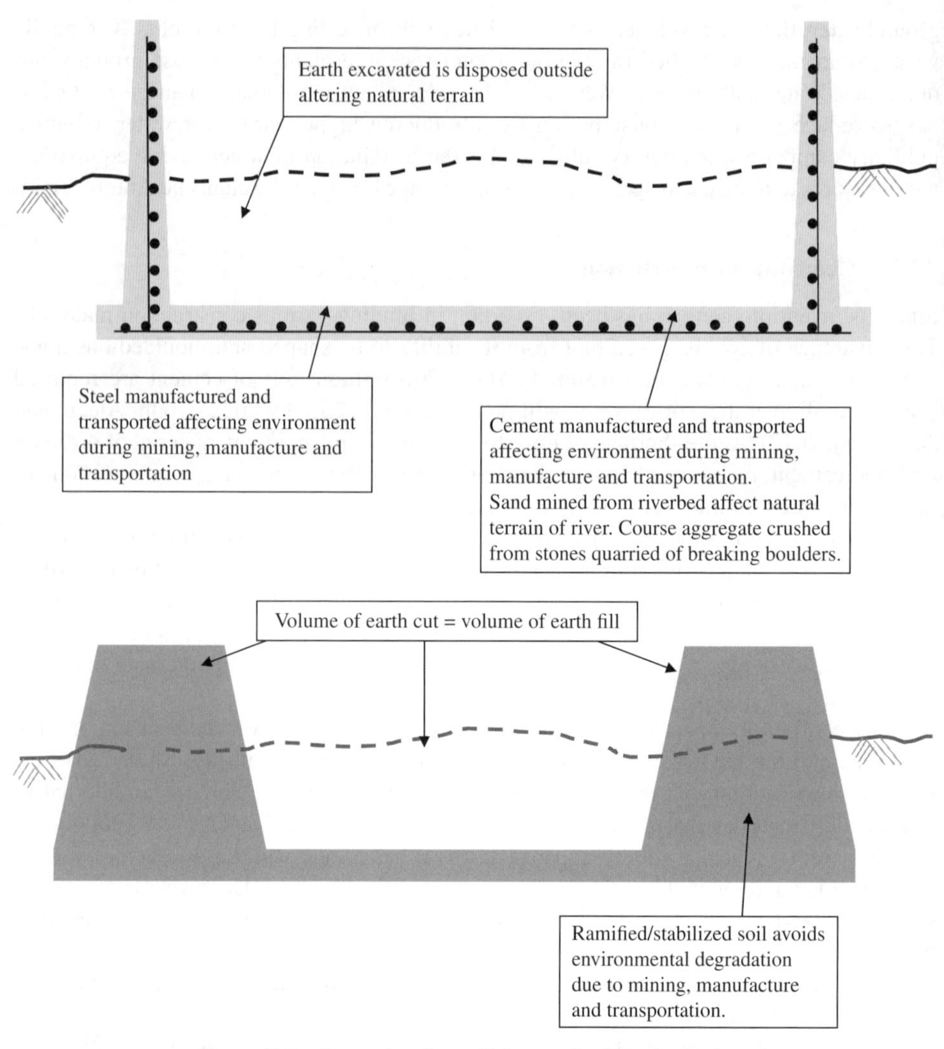

Figure 7.5 *Example of possible sustainable alternative.*

7.1.3 Choosing Eco-Friendly Construction Material

Eco-friendly constructions incorporate a wide range of concepts and strategies during the design and construction process. Eco-friendly materials (Table 7.4) are vital components of sustainable structures. They come from renewable resources and construction/demolition sites. Such materials conserve the environment and reduce the overall impact of resource extraction, transportation, processing and fabrication as well disposal of materials. They include: (i) bagasse board, (ii) building blocks from mine waste, (iii) bricks from coal washery rejects, (iv) burnt clay fly ash bricks, (v) compressed earth blocks, (vi) coir cement board, (vii) fibre flyash cement boards, (viii) eps composites and door shutters, (ix) fibre reinforced concrete precast elements, (x) fibrous gypsum plaster boards, (xi) flyash cellular

Figure 7.6 Earthen pond with liner to store leachate from landfill.

concrete/flyash cement blocks /brick, (xii) flyash lime gypsum brick, (xiii) flyash lime cellular concrete, (xiv) jute fibre polyester, (xv) insulating bricks ash of rice husk, (xvi) nonerodable mud plaster, (xvii) polytiles, (xviii) timber from fast growing trees, (xix) precastwalling roofing components, (xx) prefab brick panel system and (xxi) stabilized mud blocks (Figure 7.8).

7.1.4 Energy Saving during Construction

Energy saving during construction is something that environmental engineers often oversee. Energy saving need not be measured in lower electricity/fuel charges at a site. Transporting material from long distances often consumes unnecessary energy, so it is a waste of construction material. Adoption of local technology and material will result in a lower carbon

Figure 7.7 Inside view of wastewater treatment tanks located underground at Stockholm.

Table 7.4 Examples of eco-friendly construction material.

Sl. no.	Criteria	Description	Example
1	Reduce pollution	Use of materials with low VOC emissions	Cement paints
		Materials that prevent leaching	
		Materials that are less toxic	
		Materials that reuse waste	Slag from metallurgical industry
2	Energy conservation	Materials that require less energy during construction	Precast slabs
		Materials that help reduce the cooling loads	Aerated concrete blocks
		Products that conserve energy	Compact fluorescent (CF) lamp
		Fixtures and equipment that help conserve water	Dual flush cisterns
3	Recyclable	Reuse or recycle as different product	Steel, aluminium
4	Biodegradable	Materials that degrade biologically	Wood or earthen materials

footprint. Due to the oil crisis, energy saving in the construction and operation of buildings including effluent treatment plants has been an important issue. Energy consumption has been a significant impact on the environment due to emissions and byproducts.

Wastewater and water works are the last things to be considered for sustainable construction. Energy saving can be done by energy budgeting. Most of the energy at the construction site is required for: (i) transportation, (ii) lighting, (iii) mixing concrete, (iv) welding,

Figure 7.8 Stabilized mud blocks moulded using soil at site.

(v) earthworks, (vi) site preparation and (vii) pumping water. Energy is required off site for (i) manufacturing construction material and (ii) transportation of men and material.

7.1.5 Precautions to be Taken during Construction to save Energy during Operations

Energy consumption during operations can be reduced many times by proper planning. The major precautions to be taken during construction are: (i) installing energy-efficient pumps; (ii) installing pumps with the correct capacity; (iii) avoiding unnecessary bends and fixtures to reduce loss of pressure; (iv) submerging surface aerators optimally; (v) installing a blower or aerator with the correct capacity, (vi) placing house effluent treatment plants inside tunnels/buildings to avoid freezing and loss of heat; (vii) installing efficient membranes for reverse osmosis, membrane filtration and other membrane technology; (viii) avoiding unnecessary lighting; (ix) installing wind turbines or solar panels to provide an energy source; (x) making arrangements to feed an optimal dosage of chemicals; (xi) avoiding unnecessary pumping by proper hydraulic design; (xii) optimization of plants by choosing proper aerobic and anaerobic treatment technology; (xiii) energy recover from wastewater; (xiv) adopting nonanthropogenic energy-dependent technologies and methods like constructed wetlands, septic tanks, oxidation ponds and so on.

7.2 Intake Structures

An intake is a structure placed in a surface water body to withdraw water. Types of intake structures include intake towers, intake pipes, submerged intakes, movable intakes and shore intakes. Intake towers are used for big waterworks drawing water from reservoirs, lakes and rivers in which there is wide fluctuation in the water level or to avoid clogging.

Theoretically rainwater provides the best quality of water after few minute of scrubbing air pollutants. The best practice to maintain pure water resources is to avoid: (i) clearing of trees and vegetation, (ii) closed storage sheds, (iii) open defecation by humans and animals, (iv) fuel and chemical storage (v) application of agro chemicals, (vi) disposal of solid waste near pure water resources, (vii) restricting discharge of wastewater near pure water, (viii) septic tank installation and transportation activities. In a nutshell, there should be no human activities in the watershed area to ensure safe drinking water. In practice such a scenario is hard to achieve because most of the cities are built on fertile lands and coastal areas due to the abundance of food and jobs, which attracts many more people thereby placing pressure to improve infrastructure, which attracts further business and industries. Increased activity needs new and improved connectivity, which means engineers concentrate on passenger and goods safety rather than water issues. The increased activity generates solid waste, which will be disposed of on the outskirts of city. All developing countries have to undergo this change. While the developed countries make the best use of skilled manpower and experience, the developing counties can hardly afford such activities due to different priorities in life in the existing set up. The stress on food production increases the use of agro chemicals and regional and national policies are often made to address immediate requirement of vote banks rather than long-term water quality management. Hence, the government welcomes new industries, accepts pollution, safeguards polluters and provides subsidized agro chemicals and free/subsidized energy to farmers and the industrial community.

Reservoirs serve flood control, agricultural, navigation, recreation, water supply, hydropower generation and other purposes. More reservoirs should be built to cope with the population explosion and the increasing demands of the population. Five types of intake can control sediment, namely, intakes with sluice gates, tiered intakes, lateral intakes, bottom-grate-type intakes, bend-type intakes and combined intakes (Tan, 1996).

Water diversion works (for irrigation/water supply/flood control, etc.) usually consist of sluice gates, intake structures and other structures. The intake structure is generally located near sluice gates to ensure water flow into the intake structure. The intake may be parallel, oblique or at right angles to the river flow. Usually, a smaller angle will reduce the sediment entering an intake.

7.3 Treatment Plants

There is a tremendous shift in construction practices in the developed countries towards ecofriendly construction while the relevant technologies have yet to be picked by the developing countries. Civil engineering in the developing countries has yet to be tuned to achieve sustainable development. The reutilization of scrap is more eco-friendly that the use of steel manufactured from iron ore. The EU cement industry utilizes more than 31 million tons of fuels and secondary raw materials. Quarrying of raw materials for clay brick production in EU has been replaced to a limited extent (20%) by the use of waste materials and recycled aggregates, like paper pulp, and crushed clay brick. Although fired bricks have higher strength and water resistance than unfired clay, many situations do not require these properties and, hence, such bricks can be avoided.

Prefabrication of buildings and water tanks in the developed countries has led to significant material savings and lower quantities of construction waste. The practice has yet to be picked up in the developing countries. Off-site production is not affected by adverse weather conditions and ensures better product quality.

Prefabrication, easy handling and light materials save the environment and improve safety for workers (Van Holm *et al.*, 2011).

7.4 Water Storage and Distribution Systems

Water storage is essential for domestic, industrial and fire-fighting purposes. The type and capacity of water storage needed in a distribution system varies with the topography of the area, the size of the system, how the water system is laid and other considerations. The most common water storage structures are hydropneumatic tanks, buried reservoirs, ground-level reservoirs and elevated tanks. Water-storage tanks are used for operating storage or emergency storage. The storage tank is connected to the distribution system by a network of pipes (Figure 7.9). The elevation of the water in a storage tank is determined by the pressure required in the distribution system.

Unlike minerals and stones, a royalty is not charged for the extraction of groundwater stored in the earth in many of the countries. Groundwater is therefore considered as free resource and it is used without any plan to save it. Figure 7.10 shows the annual production of

Figure 7.9 Water pipes waiting to be laid in Bangalore, India.

different resources in 2001. At a withdrawal rate of 600–700 km^3/year, global groundwater is the world's largest extracted raw material (Zektser and Everet, 2004). Groundwater is the most important and safest source of drinking water. It is the main water supply source in many mega-cities, which include Bangalore, São Paulo, Mexico City and Bangkok. Groundwater provides nearly 70% of the water supply for the EU countries. Agriculture and industries in many countries depend on groundwater resources.

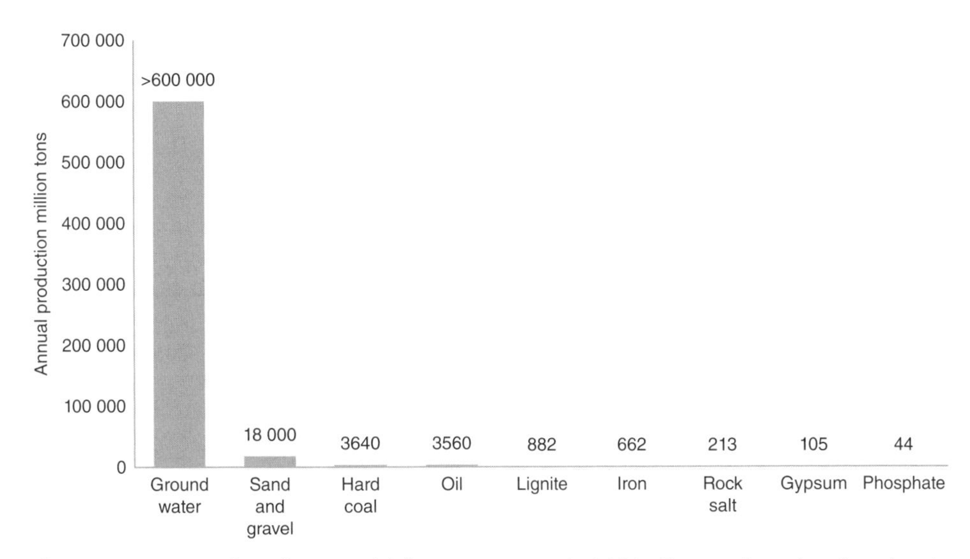

Figure 7.10 Annual production of different resources in 2001. (Source: based on Struckmeier et al., 2005.)

Table 7.5 Some of the longest pipelines supplying water to urban settlements.

From	To	Country	Length
Helena River	Kalgoorlie	Australia	530 km
Persian Gulf	Riyadh	Saudi Arabia	467 km
Fresh water wells in Oguz	Baku	Azerbaijan	265 km
Cauvery River	Bangalore	India	100 km

Groundwater depletion can lead to following problems: (i) increased pumping costs, (ii) a change of flow in rivers and other surface-water bodies as they receive a major proportion of their water from groundwater, (iii) vegetation or wetlands in the vicinity suffer notable changes, (iv) groundwater quality characteristics change, (v) there is land subsidence (Vrba and Lipponen, 2007).

Some of the longest pipelines supplying water to urban bodies are given in Table 7.5 and the approximate total pipeline length for water distribution in some of the major cities is given in Table 7.6.

Irrespective of water characteristics, pipe material or flow dynamics, biofilms unavoidably develop and persist on the inner surfaces of pipes, despite the presence of residual disinfectant (Ridgway and Olson, 1981). In addition to the likelihood of hosting pathogens, water biofilms are a source of coliforms of nonfaecal origin in distribution pipes (LeChevallier *et al.*, 1987) that lead to false positive tests and are hence cause for added uncertainty in the analysis of the water. Mature biofilms consist of microbial cell group embedded in a common glycocalyx, predominantly composed of polysaccharide materials that hamper the aesthetic properties of the water (AWWA, 1995). Drinking water pipe biofilms originate from microbial cells that contaminate the water (van der Wende *et al.*, 1989; Peyton and Characklis, 1992).

Broken water mains cause interruptions in water supply and loss of water. External corrosion, thinning of pipe walls over a period of time, frost loading due to low temperature, poor workmanship and low-quality pipes are causes of broken water pipes.

Unaccounted-for water (UFW) is an important overall indicator. Rates of UFW vary from city to city and system to system. Over the years, Singapore has lowered its UFW rate to about 5%. Rates of 20% to 40% are usual in many other countries. There is enough UFW in Mexico's water supply system to supply a city the size of Rome (Falkenmark

Table 7.6 Approximate length of total pipeline for water distribution in major cities.

City	Approximate length of total pipeline for water distribution (km)	Population (million)	Country
Toronto	5850	5.50	Canada
Tokyo	25 652	13.19	Japan
Chennai	2582	5.00	India
Fukuoka	2900	12.80	Japan
London	1400	8.00	United Kingdom
New York	10 500	19.00	United States

and Lind, 1993). Unaccounted-for water occurs due to physical losses and administrative losses. Physical losses are water lost from leaks in the system. Administrative losses are water used but not paid for.

Sustainable water solutions should aim to (i) keep costs competitive, (ii) increase water resources and (iii) manage water quality and security. The unsustainable water problem has occurred mostly because, all over the world, water has been treated as free commodity, which has meant that it has been drawn, used and discharged without sufficient consideration of the implications. The country or region as a whole suffers, resulting in conflicts, an unhealthy environment and socio-economic problems. The free supply of water from dams in many regions of the globe has resulted in unsustainable use and insufficient attention to losses. As consequences of such practice the reservoirs have gone dry due to climate change and changing rain pattern along with fragmented watersheds due to urbanization.

A sustainable approach should not just look at water treatment plants and the distribution network but should also manage water sources and the consumption pattern with intelligent software and databases. An intelligent water-management system will optimize workers' efficiency, lower energy and maintenance costs and lower work backlogs while increasing service-level quality and water sustainability. Revenue should be based on targeted pricing by user segment but, in order to attract investment, governments in the developing countries usually provide free/cheap water and land to develop the economy of the region and job opportunities, without considering long-term problems.

Sustainable water solutions should target conservation by ensuring adequate water during water stress or shortages. The solution should consider new consumption patterns and cut down conventional nonintelligent consumption patterns, especially in farming communities where conventional water consumption is high. The use of some rivers for irrigation by construction of dams in different locations has resulted in conflicts. Instead of improving water efficiency, the consumers clash with each other.

Intelligent systems should forecast future demand to help strategic planning based on the analysis of historical and current consumption data. The solution should provide information about real-time consumption. This can occur by integrating the GIS, flow patterns in rivers, groundwater depth and quantity of water in reservoirs/lakes onto a single platform.

A sustainable solution for a city or region should (i) monitor source quality/quantity to optimize the blending of water from different sources, (ii) generate and maintain maps that contain the locations of water sources, (iii) manage water-intake flow, (iv) inspecting raw water quality, (v) detect leakage, (vi) minimize energy consumption.

The supply solution should have real-time data on: (i) water leakage, (ii) efficiency of water pumps, (iii) water pressure in pipes, (iv) water quality at different locations, (v) water theft/tampering, (vi) metre outage or failure, (vii) customer segmentation, (viii) demand and supply patterns, (ix) number of water users and (x) industry/ward/area/village consumption pattern. The supply solution should not be isolated with wastewater and stormwater flow pattern in order to assess the possibility of water reuse and harvesting water from rain/snow/fog /dew. Water discharge real-time monitoring should include (i) sewer discharge, (ii) overflow, (iii) flood information, (iv) weather information (rainfall/temperature/humidity).

Over the past four decades, a number of water projects have been started in Singapore and have developed into full-scale water resource management systems. Today, Singapore has a sustainable water supply popularly recognized as the Four National Taps: (i) water from local catchments; (ii) water imported from Malaysia through pipes; (iii) high-grade reclaimed water; and (iv) desalinated water. Rainfall collected by a network of drains/canals is channelled to the reservoirs from where it is treated to make potable water. New technology called a 'variable-salinity plant' (VSP) treats collected rainwater during the rainy season and treats seawater by desalination during the dry season (National Water Agency, 2012). During the rainy season, rainwater accumulated in a nearby canal is trapped by an inflatable rubber weir, from where water is transferred to the plant. When the canal runs dry, seawater is processed in a VSP obtained from an intake feed located a few hundred metres offshore.

The city of Durban, South Africa, hosts 3.5 million citizens; it has extremely wealthy people and poor people. All of them have free, basic water provided by the government. All houses are supplied with 300 l of free water per day. In some regions of city, the water is distributed at midnight through pipes and tanks with automatic valves that have the ability to operate at low pressure. Conventionally, water pressure and capacity in urban areas is designed to allow firefighting but the higher pressure and flow results in higher losses, which is unsustainable. Lowering the pressure in the pipes reduces the flow of leaking water and the holes do not enlarge. Distribution through lower pressure results in less leakage and therefore infrastructure costs will reduce.

7.5 Wastewater Collection and Disposal System

The sustainability of wastewater collection and disposal depends on the type of wastewater collected and disposed of. In the developing countries where land use is highly fragmented due to land grabbing by influential people, it is quite common to find industries between residential areas. Often such areas are near industries, making it difficult for industries to dispose of the waste into domestic network as the toxic effluents may hinder the population of microbes in domestic sewage, affecting the effective treatment of domestic effluents. The waste varies to a great extent. If the same type of industry is scattered across a city, like electroplating units in Bangalore, collection by trucks and disposal into a common facility may provide an opportunity for sustainable disposal.

Common effluent treatment plants (CETP) have been constructed and operated throughout India because small-scale industries cannot operate wastewater treatment plants due to lack of finance for investment and lack of skill to operate effluent treatment plants. The effluents from industries are collected through pipes or in tanks, depending on the area where the CETP is located. Other advantages of CETP include reducing labour/chemical/energy costs and control of capital/operating expenditure.

Wastewater collection and disposal can be done by canals and conduits. Even though canals mix with rainwater and snow, this approach is still practised all over the world for water that overflows in the street.

Water, which is considered as the lifeblood of this planet, has supported all the species that live in it and on land. Humans have a complicated relationship with water and rely on it for far several purposes – power, transportation, agriculture and energy (IBM, 2013). Humans consume more water than any other species and change it every time they use it and dispose of it. It takes few hundred litres of water/kg for potato cultivation, a few thousand litres to manufacture one jean pant and few million litres to manufacture one automobile. Every time humans interact with water, its position is relocated from the normal water cycle and this changes its quality. Although the total quantity of water on this planet remains the same, as per the law of conservation of mass, all species are suffering due to unsustainable practices. It is the duty of the present generation to sustain and conserve water for the next generation.

By monitoring real-time water use at each building cities can dispose of wastewater sustainably and build an awareness of the need for conservation. Such an arrangement can also make it possible to stop the water supply individually and automatically once the consumption reaches the predefined water quantity allotted to consumer. Acoustic sensors with GPS can provide information about the smallest of leaks. Using wastewater, hydroponics and farming plots stacked vertically, cities can grow food without using excess water or land (IBM, 2013). Not all solutions to water and wastewater problems aim to tackle water scarcity. Around three billion people live in coastal areas and other low-lying area vulnerable to hooding and storm surges, which are augmented by global warming. Nearly 60% of the citizens of the Netherlands live at or below sea level, largely on land retrieved from the deltas of the Maas, the Rhine and the Scheldt and, hence, for 2000 years, the country has been developing and refining the management of an overabundance of water.

Figure 7.11 shows the water cycle and human intervention. Unsustainable practices have changed the rhythm of the water cycle and the environment, which was formed over millions of years. Anthropogenic activity has fragmented natural water streams at all levels (first/second/third order) by construction of dams, roads, buildings, slums, airports, canals. Many of the lakes and water bodies have been closed to constructed urban infrastructure. Such practices have resulted in drought/flood and other disasters.

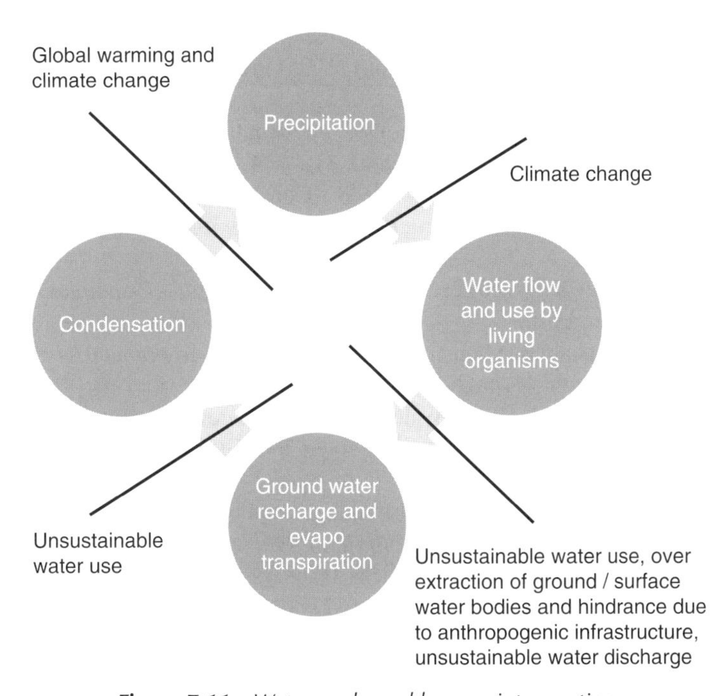

Figure 7.11 *Water cycle and human intervention.*

Durban has provided more than 60 000 urine deviation toilets, which have two chambers that separate urine that can be collected and used for fertilizing crops. The city services around 60 000 ventilation improved pit latrines that store black waste up to five years before emptying, which can be used for urban agro forestry.

About 80% of the freshwater resources in world are used for agriculture. Farmers are usually given priority over municipal water supplies, which are usually subsidized by governments, resulting in careless handling and nearly 60% waste. Since 1964, Israeli farmers have increased their water consumption by 3% while producing more than nine times the quantity of food (IBM, 2013).

The greater Tokyo region includes five river basins providing shelter to about 27 million people, covering an area of around 22 600 km^2. Due to anthropogenic activities, numerous water-related problems have emerged, increasing the need for better water quality, protection, diversification and enhancement of the environment. The water resources required to supply the urban area and maintain the safety against drought are tough to manage. Further, groundwater withdrawal is causing land subsidence. Changes in land use and the changing climate have increased flooding in recent years. The quality of water has deteriorated in greater Tokyo. Despite efforts to prevent it, the pollution level is high in some water bodies. Due to increase in imported non-native species of plants and fish the region is facing ecological problem.

References

Abdul Aleem, M.I. and Arumairaj, P.D. (2012) Geopolymer concrete – a review. *International Journal of Engineering Sciences and Emerging Technologies*, **1** (2), 118–122.

AWWA (American Water Works Association) (1995) Problem Organisms in Water: Identification and Treatment, AWWA, Denver, CO.

Bon, R. and Hutchinson, K. (2000) Sustainable construction: some economic challenges. *Building Research and Information* **28** (5/6), 310–314.

Bossink, B.A.G. and Brouwersz, H.J.H. (1996) Construction waste: quantification and source evaluation. *Journal of Construction Engineering and Management* **122** (I), 55–60.

Chandrappa, R. and Das, D.B. (2012) *Solid Waste Management – Principles and Practice*, Springer Verlag, New York.

Ding, G.K.C. (2008) Sustainable construction – the role of environmental assessment tools. *Journal of Environmental Management* **86**, 451–464.

Falkenmark, M. and Lindh, G. (1993) Water and economic development, in *Water in Crisis: A Guide to the World's Fresh Water Resources* (ed. P.H. Gleick), Oxford University Press, New York.

Freidin, C. (2002) Stability of cementless building units based on oil shale fly ash binder in various conditions. *Construction and Building Materials* **16**, 23–28.

Freidin, C. (2005) Cementless pressed blocks from waste products of coal-firing power station. *Construction and Building Materials* **21**, 12–18.

IBM (2013) *Water: a Global Innovation Outlook Report*, http://www.ibm.com/ibm/gio/media/pdf/ibm_gio_water_report.pdf (accessed 8 January 2014).

Indiaonlinepages (2013) *Population of Delhi 2012*, http://www.indiaonlinepages.com/population/delhi-population.html (accessed 8 January 2014).

India Today (2012) Delhi ground water reserves depleting at alarming rate, finds CGB study, http://indiatoday.intoday.in/story/delhi-groundwater-reserves-depleting-fast-cgb-study/1/200575.html, 14 June 2012 (accessed 8 January 2014).

Khan, N. (2005) The development of the emerging technologies sustainability assessment and its application in the design of a bioprocess for the treatment of wine distillery effluent. Master of Science thesis. Rhodes University.

LeChevallier, M.W., Babcock, T.M. and Lee, R.G. (1987) Examination and characterization of distribution system biofilms. *Applied and Environmental Microbiology* **53**, 2714–2724.

Miyatake, Y. (1996) Technology development and sustainable construction. *Journal of Management in Engineering* **12**, 23–27.

National Water Agency (2012) Innovation in Water, Singapore.

Peyton, B.M. and Characklis, W.G. (1992) Kinetics of biofilm detachment. *Water Science and Technology*, **26** (9–11), 1995–1998.

Rashad, A.M. (2013) Metakaolin as cementitious material: history, scours, production and composition – a comprehensive overview. *Construction and Building Materials*, **41**, 303–318.

Ridgway, H.F. and Olson, B.H. (1981) Scanning electron microscope evidence for bacterial colonization of a drinking water distribution system. *Applied and Environmental Microbiology*, **41**, 274–287.

Salian, P. and Anton, B. (2013) Making Urban Water Management More Sustainable: Achievements in Berlin, ICLEI European Secretariat., http://www.switchurbanwater.eu/outputs/pdfs/w6-1_gen_dem_d6.1.6_case_study_-berlin.pdf (accessed 14 January 2014).

Struckmeier, W., Rubin, Y. and Jones, J.A.A. (2005) Groundwater – Reservoir for A Thirsty Planet? The Year of Planet Earth Project, Leiden.

Tan, Y. (1996) Design of silt related hydraulic structures. Proceedings of the International Conference on Reservoir Sedimentation, Fort Collins, Colorado, pp. 675–731.

van der Wende, E., Characklis, W.G. and Smith, D.B. (1989) Biofilms and bacterial drinking water quality. *Water Research* **23**, 1313–1322.

Van Holm, M., Simões da Silva, L., Revel, G.M. *et al.* (2011) Scientific Assessment in support of the Materials Roadmap enabling Low Carbon Energy Technologies, European Commission, Joint Research Centre, Institute for Energy and Transport.

Vrba, J. and Lipponen, A. (2007) *Groundwater Resources Sustainability Indicators*, UNESCO, Paris.

Yang, K.-H., Hwang, H.-Z., Kim, S.-Y. and Song, J.-K. (2007) *Building and Environment* **42**, 3717–3725.

Yang, K.-H., Mun, J.-H., Lee, K.-S. and Song, J.-K. (2011) Tests on cementless alkali-activated slag concrete using lightweight aggregates. *International Journal of Concrete Structures and Materials* **5** (2), 125–131.

Yang, K.-H., Song, J.-K., Ashour, A.F. and Lee, E.-T. (2008) Properties of cementless mortars activated by sodium silicate. *Construction and Building Materials* **22**, 1981–1989.

Zektser, I.S. and Everet, L.G. (eds) (2004) Groundwater Resources of the World and their Use, UNESCO, Paris.

8

Safety Issues in Sustainable Water Management

Water is an essential resource for human and economic wellbeing. Modern society depends on a complex water infrastructure, which is vulnerable to intentional or accidental disruption due to war, intrastate violence and terrorism (Gleick, 2006).

Sustainable development is more than just protecting the environment. World Commission on Environment and Development (1987) stated that sustainable development is 'development that meets the needs of the present without compromising the ability of future generations to meet their own needs.' According to IUCN, UNEP, WWF (1991), sustainability means 'improving the quality of human life while living within the carrying capacity of supporting ecosystems.' According to the ICLEI (1996), sustainability means 'development that delivers basic environmental, social and economic services to all residences of a community without threatening the viability of natural, built and social systems upon which the delivery of those systems depends.'

Protection of the environment and meeting the needs of the future generation should not jeopardize safety, health and the lives of the current generation. The sustainable health and safety idea approaches both the concept and the execution of health and safety measures from a special perspective (Rajendran, 2006).

Risks apply to all individuals irrespective of nationality or wealth. Hazards can occur anywhere if they are not properly managed. All human activities carry risk. People working in the water sector should think not only of their own safety but also of that of their customers. The simplest way to achieve safety is: (i) identify hazards, (ii) evaluate the recognized hazards and (iii) control hazards.

Combined sewer systems that include stormwater that carries oils, salts, metals and asbestos are associated with major risks. Some of sewers receive infiltration that carries pesticides and herbicides. For many years, due to deaths associated with confined space entry, sewer works have been considered as hazardous. Some chemically related health hazards are acute in nature, whereas other problems are chronic. Wastewater treatment may

Sustainable Water Engineering: Theory and Practice, First Edition. Ramesha Chandrappa and Diganta B. Das.

generate aerosols contain microbiological and chemical constituents. A sewage treatment plant comprises open tanks and basins. Volatile organics in wastewater may be transferred from the liquid into an air stream during treatment due to dispersion or vaporization (Brown, 1997). Sewage consists of carcinogens and/or mutagens, exposing sewage workers to cancer or serious ailments. Infections from exposure to pathogens may be subclinical or may appear as diseases in wastewater workers who have reported nausea, indigestion, diarrhoea, vomiting and flu-like complaints. Even though several years of exposure generate eventual immunity for workers to some extent, new workers become sick more often than experienced workers.

Hazards can occur anywhere, which includes at wastewater treatment plants. Industrial effluent treatment plants may pose safety-related risks due to the range of chemicals in the influent and sludges. Water/wastewater treatment plant labour becomes exposed to microbes or chemicals by contact with wastewater/sludges, or by inhalation of particles, gases, aerosols, vapours or droplets

Water and waste management have become complex matters. Effects on human health at treatment places and at the point of consumption/disposal are causes of concern. All the activities involve risk of water/wastewater treatment, distribution, transportation and use.

Figure 8.1 shows a relationship between water/wastewater treatment and risk and Figure 8.2 shows a relationship between water/wastewater treatment, sustainability and risk. In the water/wastewater sector, safety is important because failure to control risks could (i) affect the wages of the personnel who are ill or injured, (ii) lead to a loss of productivity, (iii) cause damage to equipment, (iv) incur costs for investigation so that the problem can be corrected, (v) lead to consumer dissatisfaction and (v) result in fines and legal costs if there are prosecutions.

The risk associated with water depends upon the source of water/wastewater and the recipient. Falling objects, bending, turning, twisting and tumbling of improperly stacked materials are common workplace issues. Wastewater treatment-facility operators can be exposed to chemical, physical and biological hazards like falling into tanks, confined space entries, inhalation of VOCs, bio aerosols, methane, contact with pathogens/vectors and

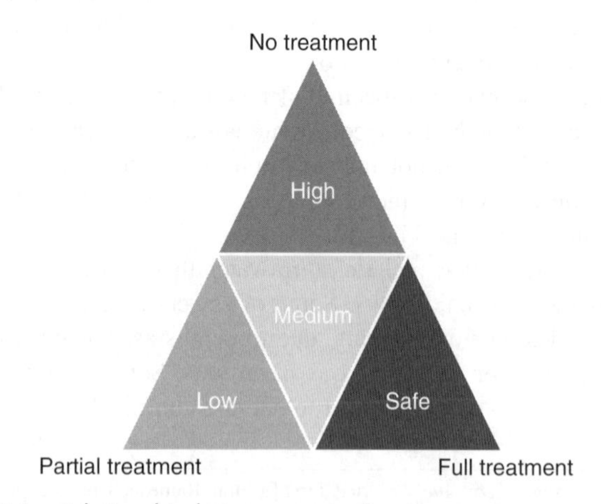

Figure 8.1 *Relationship between water/wastewater treatment and risk.*

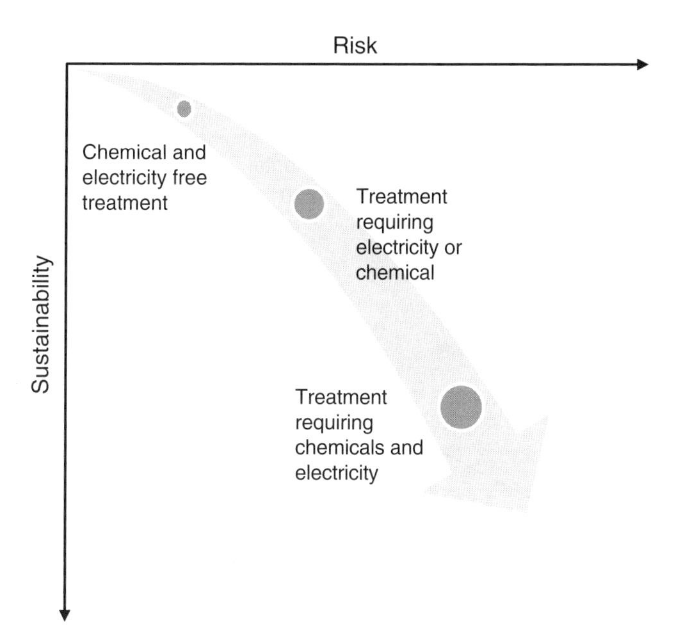

Figure 8.2 *Relationship between treatment, sustainability and risk.*

exposure to potentially hazardous chemicals (like chlorine, sodium or calcium hypochlorite and ammonia) (IFC, 2007).

In order to keep track of safety performance and to take corrective action it is essential that accident records be maintained. Records should include (i) an accident report, which should include a description of the accident and accident analysis, (ii) a doctor's report and (iii) action taken.

Hazards can be avoided at workplaces to a great extent and some methods are: (i) to examine records of what has caused accidents in the past and take corrective action, (ii) conduct a daily audit in the treatment plant, (iii) provide proper training and personal protective equipments (PPEs), (iv) supervise risky operations so that an alarm can be raised in case of accidents, (v) keep all emergency equipment like fire extinguishers and emergency baths in good condition, (vi) provide closed-circuit cameras with backup arrangements, (vii) provide fencing and (viii) guard the treatment plants to avoid entry of untrained/unauthorized persons.

Figure 8.3 shows a risk map for health, safety and sustainability. Figure 8.4 shows a risk map of safety of construction workers. Health risks from wastewater depend on (i) the composition of the water/wastewater, (ii) the products and byproducts during waste reaction/decomposition, (iii) work conditions, (iv) energy use, (v) the health of individuals, (vi) the methodology/technology adopted for handling/treating of water/wastewater, (vii) disasters in the area, (viii) the climatological/environmental setup of the location, (ix) personal hygiene practices, (x) personal protective equipment, (xi) safety regulations and (xii) the efficiency of regulators.

Safety issues with respect to water should be considered together with climate change. Change in climate can change the probability of risk occurrence. One of the striking effects

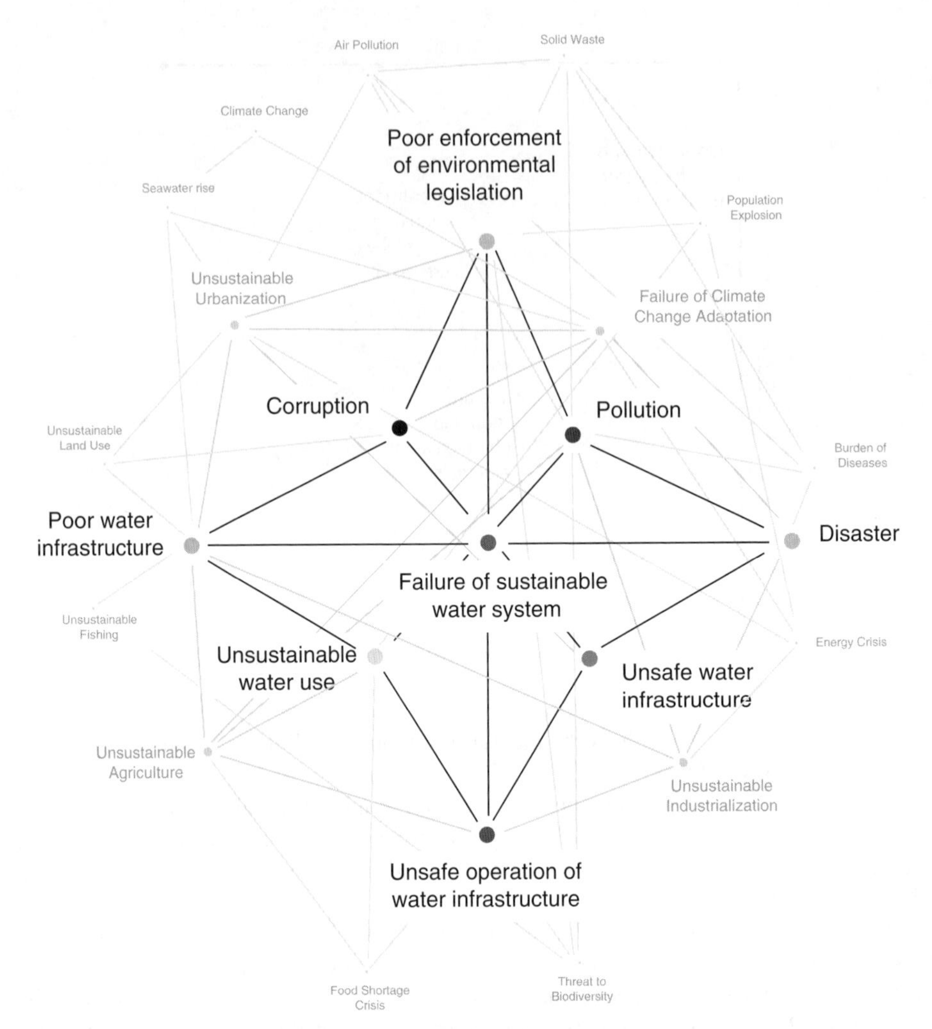

Figure 8.3 *Risk map for health, safety and sustainability. (For a colour version of this figure, see the colour plate section.)*

of change in climate is melting of ice resulting in: (i) variation in runoff due to the melting of glaciers and (ii) a rise in sea levels. The increase in sea level leads to: (i) changes in overturning circulation in deep sea; (ii) a change in ocean currents; (iii) damage to terrestrial and aquatic flora and fauna; (iv) the destruction of property in coastal areas and (v) seawater intrusion into surface water and groundwater near coasts. River deltas and cities near the sea are vulnerable to increases in sea level together with land subsidence.

8.1 Health, Safety and Sustainability

All water works are linked to health and prosperity. The dams, irrigation water structures, water supply, wastewater collection, water/wastewater treatment, urban drainage, industrial

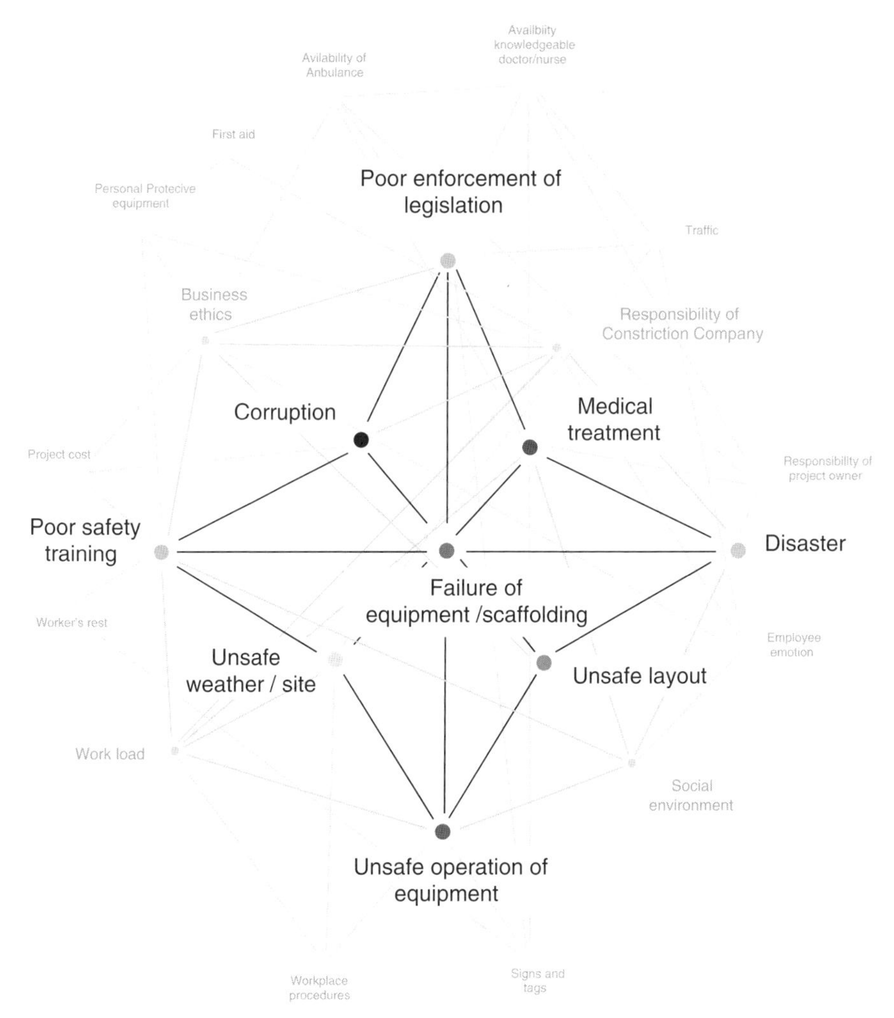

Figure 8.4 *Risk map of safety of construction workers. (For a colour version of this figure, see the colour plate section.)*

water use/management are linked with health, safety and sustainability. Health can be affected by (i) poor quality of water, (ii) exposure to water for an extended period during disasters, (iii) immersion of victims in water, (iv) vector-borne diseases, (v) food poisoning, (vi) reaction of water with chemicals, (vii) explosion of steam-generating/transporting systems and (viii) the flow of electricity in water.

Water bodies and deep water-holding structures are safety concerns due to the possibility of immersion of humans and animals.

Safety has been affected by many theories, including the domino theory developed by Heinrich in 1930 (Heinrich *et al.*, 1980; Howell *et al.*, 2002). His early work was criticized and was updated to focus on the responsibility of management for accidents. Other models can be classified as epidemiological, human factor, behaviour, systems and decision models

(Heinrich *et al.*, 1980) but other issues, indicated in Figure 8.4, cannot be neglected. People who die in the workplace are not people in top-level management but people who really work on site. It is quite possible that the owners of construction companies would neglect safety as they are not actually involved in hazardous activity themselves. Hence in many workplaces they are not concerned about safety equipment, rest, proper work environment, PPE, first aid, water and sanitation and so on.

In order to overcome the risk associated with water, Queensland, the second largest state in Australia, has passed legislation. It offers the framework to achieve sustainable water planning, allocation, management and supply processes and to ensure better security for water resources. It requires councils and authorities that own/operate water supply systems in Queensland to register and make strategic asset-management plans. The aim of the legislation is to ensure a safe and reliable water supply. This is achieved by providing a regulatory framework for water and sewerage services; recycled water and drinking water quality; the regulation of responsibilities for referable dams and flood mitigation and to protect the interests of customers. Such legislation is yet to be adopted in the other parts of the world. Since most of the governments in the world are responsible for providing water supplies, passing such legislation may not happen across the globe in immediate future as the government authorities usually resist any legislation regulating them.

Human development can be derailed by climate-related disasters (UNDP, 2007). Besides the cost of recovery, disasters can take away lives and disrupt livelihoods. Disaster-related shocks are likely to follow people throughout life. The Tokyo metropolitan area, which has an extremely high population density, is subject to seasonal floods and other hazards like droughts and earthquakes but Japan has the means and the skills to administer these risks by creating infrastructure (such as levees, dams and underground floodways), public awareness and disaster preparedness. The country has developed Internet- and GIS-based early warning systems and hazard mapping. The country has shelters for people to take refuge. Such efforts have ensured the safety of the population in the area but such preparedness is lacking in other parts of the world. The flood in Uttarkhand, India, in June 2013, resulted in the deaths of thousands of people and many tourists were reportedly raped, murdered and looted during the floods there (Singh, 2013).

Floods result from heavy rainfall, tsunamis, unusual high tides, severe winds over water, rapid melting of snow, or failure of levees/retention ponds/dams/structures that retain the water. Periodic floods form flood plains in the surrounding regions. Flooding damages property and, endangers the lives of people and other species. Water runoff results in soil erosion contaminating the water. The spawning places for fishes can become polluted/destroyed. Flood-control methods include planting vegetation, terracing hillsides, construction of floodways, reservoirs (retention ponds), levees, lakes, weirs, dams, sandbags or portable inflatable tubes to hold water during times of flooding. The self-closing flood barrier is another flood-defence system built to protect properties and communities. Coastal flooding defences include beach nourishment, sea walls and barrier islands. In some places tide gates are used with culverts and dykes that are placed where an estuary begins, at the mouth of a stream, where tributary drainage/streams-ditches connect to sloughs. Tide gates open during outgoing tides and close during incoming tides.

It is often thought that bottled water is the safest water anybody can think of. Studies conducted by Reimann *et al.* (2010) for 57 chemical elements in bottled water sold in the European Union demonstrated significant differences in median concentrations for Al, Bi,

Ce, Cr, Cu, Er, Dy, Fe, Gd, Lu, La, Nb, Nd, Pr, Pb, Sb, Sn, Tb, Ti, Th, Sm, Tm, Yb, Y, Zn and Zr. Antimony had a 21-times higher median value with respect to water sold in polyethylene terephthalate (PET) bottles. Water in glass bottles had more Ce, Pb, Al, Zr, Th, La, Tb, Cu, Ti, Pr, Fe, Zn, Nd, Sn, Er, Cr, Gd, Sm, Bi, Y, Yb, Lu, Tm and Nb. Testing the same water sold in green and clear glass bottles demonstrated higher concentrations of Cr, Th, La, Zr, Nd, Ce, Pr, Nb, Ti, Fe, Co and Er in green bottles.

Studies conducted by Shotyk and Krachler (2007) on 132 brands of bottled water in 28 countries showed that antimony concentrations in water were mainly due to the use of Sb_2O_3 as the catalyst during the production of PET. Two of the brands had more than the maximum allowable concentration of Sb in drinking water in Japan (2 μg/L). Studies on 14 brands of bottled water in Canada showed Sb concentrations increased on average 19% during six months' storage, whereas 48 brands of water in 11 EU countries increased on average 90% in identical conditions.

8.2 Safety of Consumer versus Operator

Plant operators are most likely to encounter hazards due to confined spaces, dust, fumes, fire, electrical shock, gases, physical injuries, noise, stored energy, toxic chemicals, vapours and so on. The safety of operators can be ensured with proper construction principles, safety instruction and equipment (Figures 8.5–8.8). The safety of water consumers depends on the precautions taken right from sourcing of water up to the point of delivery.

To ensure the safety of operators it is important to communicate hazards at work sites. A variety of chemicals are stored and used at water/wastewater treatment plants. Employees must be provided with information about hazards associated with these chemicals, which includes (i) a list of the hazardous chemicals at the workplace, (ii) information on the

Figure 8.5 Wastewater treatment plant with proper guarding.

Figure 8.6 *Flares for burning methane from wastewater to energy plant.*

labelling used to identify chemicals at the facility, (iii) location and use of the material safety data sheet (MSDS) of chemicals used/stored at the workplace and (iv) methods to detect the presence/release of hazardous materials.

It is the responsibility of the employer to provide written hazard communication to the operator to ensure the safety of the operator. In the past it was often thought that it is the responsibility of employee to ensure his own safety but the employer should act to ensure

Figure 8.7 *Safety signs and handrails on stairs of sedimentation tank.*

Figure 8.8 *Lifebuoy at outlet of wastewater treatment plant.*

the safety of the operator. It is also safety of water supplier and wastewater collector to ensure safety of consumer and others who may be exposed to risk.

All chemical containers in the workplace must be identified with a clearly understood label. These materials should never be placed in containers that may be associated with food or drink. Water/wastewater treatment facilities should communicate the following information: (i) chemical identity (Table 8.1), (ii) fire and explosion hazard data (Table 8.2), (iii) reactivity data (Table 8.3), (iv) precautions for safe handling and use (Table 8.4), (v) health hazard data (Table 8.5) and (vi) control measures (Table 8.6).

In order to manage accidents it is essential that the water/wastewater treatment facility be prepared for the administration of first aid. First aid is care given for sudden illness or injury prior to medical treatment. The first-aid provider should be trained in medical emergency procedures, prior to shifting the affected person to hospital or while awaiting medical service personnel. A first-aid programme includes: (i) management leadership and employee involvement, (ii) safety and health training, (iii) hazard prevention and control, (iv) worksite analysis (US Department of Labor, 2006). It is recommended there should be a vehicle available for transferring victims to hospital as most of the water and wastewater treatment plants are located in the outskirts of cities. Apart from vehicles, the plants should have first-aid boxes located in strategic places with easy access.

There are many instances of water contamination that are not recorded to safeguard the officials and water supply companies, although there is some literature that helps planners to opt for better precautionary measures. The services to 75 apartments housing about 300

Table 8.1 *Chemical identity (sample format).*

Name of chemical:	Manufacturer name:
Address:	Web site for information:
Date prepared:	

Table 8.2 *Fire and explosion hazard data (sample format).*

Name of substance/chemical		
Flash point	Lower explosive limit	Upper explosive limit
Extinguishing media		
Special firefighting procedure		
Unusual fire and explosion hazard		

Table 8.3 *Reactivity data (sample format).*

Unstable		Conditions to be avoided
Stable		
Incompatible materials		
Hazardous decomposition/byproducts		
Hazardous polymerization	☐ May occur ☐ Will not occur	
Conditions to avoid		

Table 8.4 *Precautions for safe handling and use (sample format).*

What to do when material is released/spilled
Waste disposal method
Handling and storage precautions
Other precautions

Table 8.5 *Health hazard data (Sample format).*

Routes of entry	☐ Ingestion ☐ Skin ☐ Inhalation
Health hazards	
Carcinogenicity	
Signs and symptoms of exposure	
Emergency first-aid procedures	

Table 8.6 *Control measures (sample format).*

Respiratory protection:
Ventilation required:
PPE required:
Work/hygienic practices:

people were contaminated with chlordane and heptachlor in a city in Pennsylvania during December, 1980. Insecticides entered into the water supply system when they were being used against termites (USEPA, 2003).

An hydraulic aspirator to drain fluids in a funeral home was the source of blood in the public water supply in one of the episodes in the United States. In another episode, sodium hydroxide back siphoned into the water main when the water main broke (USEPA, 2003).

Apart from contamination due to negligence or ignorance, water may be contaminated due to disasters. Large quantities of radionuclides were released into the Pacific Ocean and the atmosphere from the Fukushima Dai-chi Nuclear Power Plant (FDNPP) due to the damage caused by the tsunami and earthquake in March 2011 (Kinoshita *et al.*, 2011, Ueda *et al.*, 2013), resulting in radioactive substances in tap water throughout eastern Japan (Kamei *et al.*, 2012).

Based on studies after an accident occurred in a storage tank in wastewater treatment plant of a pharmaceutical company, Sasso *et al.* (2012) concluded that there had been a lack of methodical planning of the tank's draining operation; this was necessary to avoid biochemical reactions due to uncontrolled accumulation of waste fluids.

Climate and environmental change have increased the occurrence and magnitude of hydro meteorological risks, which are closely connected with each other and are sometimes induced/amplified/triggered by anthropogenic activity activities. There are huge variations in water availability across the globe, as some places receive heavy rain whereas other places face drought. There has been a reduction in groundwater availability due to water extraction, which has altered water quality to a great extent. Millions of people worldwide consume excessive arsenic through food and drinking water. The consumption of seafood is the main arsenic exposure route in people; it can lead to arsenic-related sickness including cancers (Chen *et al.*, 2005, 2010; Hata *et al.*, 2007; Lindberg *et al.*, 2008). Hexachloro-bicycloheptadiene (HEX-BCH) was detected more often in the urine from workers of a municipal wastewater treatment plant that received waste from a pesticide manufacturer than in the urine of workers at another wastewater treatment plant in the same city. These are only a few of the many similar episodes that go unnoticed worldwide.

The entry into water of pathogens and chemicals occurs all over the world but it is considered serious in the developed world. Unlike industrial safety, where the main objective is the safety of people working in the industry, water works should consider the safety both of consumers and operators. Working with wastewater may expose staff to blood-borne pathogens, which include, but are not limited to hepatitis B (HBV), hepatitis C, hepatitis D, hepatitis G, malaria, human immunodeficiency virus (HIV), syphilis. Other potentially infectious materials (PIM) may also pose a risk in the wastewater work place. Potentially infectious materials include, but are not limited to, amniotic fluid, cerebro-spinal fluid,

Table 8.7 *Safety risk of worker, consumer and general public/animals who are neither workers/consumers but come into contact with water.*

Particular	Worker	Consumer	General public/ animal who is neither worker/ consumer but come into contact with water
No protective measures	High	High	High
Human exposure control	Low	Low	Low
Partial treatment	Low	Low	Low
Partial treatment, crop restriction and human exposure control	Low	Safe	Low
Full treatment	Safe	Safe	Safe

semen, mucous membrane secretions, nose/sinus drainage, peritoneal fluid, saliva in dental procedures, synovial fluid, urine, vaginal secretions and vomit. Accident situations include injury by syringes, splashed wastewater entering eyes, pathogens entering cuts and wounds and so on. The PIM causes infection when transmitted into the victim's bloodstream, which means the PIM needs to penetrate the victim's skin or mucous membranes (eyes, mouth and nose). Hence, it is essential that any visitor or worker at the site should cover all skin breaks and use appropriate PPE. Wastewater workplace should also have controls preventing eating, smoking, handling contact lenses, applying makeup and drinking (Alan and Tucci, 2010).

Figures 8.5, 8.6, 8.7 and 8.8 shows some precautionary measures taken to safeguard operators and Table 8.7 shows the safety risks of workers, consumers, animals and the general public (who are not workers or consumers but come into contact with water). Table 8.8 shows some possible hazards during wastewater treatment. Figure 8.9 shows a slippery floor, which demands proper PPEs, Figure 8.10 shows safety issues with respect to water distribution and Figure 8.11 shows some important safety issues with respect to water/wastewater treatment.

The safety of consumers and the personnel of water works can become affected due to inadequate funding. Other causes include: (i) disasters, which could break the pipe network, (ii) failure of tanks, (iii) electric short circuits. Electrical hazards could affect personnel operating pumps and other electrical equipment. Suboptimal loading and overloading of water treatment plants can affect performance of the plant and, hence, can affect water quality and the safety of consumers. Groundwater depletion and contamination can also affect the safety of consumers/operators.

Recreation has a considerable role in life all over the world and people tend to combine it with water. Recreational water illnesses (RWIs) have been a serious issue all over the world, even though they are neglected most of the time. They refer to sicknesses associated with recreational water locations such as hot tubs, swimming pools (Figure 8.12), water parks, beaches and the ocean. They comprise an array of illnesses that includes infections of the skin, eye, respiratory organs, ears, neurological and gastrointestinal systems. Infections with certain pathogens will have serious effects and life-threatening consequences. High-risk groups with respect to RWIs are immunosuppressed, pregnant women, elderly and

Table 8.8 *Possible hazards during wastewater treatment.*

		Possible hazard				
Unit operation	Classification	Electrical	Chemical	Biological	Physical	Radiological
Screen	Manual cleaning			✓	✓	
	Auto cleaning	✓		✓	✓	
Grit chamber	Manual cleaning	✓		✓	✓	
	Cleaned with rakes operated by motors	✓		✓	✓	
Commenting		✓		✓	✓	
Sedimentation	With manual cleaning			✓	✓	
	Cleaned with rakes operated by motors	✓				

		Possible hazard				
Unit process	Classification	Electrical	Chemical	Biological	Physical	Radiological
Activated sludge process		✓		✓	✓	
Ponds				✓	✓	
Lagoons		✓		✓	✓	
Trickling filter		✓		✓	✓	
Rotating biological contactors		✓		✓	✓	
Filter	Rapid sand filter	✓		✓	✓	
	Pressure sand filter	✓		✓	✓	
	Activated carbon filter	✓		✓	✓	
Coagulation		✓		✓	✓	
Up flow anaerobic sludge blanket reactor		✓	✓	✓	✓	
Anaerobic filter		✓	✓	✓	✓	
Air stripping		✓		✓	✓	

(continued)

Table 8.8 *(Continued)*

Unit process	Classification	Possible hazard				
		Electrical	Chemical	Biological	Physical	Radiological
Chemical precipitation		✓	✓	✓	✓	
Softening					✓	
Membrane technology	Ultrafiltration				✓	
	Reverse osmosis				✓	
	Forward osmosis	✓		✓	✓	
Disinfection	Chlorination	✓		✓	✓	
	Ozone – disinfection	✓		✓	✓	
	Pasteurization	✓		✓	✓	
	Solar pasteurization			✓	✓	
	UV-disinfection	✓		✓	✓	✓
Nanotechnology		✓		✓	✓	✓
Sono-photo-penton		✓		✓	✓	
Advanced oxidation process		✓		✓	✓	
Photocatalytic degradation		✓		✓	✓	
Electrochemical oxidation		✓		✓	✓	
Adsorption process using macroporous resin		✓		✓	✓	
Constructed wetlands		✓		✓	✓	
Vermi compost		✓		✓	✓	
Wastewater incineration		✓		✓	✓	
Sludge treatment	Sludge thickening	✓		✓	✓	
	Sludge digestion	✓		✓	✓	
	Sludge drying – solar			✓	✓	
	Air drying	✓		✓	✓	
	Incineration	✓		✓	✓	

Figure 8.9 *View of slippery floor.*

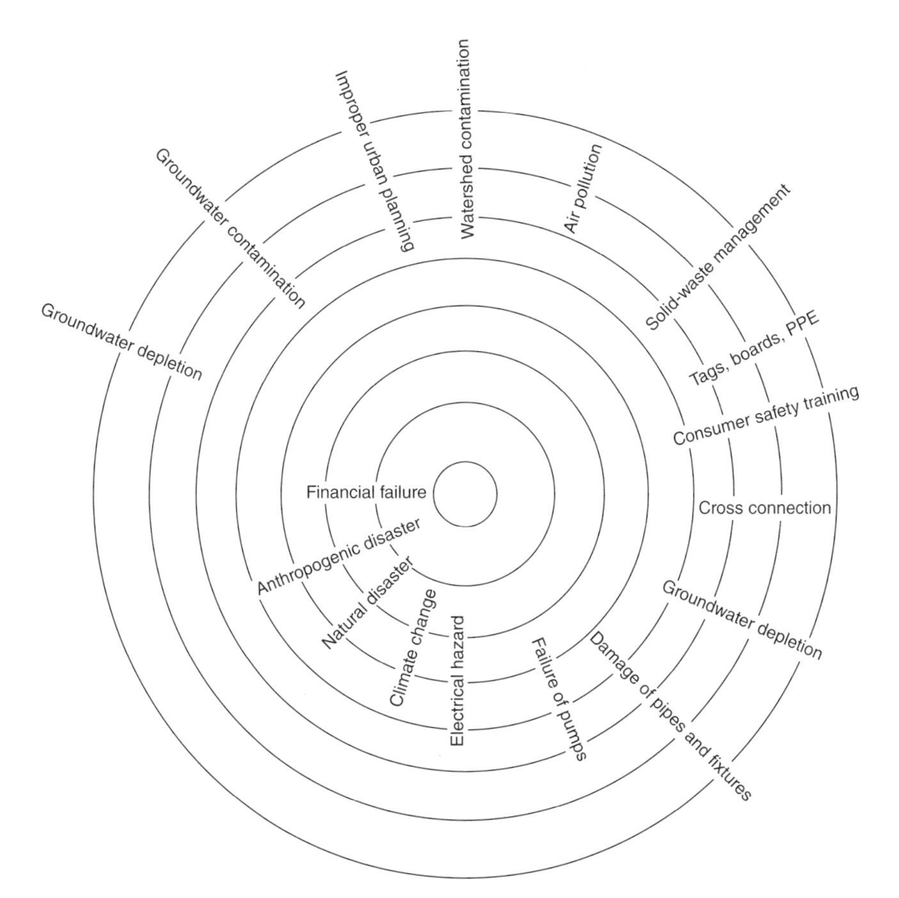

Figure 8.10 *Safety issues with respect to water distribution.*

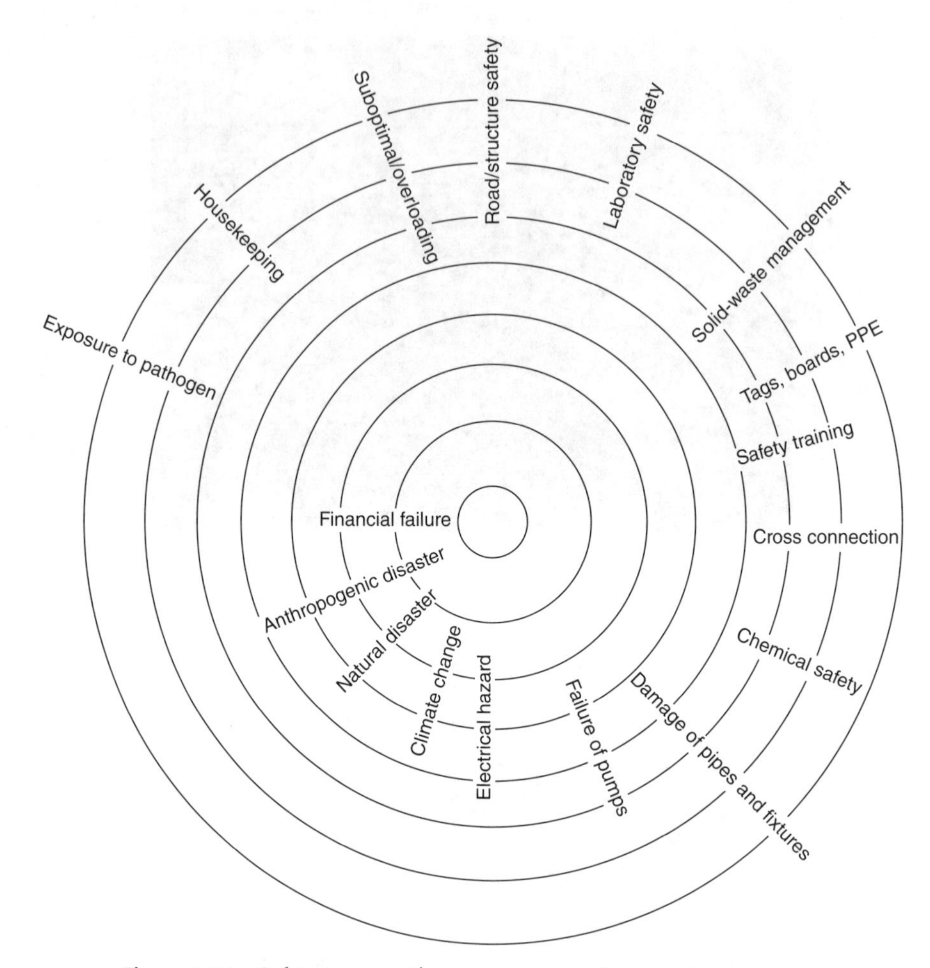

Figure 8.11 *Safety issues with respect to water/wastewater treatment.*

young people. Most waterborne pathogens in recreational facilities are spread faecal-orally. Depending on the source, pathogens can be of animal origin (zoonotic) or can be human pathogens originating from another bather. Pathogens of animal origin can arise from infected animals close to the recreational facility (Barna and Kádár, 2012). In addition to pathogens, unsafe storage of chemicals (Figure 8.13) and improper water treatment also pose a threat to people's health and lives.

Safety issues with respect to residue management also need to be considered while designing the plant. With the high level of dryness and fine particles, a fire/explosion hazard may exist when dried sludge is conveyed/stored in dryers using heat from fuel combustion. The heat of combustion can lead to an explosion due to the increase in volume of hot combustion products.

Fast-growing populations and urban migration in the developing countries has resulted in the establishment of centralized water systems. Ageing, stressed and poorly maintained distribution systems reduce the quality of piped drinking water and pose serious health

Figure 8.12 *Poor water quality in swimming pool and amusement parks is a cause for diseases among users.*

risks. Deficiencies in public water distribution are caused by: (i) failure to disinfect, (ii) low pipeline water pressure, (iii) inadequate sewage disposal, (iv) excessive network leakages, (v) intermittent services and (vi) corrosion of parts (Lee and Schwab, 2005). Improved monitoring, research and understanding of distribution system deficiencies can improve public health and reduce the global disease burden.

Figure 8.13 *Unsafe chemical storage practice.*

Unlike the developed countries where failures of water treatment are relatively rare, failure of treatment has become the norm in many developing countries where many systems operate at a part of their capacity resulting in poor water quality. The developing countries experience deterioration of infrastructure, failure of distribution systems and a larger magnitude of problems related to the water distribution system (WHO and UNICEF, 2000).

An upsurge in the total coliforms and thermotolerant coliforms was observed as residual concentrations of residual chlorine were reduced at delivery point to 0.2 ppm at residences at San Fernando, Trinidad (Agard *et al.*, 2002). Similarly, the concentration of free chlorine reduced as the distance from the treatment plant increased resulting in coliform at Pietermaritzburg, South Africa (Bailey and Thompson, 1995). Due to shortages of the required chemicals, the correct chlorine dosage at a plant at Phnom Penh, Cambodia, was not achieved, resulting in lower free chlorine concentrations than the recommended 0.2 mg/l (Dany *et al.*, 2000). Gastrointestinal illness increased due to a decline in residual chlorine concentrations at Cherepovets, Russia (Egorov *et al.*, 2002). Improper roads during the monsoon interrupted the supply of chlorine, causing cholera outbreaks in India (Gadgil, 1998). Chlorine concentrations of 0.2 mg/l were detected in only 79% of water samples in Panaji City, India. The urban population coverage of disinfected water varies from 20% in Haiti to 100% in a number of Latin American countries in the Caribbean and Latin America. It is not safe for people to be supplied with water that is not disinfected. A hepatitis E outbreak occurred during a major repair of a treatment plant in Islamabad, Pakistan. Similarly, an increased incidence of diarrhoea occurred in Nukus, Uzbekistan, due to low pressure and inadequate chlorination within the distribution system (Semenza *et al.*, 1998). Cross-contamination of the water supply occurred due to negative pressure and back-siphonage in Riohacha, Colombia. Inadequate system pressure from overhead tanks and pumphouses resulted in zero pressure in Bangladesh, making the system vulnerable to contamination. Similarly, water pressure was deemed inadequate as pumps were not functioning well in Phnom Penh, Cambodia (Dany *et al.*, 2000). Fluctuations in pressure led to the contamination of water in Cebu, Philippines (Moe *et al.*, 1991). Frequent power cuts resulted in the inoperability of water pumps, leading to an unreliable water supply in Iganga, Uganda (Thompson *et al.*, 2000). Since 1920 there have been about 1884 waterborne outbreaks, in the United States, resulting in 882 144 cases of illness and 1169 deaths. The largest outbreak reported in the United States was in Milwaukee, Wisconsin, in 1993, which resulted in 403 000 cases of diarrhoeal and 50–100 deaths (Gallagher, 2010).

8.3 Safety of People and Animals other than Consumers and Operators

Accidents and loss of property can occur at: (i) the point of operation of a water/wastewater treatment plant, (ii) along the line of distribution/collection, (iii) at the point of discharge/intake due to unsafe working condition. In some cases, hazards can go beyond the water infrastructure. Figure 8.14 shows a schematic diagram of the linkage between a cause and a loss. Figure 8.15 shows some effluents from operations with poor safety precautions and Figure 8.16 shows an example of bad safety procedures that still exist in the developing countries.

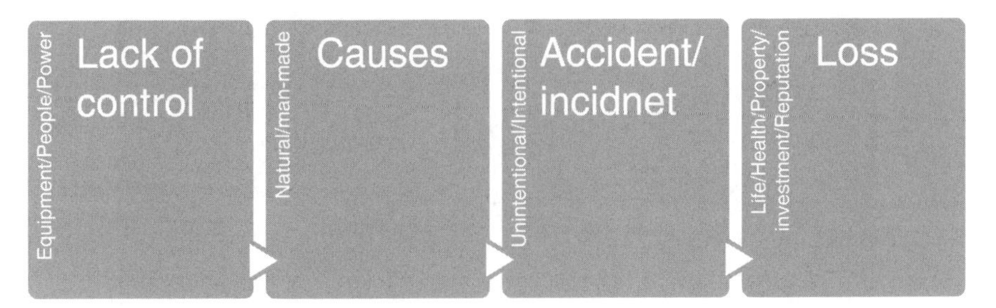

Figure 8.14 Schematic diagram of linkage between cause and loss.

The effects of accidents in water/wastewater treatment plants vary hugely. Consumers' health can be at risk, despite measures being taken to ensure safety for everything and everybody at a water/wastewater treatment plant. Contamination can occur due to pipes or fittings breaking, or intentionally due to deliberate acts of terrorism.

The importance of freshwater and water infrastructure to humans makes water and water systems targets of terrorism. There is a real possibility that terrorists may strike at water systems and there is a long history of terrorist attacks. Water infrastructure can be contaminated by the introduction of disease-causing agents or poison. The damage is done by rendering water unusable, hurting people or destroying purification and supply infrastructure (Gleick, 2006). Even though disaster-mitigation planning is well advanced in theory, we fail when actual disasters occur and we are vulnerable to eco-terrorism and environmental terrorism.

Figure 8.15 Effluent from operations with poor safety precautions.

Figure 8.16 Example of bad safety procedures that still exist in developing countries.

Box 8.1 Eco Terrorism and Environmental Terrorism

The term 'environmental terrorism' refers to the illegal use of force against environmental resources/systems with the intention to harm people or property. 'Eco-terrorism' refers to the unlawful use of force against people/property with the intention of saving the environment.

An explosion on 1 November 1986 in the Sandos Chemical factory in Basel, Switzerland, led to the pollution of the River Rhine. Fire-extinguishing activity resulted in the entry of 10 000 to 15 000 m^3 of water with organic mercury compounds, herbicides, insecticides, fungicides and other agrichemicals into the Rhine. They made their way for 900 km, through six sovereign states, before entering the Baltic Sea, killing thousands of fish and waterfowl.

Rapid industrialization in China has resulted in frequent water-pollution accidents (Xiao *et al.*, 2011). The country witnessed 6677 water-pollution accidents between 2000 and 2008 (National Bureau of Statistics of China, 2007, 2008). An accident in a petrochemical plant in Jilin Province resulted in the release of around 100 tons of benzene, nitrobenzene and aniline into the Songhua River. A chemical spill in Shanghai, confirmed on 11 January 2013, was the cause of water contamination in villages and more than 400 factories/companies due to an illegal chemical waste dump (Li, 2013).

Waterborne disease is a concern for wastewater workers. Regular physical examinations are a necessity for workers who are exposed to pathogens and hazardous materials. Transportation safety within the wastewater treatment plant is important and proper signs for speed limits, turnings and construction activities need to be put in place. Electrocution or mechanical hazards during equipment repair or servicing can lead to loss of life and property. Hence a 'lockout/tagout' procedure needs to be in place as it prevents accidents. Valves

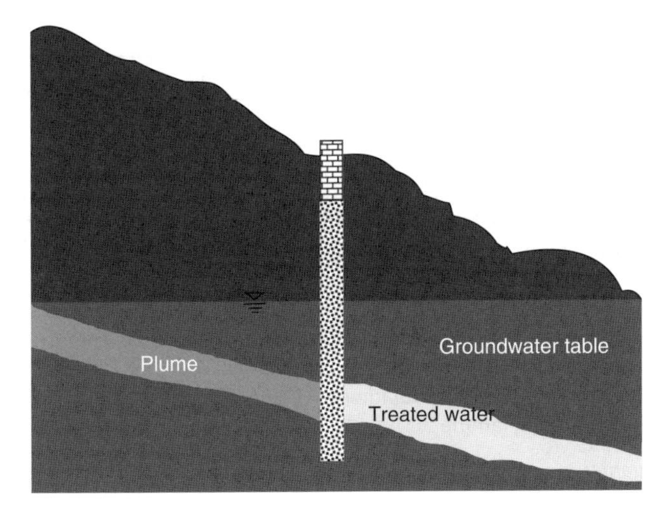

Figure 8.17 *Schematic diagram of a permeable reactive barrier.*

to process tanks should be locked to avoid accidental flooding during maintenance. The wastewater laboratory can be a hazardous place as acids or poisonous reagents need to be handled and stored in the laboratory. Use of personal protective equipment, avoiding eating, storing food, smoking or drinking should become workplace policies. Labelling chemicals and avoiding positioning incompatible chemicals next to each other should become part of standard operational procedure. Facilities should have plans on how to deal with emergencies and an emergency preparedness plan with respect to fire, explosion, chemical releases, severe weather, floods, spills, medical emergencies or other disasters.

In addition to concerns about safety in the water infrastructure, groundwater contamination has been attracting the attention of experts for several years. Permeable reactive barriers (PRBs) are widely used to control organic contamination in groundwater (Das, 2002; Das, 2005) (Figure 8.17). They have been effective in reducing groundwater contamination resulting from migration of accidentally or intentionally released pollutants into groundwater. A reactive material is placed underground to interrupt and react with a contaminant in groundwater (Das and Nassehi, 2002).

This technology is more cost-effective than pump-and-treat methods. A variety of materials can be used for permeable reactive barriers (Table 8.9). Apatite II, a waste product from the fish industry, is a good buffer for neutralizing the acidity and is one of the good nonspecific metal sorbents. It is used as reactive media in PRB at the Mine and Mill site in northern Idaho (Geranio, 2007). Typically PRBs are usually built in two configurations: (i) the continuous PRB and (ii) funnel-and-gate. But reactive barriers can also be placed in series of bore wells. Permeable reactive barriers measure about 15 to 20 m below ground level. In funnel-and-gate PRBs, walls are used to direct the contaminant plume towards reactive media, whereas a continuous PRB transects the plume. Reactive media can also be installed above the ground and are referred to as groundwater treatment cells. The targeted contaminant is removed by adsorption, mineral precipitation, or degradation to an innocuous compound. They do not incorporate mechanical devices and are the most promising passive treatment technologies (NTUA, 2000).

Table 8.9 Reactive material used in PRB.

Sl. no.	Contaminants	Reactive material
1	*Organic*: 1,1,1-trichloroethane;1,2-dibromoethane; 1,1,2-trichloroethane; dichloromethane; 1,1-dichloroethane; tetrachloromethane; trichloromethane; tetrachloroethylene; cis-1,2-dichloroethene;trichloroethene. vinyl chloride; 1,1-dichloroethene; 1,2-dichloropropane; benzene; Freon 113; toluene;trans-1,2-dichloroethene; ethylbenzene; hexachlorobutadiene; N-nitrosodimethylamine	Fe^0 Iron sponge Cu coated Fe Pd, Ni, Fe^0, sand, concrete mixture Zero-valent iron pellets (Figure 8.19)
	Inorganic: As, Cd, Cr, Co, Cu, Hg, Fe, Pb, Ni, U, Tc, Mn, Se, SO_4, NO_3, PO_4, Zn,	
2	Cd, Co, Cu, NO_3, SO_4, Pb, Ni, Zn	Materials with organic carbon composted leaf mulch, pine mulch, manure, sewage sludge, sawdust, leaf, peat, pine bark, wood waste.
3	As, Cr, U, Mo, PO_4, Se	Limestone, hydrated lime
4	As, Mo,U,	Phosphates
5	As, Cr, Mo, U,	Ferrous sulfate
6	Ba, Cr, PCE, Sr,	Natural zeolites: mordenite Surfactant modified zeolite (SMZ), clinoptilolite,
7	As, PO_4, Sr, U	Basic oxygen furnace oxide (BOF), Amorphous ferric oxide (AFO), Iron oxide,
8	As, PO_4, Sr	Activated alumina
9	Halogenated hydrocarbons; aromatic compounds	Organic polymers: Cyclophane I, II
10	Cr	Sodium dithionite
11	Alpha-hexachlorobenzenes (HCB), DDD, ethylbenzene, DDT, lindane, beta-HCB, methyl parathion xylene,	Activated carbon
12	Metals, sulfate and nitrate.	Apatite, apatite II
13	HCB, DDD, beta-HCB, DDT, ethylbenzene, methyl parathion, lindane, xylene	Micro-organisms: A. Putrefaciens, G. matallireducens,

Apart from PRB, slurry walls, grout walls, sheet piling and groundwater pumping are used to protect groundwater quality.

Slurry walls: a slurry trench is excavated to the impervious layer. The trench is normally 1 to 1.5 m in width. Slurry of a 4% to 7% bentonite clay suspension in water is mixed with excavated soil and other suitable soil, along with additives to form a very low permeable wall. An HDPE membrane can be used to line the excavated trench or as a curtain at the centre of the trench.

Grout curtains: in this method, suspension grouts of bentonite and/or Portland cement are injected in single, double or multiple lines of holes drilled into the earth in such a way that the grout from adjacent holes merges to form a continuous barrier.

Sheet piling cutoff walls: in this method, wood, precast concrete, sheet metal or any material is use to form a cutoff wall.

Groundwater pumping: the groundwater pumping at a contamination site is done to (i) lower the water table (Figure 8.18), (ii) contain a plume, (iii) pump contaminated groundwater for treatment.

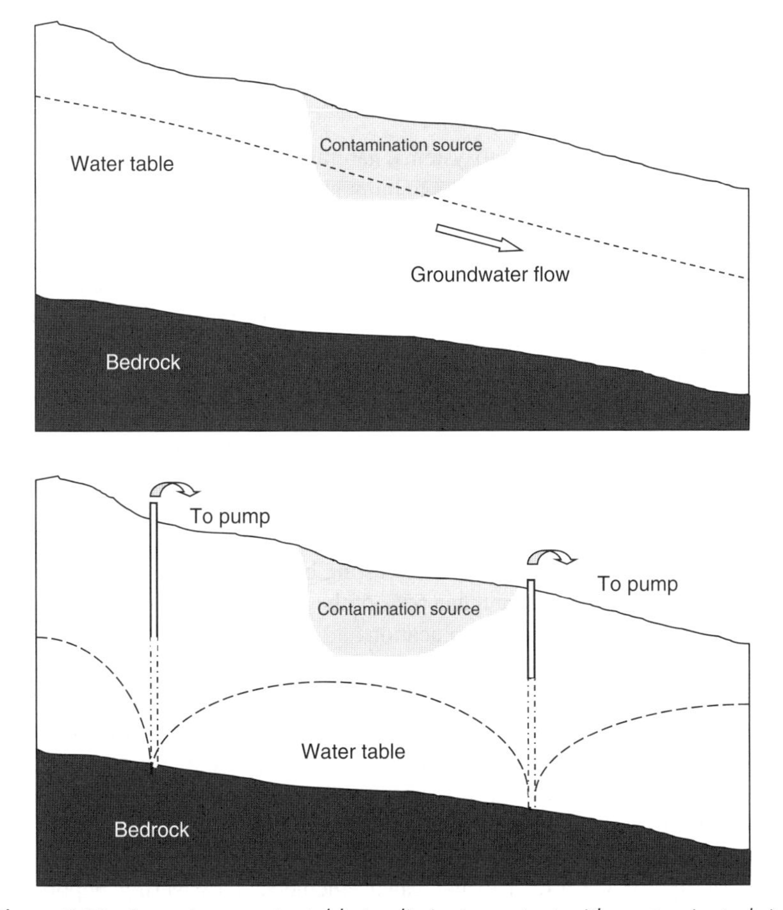

Figure 8.18 *Lowering a water table to eliminate contact with contaminated site.*

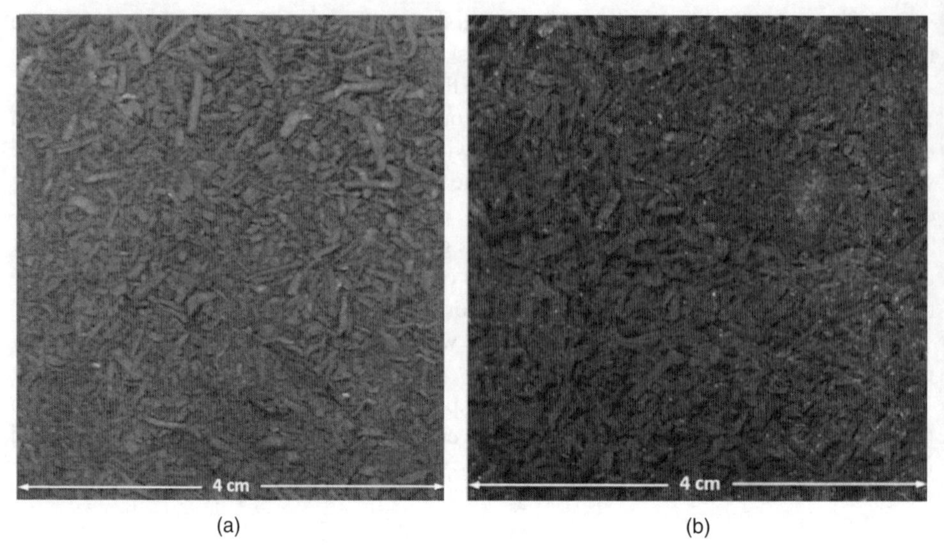

Figure 8.19 *Zero valent iron (a) before adsorption (b) after adsorption.*

Disasters can cause a threat to ecosystems, natural habitats, life and property. They can cause physical and psychological damage to humans. During the latter part of the twentieth century, many water bodies, estuaries, continental shelves, bays and seas experienced a significant oxygen deficiency at the bottom, referred as dead zones. These are areas in water bodies where the concentration of oxygen is below 2 mg/l, which is too low to support aquatic animals that need oxygen. The noted dead zones are the Gulf of Mexico, the deep hypoxic zone of Baltic Sea and Chesapeake Bay. Around 169 hypoxic zones have been documented and they are increasing (Selman *et al.*, 2008).

The most important fishing grounds are on and along continental shelves located less than 200 nautical miles from sea shores. Distribution of these fishing grounds is very localized. Primary fishing areas are likely to become more and more infested by invasive species, most of them introduced from ship ballast water (Nellemann *et al.*, 2008). Each of the big five stressors, (i) climate change, (ii) pollution, (iii) fragmentation and habitat loss, (iv) invasive species infestations and (v) overharvesting by fisheries, results in severe impacts on the biological environment. More than 65% of the global sea grass communities are lost due to land reclamation, disease, eutrophication and unsustainable fishing practices (Lotze *et al.*, 2006).

While the physical footprints of roads, canals and water works are relatively minute, the *ecological* footprint of these infrastructures extends further. Effects of these infrastructures include habitat loss and fragmentation. The magnitude of the effects depends on the characteristics of the ecology. In deserts, some species use canals as a water source. But bighorn sheep, Sonoran pronghorn and desert mule deer have drowned in canals (Rautenstrauch and Krausman, 1989).

Ensuring safety for wildlife in water canals includes: (i) ensuring opportunities for animals/birds to cross the canal, (ii) installing fencing where crossing structures are absent,

(iii) providing alternative sources of water adjoining crossing structures, (iv) providing escape structures for animals along the canal, which do not have a fencing or crossing structure.

8.4 Safety Issues during Construction

Safety issues on construction sites vary with place, season and time. In 2009, construction workers in private industry had a fatal injury rate of about three times that of all workers in the United States. Cases of lethal injuries in construction reduced to 816 in 2009 from 975 in 2008. The proportion of fatalities amongst labourers rose from 22% in 2008 to 25% in 2009 and 48% of all fatal falls that resulted in death in private industry in 2009 were from construction activities followed by transportation (25%) in construction, contact with equipment/objects (19%) and exposure to hazardous substances or environments (16%) (USBLS, 2011).

Working and living conditions of construction labourers were poor in north-eastern Thailand (Thinkhamrop *et al.*, 1997). The camp sites studied were unhygienic and 93% of workers never used helmets. Less than 10% of the workers reported that safety belts, gloves, shoes and helmets were provided by the companies. Thirty-eight per cent of workers from the bigger sites had at least one day away from work per month as compared to 29% in smaller sites.

The water bodies were historically used by large number of people with vested water rights, granting the right to use water for public and private purposes. With the increase in disputes on water rights, legal systems have evolved in many parts of the world that have set up new official bodies for administering river water. Planning construction site layouts is one of the most significant aspects to safeguard people and the environment. Site management (Figure 8.20) includes identifying the places for temporary facilities like storage areas for material, security fences, access roads, equipment, site offices, factory shops, soil heaps present on the site after excavation and concrete-mixing plants (Yeh, 1995; Hegazy and Elbeltagi, 1999). Proper site layout can (i) reduce the cost of materials handling, (ii) lower travel times of material, labour and equipment on site, (iii) improve construction productivity and (iv) improve safety and quality (Tommelein *et al.*, 1992; Anumba and Bishop, 1997). Preparing good site layout is a significant and vital task that should be performed properly during the planning and construction phases.

Common precautions to be taken in construction sites are:

(A) Platform and scaffolding
   (a) Proper flooring with preclusion to avoid skidding.
   (b) Safe access to movable platforms.
   (c) Rigid footing for scaffolding.
   (d) Proper jacks and levelling screws.
(B) Roadways
   (a) Surfaces are in good condition without potholes and humps.
   (b) Skid proof.
   (c) Standard signs and marks.
   (d) Proper preparation for seasonal weather extremes such as snow, rain and so on.

Figure 8.20 *Building under construction covered completely to avoid dust nuisance.*

(C) Signs and tags
   (a) Hazard warning.
   (b) Tags attached to defective equipment.
   (c) Racks/platform load limits are posted.
(D) Stacking and storage
   (a) Aisles and access paths are unobstructed.
   (b) All stalks are stable against sliding collapsing.
   (c) Proper drainage in storage area.
   (d) Proper housekeeping.
   (e) Precautions to be taken to check presence of snakes/scorpions and insects.
(E) Stairs
   (a) Treads are slip resistant.
   (b) Handrails are provided on open sides.
   (c) Adequately lighted.
   (d) Long flights are provided with rest platforms at regular intervals.
   (e) Unobstructed.
(F) Valves and mechanical control
   (a) Colour coded and labelled.
   (b) Readily accessible.

(G) Ventilation
  (a) Enclosures are provided with continuous inflow of air.
  (b) Adequate ventilation is provided.
(H) Warning system
  (a) Fire/emergency system is in place and functional.
  (b) Appropriate warning system is provided on vehicles.
(I) Nets
  (a) Proper nets are provided to avoid injuries due to falling objects.
  (b) Nets with proper opening are provided with adequate strength.

Cranes are used for handling heavy objects and equipment on construction sites. Available statistics suggest that cranes and falling objects are major causes of construction accidents (Anumba and Bishop, 1997; NIOSH, 2000; Hiller and Schneider, 2001; El-Rayes and Khalafallah, 2005). Crane operations caused 151 safety violations during the recovery efforts after the collapse of the World Trade Center in New York on 11 September 2001(OSHA, 2003a).

An average of 71 deaths occurs every year in the United States due to crane accidents (OSHA, 2003b). Hazardous materials and equipment are used and located on construction locations, exposing construction labourers to risk.

Trenching is a common construction activity that is carried out at construction sites for (i) water treatment plants, (ii) wastewater collection, (iii) water distribution, (iv) wastewater treatment plants. A trench is confined and dangerous as both walls can collapse, killing the workers. A cave-in can crush workers. Soil conditions are important factors that affect the stability of soil.

Conditions that can be causes of trench failure are (Merritt, 1986; Asfahl, 1990; Fang, 1991): (i) previously excavated areas, (ii) intersecting trenches, (iii) wet soils, (iv) clay soil, (v) layers of different soils, (vi) loose soil, (vii) cracks in rocks.

Hazard-increasing factors in trenches are: (i) rainy weather, (ii) vibrations caused by vehicle equipment, (iii) equipment or material too near the edge, (iv) load from existing structures

Safety precautions for trenching include (Hammer, 1981; Jannadi, 2008; Helander, 1981; Ridley, 1983): (i) studying soil conditions, (ii) providing perimeter guarding at the job site, (iii) providing training in safe practices for all workers, (iii) providing PPE to people involved in trenching operation, (iv) good housekeeping, (v) providing barricades around the trench, flagging, warning signs and lights, traffic signs and proper bridges across the trench for people and traffic if needed, (vi) providing first-aid facilities, (vii) keeping equipment in good operating condition, (viii) protection for existing utility lines.

Electric shock occurs when an animal/human becomes part of a path of an electric current. Injury or death can occur as a result. The expression is typically used to explain an injurious exposure to electric current (Patrick, 1998). Even though the electric current may be well below the strength needed to cause noticeable injury, a person's reaction can result in a fall or movement into machinery leading to injury or death. Burns can occur when a person touches an electrical wire or equipment that is improperly used/maintained. Usually such injuries/burns occur on the hands. Arc-blasts occur due to high-amperage electric currents arcing through air. Explosions occur when electricity ignites a mixture of explosives in the atmosphere. Electricity is one of the causes of fire due to defective or misused electrical equipment. High voltage results in destruction of tissue.

The severity of the effect on the body due to electricity depends on: (i) the amount of electric current through the body, (ii) the path of the current through the body, (iii) the duration of exposure to the electric current and (iv) the current's frequency. The current felt by humans depends on whether it is AC or DC and its frequency. Generally, 1 mA is not perceptible. A person can feel 1 mA of AC at 60 Hz, whereas at least 5 mA is required for same effect for DC. A current of 5 mA is painful. Men can be thrown if extensor muscles are stimulated at a current of 9–30 mA. A current of 6–25 mA leads to painful shock and loss of muscular control for women; 50–150 mA results in extreme pain, severe muscular contractions and respiratory arrest resulting in possible death. A current of magnitude 1000–4300 mA can stop rhythmic heart pumping, damage nerve and causes muscular contraction. An electric current of 10 000 mA results in cardiac arrest, severe burns and probable death (Kouwenhoven, 1998).

8.4.1 Electrical Protective Devices

Electrical protective devices include circuit breakers, fuses and ground-fault circuit-interrupters (GFCIs). These devices should be of the correct size and type to interrupt the flow of current when it surpasses the capacity of the conductor. Overcurrent can occur due to overload and when insulation fails in a circuit. A short circuit can occur between: (i) an enclosure and conductor or (ii) between two conductors. A fuse is a device that opens a circuit when the current through it exceeds its rating. The fuse contains a special metal wire/strip designed to melt and blow out when the current through it exceeds its rating.

Circuit breakers are protective devices that open circuits during excessive current to avoid the inconvenience of changing fuses.

A ground-fault circuit-interrupter is used to open a circuit when the current through the circuit does not return by the prescribed route.

8.5 Chemical Handling and Storage

Chemicals are used extensively in wastewater treatment. Proper storage and handling should be followed to eliminate risks associated with chemicals. Chemicals should be labelled properly and stored as per information specified in the MSDS. The eye wash, first-aid supplies and shower facilities, emergency phone numbers, spill-cleanup supplies, PPE and fire extinguishers should be readily available.

Many chemicals react violently when mixed. Mixing/storing the following chemical classes together would lead to dangerous reactions: (i) bleaches, (ii) acids and alkalies, (iii) reducing agents, (iv) oxidizing agents and (v) solvents and flammables.

Some major incompatible chemicals that result in hazards are: (i) nitric acid with sulfuric acid, (ii) ammonia with hypochlorite bleach, (iii) nitric acid with acetic acid, (iv) 1-butanol with strong mineral acids, (v) n–n-dimethyl form amide with halogenated hydrocarbons, (vi) n-butyl amine with copper and copper alloys, (vii) ethylene glycol with sulfuric acid, (viii) ethyl acetate with strong alkalies, (ix) ethylene dichloride with oxidizing materials, (x) 1,1,1 trichloroethane with caustic soda and caustic potash, (xi) MEK peroxide with anything flammable.

Corrosive, oxidizing and reactive chemicals create hazards and, hence, require precautions. Inadvertent mixing may cause adverse reactions that can lead to the release of toxic/flammable substances resulting in fires and/or explosions. The following measures need to be observed while handling such chemicals:

- Corrosive/oxidizing/reactive chemicals must be separated from flammable/incompatible chemicals to minimize intermixing during spills.
- Workers handling corrosive, oxidizing, or reactive waste need to be provided with specialized training.
- People handling corrosive/oxidizing/reactive chemicals should wear, appropriate PPEs.
- Where corrosive/oxidizing/reactive waste is handled/stored, qualified first-aid should be ensured.
- Asbestos-containing materials should be handled by specially trained personnel.

Practising safety in the laboratory also calls for safety procedures to safeguard people working there.

General checkpoints for chemical storage (Figure 8.21) are:

- tanks/drums of appropriate material;
- onsite storage based on safety requirement and emergency preparedness;
- storage places adequately vented;
- pressure-relief valves provided on tanks;
- adequate provision to contain spills;
- provision for proper spill absorbance and drainage;
- container corrosion prevention;
- adequate fire-resistant storage cabinet with venting;
- prohibition of smoking and postage of no-smoking boards;
- approved safety containers.

Figure 8.21 *Chemicals stored at a wastewater treatment plant site.*

General checkpoints for compressed gases are:

- they should be stored upright and secured against failing over;
- they should be legibly marked;
- caps should be in place and hand tight;
- protection against rust/corrosion;
- they should be stored away from heat sources;
- they should be stored away from stairs, elevators and exit routes;
- inspection for denting, corrosion, test records;
- there should be sufficient ventilation.

8.5.1 Chlorine

Chlorine gas is relatively stable; however, precautions should be taken as a large volume of gas can be released in a short duration of time. Chlorine cylinders should be stored in a proper enclosure to avoid unauthorized tampering (Figure 8.22). Cylinders should be chained to a wall to prevent tipping. Open the cylinder valve only as much as needed. Check for leakages of chlorine using an ammonia bottle after replacing cylinders.

Dry chlorine is available as: granular chlorine, liquid (household bleach), tablet chlorine (Sanuril) and gas. Dry chlorine, whether tablet or powder, is volatile when mixed with other substances; hence, it should be kept in a dry container with a seal to prevent contact with moisture. Dry chlorine should be handled with clean containers that should only be used for handling it. It should not be stored in an environment with electrical controls as it can cause corrosion of electrical wiring and controls if left open. A chlorine gas mask should be available in the plant in case of emergency.

Figure 8.22 Chlorine stored at wastewater treatment plant site. (For a colour version of this figure, see the colour plate section.)

8.5.2 Herbicides/Pesticides

Herbicides/pesticides are used in WWTP to control weeds and to control algae in lagoons. Herbicides/pesticide should be added in a clean, disposable container wearing appropriate PPE. Herbicides/pesticides should be mixed in the proper dosages using recommended application procedures. Unused substances and packaging materials should be disposed of properly.

8.6 Safety during Water/Wastewater Treatment Plant Operation

Employees in water/wastewater work are exposed to personal injuries due to slippery ladders, steps and walks; exposed machinery; fast-flowing water and deep tanks; electric currents; heavy lifting and explosive and asphyxiating gases. Accidents are normally caused by unsafe acts by people and/or due to hazardous conditions. Safety issues during winter include slippery workplaces and increased flows in streams (Figure 8.23).

Water-insoluble compounds are released into the atmosphere during aeration, splashing and so on. Emission may occur from the primary clarifier weir trickling filteret. Metals tend to pass through into the receiving water or accumulate in sludge. Mercury is a possible exception as it may be air-stripped if it is present in the effluent in sufficient concentrations. Mercury metal removal by activated sludge shows whether the plant is acclimated to the metal. The metals content of activated sludge will be about two to five times that of primary sludge (Brown, 1997).

Many accidents can be avoided by proper safety procedures. Safe wastewater treatment plants adopting conventional treatment should protect pipes and fixtures against corrosion and harsh environments. Treatment plants in the developed countries use PVC-coated conduits and fittings to provide superior protection against corrosion and liquid ingress. Such conduits and fittings may not be easily available in many parts of the world, thus making conventional treatments unsustainable.

If we fail to identify/evaluate/control hazards, we will be killed/injured. The simplest way to achieve safe living is:

1. Identify hazards. The first and foremost step of the safety model is to recognize the hazards. Hence we should plan and discuss hazard recognition with co-workers rather than people who will never work in the place or who worked in similar places. Constant feedback is essential as safety depends on the reality on the ground, which varies from place to place and depends on the personnel involved.
2. Evaluate hazards. It is necessary to recognize all possible hazards, considering the background of the personnel, the plant setup, climate, seismic zone and location with respect to water bodies. It is not prudent to make assumptions about the risk until the hazard is fully evaluated.
3. Control hazards. Once hazards are identified and evaluated, action should be taken to prevent the risk and arrangement should be made for emergency preparedness.

Some of the costs involved in safety are (i) wages of the sick/injured people and costs of covering their jobs, (ii) lost productivity and inexperienced replacements, (iii) damage to products, equipment and property, (iv) costs towards correcting the problem, (v) fines and legal costs, (vi) impact on the environment, (vii) loss of confidence amongst customers and investors and (viii) loss of reputation.

Figure 8.23 *Safety issues during winter include slippery workplace and increased flows in streams.*

Insurance premiums will depend on the precautionary measures taken at the site. Plant layout is a compromise between numerous factors like: (i) distances for transfer of materials, (ii) interaction with existing/planned facilities with respect to roadways, drainage and utilities, (iii) the geographical limitations at the site, (iv) interaction with neighbouring activity, (v) operability and maintainability of facility, (vi) people living in the neighbourhood, (vii) ecological setup, (viii) materials used/handled/transported/stored in the vicinity of the facility, (ix) access for emergency services, (x) emergency escape routes for personnel.

The important factors of plant layout are: (i) the need to prevent, limit and/or mitigate acceleration of adjacent events (domino effects); (ii) the control of access of unauthorized personnel; (iii) provision of access for emergency services and provision of escape routes.

The major principles inherent in safety are: (i) intensification to reduce inventories; (ii) substitution of hazardous material with less hazardous alternatives; (iii) reduction of hazardous processes/conditions; (iv) simpler systems/processes, which have the potential to reduce hazardous events; (v) fail-safe design

Piping systems are a commonly used method of conveying water/wastewater. Pipework is made up of several components that include pipes, flanges, supports, valves, strainers, gaskets, bolts, flexible and expansion joints. Safety and failure of each component needs to be considered prior to finalization of the project.

Consumers and people in the locality of water/wastewater treatment plant can be subject to risk due to maintenance procedure. Risk include contamination of water rearouses, contamination of drinking water, death/injury to workers and public. Risk associated with maintenance procedire includes: (i) human factors, (ii) poorly skilled personnel, (iii) unconscious and conscious incompetence, (iv) knowledge of failure rate and maintainability, (v) good maintainability principles and (vi) clear criteria for identification of faults and marginal performance.

The following issues constitute major hazards: (i) failure of safety due to absence of maintenance, (ii) sparks during maintenance, (iii) human error during maintenance, (iv) incompetence of maintenance staff, (v) poor communication, (vi) a lack of spares, (vii) failure to drain and/or isolate hazardous areas/material, which could lead to a release of flammable or toxic material, (viii) scheduled maintenance not being undertaken, (ix) lack of knowledge amongst the staff about where the maintenance is being carried out; (x) failure to recommission facilities correctly after maintenance, and (xi) unauthorized staff performing maintenance.

The lighting system should ensure that accidents are avoided in an area. Safe lighting requirements are given in Table 8.10.

Even though many of the wastewater treatment plants are located in the open, some of the units are being built underground, especially in cities where land is costly. In some other circumstances parts of the water/wastewater treatment unit/operation are located inside a building, making it necessary to have proper ventilation. Ventilation is defined as a method of controlling exposure of people inside to airborne toxic chemicals or flammable vapours.

The design, erection, commissioning and repair of ventilation systems should be done by qualified ventilation engineers or organizations specializing in this field. Workforce ventilation can be classified into: (i) indoor air-quality ventilation, which is used to supply fresh, heated or cooled air to buildings, (ii) dilution ventilation, which dilutes contaminated air in the building or room by exhausting some dirty air and blowing in clean air, (iii) local exhaust ventilation, which captures contaminated air near the source.

Table 8.10 *Safe lighting requirement.*

Activity	Types of work/location	Average illuminance, lux	Minimum measured illuminance, lux
Movement of vehicles/people.	Parking area, internal routes, corridors	20	5
Movement of vehicles/ people in hazardous places; rough work that does not require any viewing of details.	Construction site	50	20
Work that requires limited viewing of details	Machine house	100	50
Work that require perception of details	Offices	200	100
Work that require perception of fine details	Control room	500	200

Housekeeping plays a major role in preventing accidents and ensuring safety. Some of the accidents due to bad housekeeping are: (i) tripping over loose objects, (ii) articles dropping from above, (iii) striking against projecting/poorly stacked/misplaced material, (iv) slipping on greasy/wet/dirty surfaces, (v) tearing parts of the body by projecting nails/wire/steel strapping and so on. Typical examples of poor housekeeping are: (i) excessive material/waste/chips in the working area, (ii) congested aisles, (iii) tools left on machines, (iv) overflowing waste containers, (v) chemicals in open containers, (vi) broken glass/sharp objects, (vii) electric wires across aisles, (viii) dirty light fittings, windows/skylights. The following is good practice for housekeeping: (i) keep aisles clear, (ii) improve storage facilities, (iii) keep floors clean, (iv) paint the walls, (v) maintain the light fittings, (vi) clean the windows, (vii) dispose of solid waste and prevent spillage, (viii) get rid of dust and dirt, (ix) maintain a high standard in dining/rest rooms, (x) keep tools tidy, (xi) keep first-aid kit in proper condition, (xii) inspect fire-control equipment, (xiii) attend regularly to maintenance, (xiv) assign responsibility for cleaning

Management should accept responsibility for good housekeeping. Management must plan consistently and implement and enforce the measures decided upon.

General records to be kept near water and wastewater treatment plants pertaining to safety are: (i) assessment of risk, (ii) risk monitoring (airborne contaminants, noise, lighting, ventilation systems, etc.), (iii) information, instruction and training, (iv) health surveillance, (v) access and egress, (vi) fire control equipment and (vii) residual current devices. The water/wastewater treatment should be inspected and report be documented with corrective action taken.

It is recommended to conduct a walk-through inspection of large facilities every day. A typical safety inspection form is given in Table 8.11 and an accident report form is given in Table 8.12.

Table 8.11 *Inspection report form.*

	Form No:———
	Basic OHS Workplace Inspection Checklist
	Version: Date: / Next Review:

Inspection place:

Inspection date:

Inspection team:

Name	Designation

Work areas	Satisfactory	Unsatisfactory	Comments
Work surfaces located at proper height for job undertaken			
Sufficient rest breaks are taken during work			
Layout of work place reduces twisting/bending/overreaching			
Heavy/frequently used items are stored at proper height			
Free-standing fittings are secure and stable			
People are protected from sharp objects			
Serviceability of equipment			
Windows are safe			
Objects are stored at proper height			
Manual Handling			
All hazards are identified			
Risk assessments are done for all risks			
Appropriate equipment is provided			
Staff are trained for proper material handling			

(continued)

Table 8.11 (Continued)

Work areas	Satisfactory	Unsatisfactory	Comments
Repetitive actions are minimized			
Staff use correct manual handling techniques			
Rest breaks are given			
Preparatory exercises undertaken			
Adequate space is available for manual handling and use of mechanical aids			
Lifting devices/trolleys/stacking aids/handcarts etc. are adequate			
Housekeeping			
Floors, aisles, passageways and landings are clean			
Access/egress points are kept clear			
Storage areas are clean and tidy			
Work areas are clean and tidy			
Rest areas and dining areas are clean and tidy			
Waste disposal arrangements are adequate			
Indoor environment			
Ventilation/airflow is adequate			
Lighting is adequate			
Temperature is comfortable are adequate			
Glare levels satisfactory for tasks			
Noise levels are within the standard			
No smoking policy is adopted			
Access/egress			
Exits are adequately lit, accessible, signposted, not locked			
Passageways are free from trip hazards			
Traffic flow is safe			
Condition of floors, stair treads, carpets, landings, handrails, and so on			
Fire Safety			
Fire equipment is adequate			
Fire equipment is accessible			
Flammable materials and chemicals are stored properly			
Fire-escape evacuation plans are adequate			

Table 8.11 *(Continued)*

Work areas	Satisfactory	Unsatisfactory	Comments
First aid kit is adequate and accessible			
Emergency procedures			
Site emergency plan readily available			
Emergency checklists readily available			
Emergency evacuations/drills practice			
Availability of emergency equipment			
Audibility of sirens and alarm signals			
Electrical safety			
All electric fittings are safe and serviceable			
Serviceability of power outlets			
Tags			
Records			
Labelling			
Circuit breakers, main switches and fuses for power isolation are adequate			
Hand tools			
Correct types being used			
Serviceability and condition			
Condition of air, electrical lines and fittings			
Adequacy of instruction and training			
Availability of risk assessment documentation			
SOP			
SOPs are available and adequate, up to date and readily accessible			
Plant/machinery			
Risk assessment documentation is available			
Guards are adequate			
All machinery is in good condition and stable			
Operating controls are protected from inadvertent operation			

(continued)

Table 8.11 (Continued)

Work areas	Satisfactory	Unsatisfactory	Comments
Hazard areas defined clearly			
Adequacy of instruction and training provided is adequate			
Adequate signage			
Noise levels and vibration are within limits			
Personal protective equipment (PPE)			
Face shields			
Hearing protection			
Eye protection			
Gloves			
Protective clothing			
Safety shoes			
Respirators			
Storage areas			
Defined areas			
Accessibility and layout			
Labelling of substances			
Availability of MSDS			
Ventilation			
Lifting devices			
Availability of risk assessment documentation			
Record licenses and books			
SOPs			
Trolleys/handcarts etc.			
Labelling of load rating			
General			
Policies and procedures			
Risk assessment documentation			
SOPs			
Adherence to speed limitations			

Summary of unsatisfactory aspects

Unsatisfactory aspect	Observation/remarks

Inspection team leader: … … … … … … … … … … … … … … …
Signature: … … … … … … … … … … …

Table 8.12 Accident investigation form.

Sl. no.	Aspect	Description
1	Person who saw the incident	
2	People working with person injured	
3	Name of the supervisor of injured person	
4	Name of the person who instructed the injured person	
5	Name of other people involved with accident	
6	Description of incidents immediately prior to incident	
7	Description of incident	
8	Description of injury and injuries	
9	Description of damages to property	
10	Cost of damage	
11	Description of machines machine/tool/equipment involved in accident	
12	Precautions required to take	
13	Precautions taken	
14	PPE required	
15	PPE used	
16	Details of training needed	
17	Description given by eye witness	
19	Description of safety rules violated	
20	Time at which accident occurred	
21	Whether natural disaster or manmade disaster was associated with accident	
22	Details of last maintenance of machines associated with accident	
23	Description of instructions given	
24	Whether there was supervisor at the time of accident	
25	Description of photo graphs and video taken at the accident site	
26	Details of samples taken at the accident spot	
27	Details of past accidents in any at the site	
28	Details of first aid given	
29	Detailed of postaccident indents and action taken to admit injured person to hospital	

Radioactive elements with short half-lives are more hazardous to the worker as the element will undergo decay in the plant. Little decay will occur in material with a long half-life while that material is in the wastewater/sludge. Radium appears to be removed effectively by the activated sludge process, trickling filters or RBCs (Brown, 1997) and volatile radioactive compounds may enter the air.

Common methods to control hazards at wastewater treatment plants are as follows:

(A) Air contaminates
- (a) Substitution or replacement.
- (b) Isolation of operation.
- (c) Elimination of toxic chemicals.
- (d) Change of process/operation.
- (e) Local exhaust.

(f) Proper ventilation.
(g) Air pollution control equipment.
(h) Wetting-down methods.
(i) Housekeeping.
(j) Personal protective equipment.
(k) Personal hygiene.
(l) Air monitoring.
(m) Plant trees around the WWTP to capture the droplets/particles.
(n) Reduce the amount of air-stripping and aerosol creation by use of finer bubbles for aeration.
(o) Reduce aerosols by using diffused aeration instead of mechanical aeration.
(p) Consider floating covers on aeration basin or use biodegradable oils, ping-pong balls floating on the surface, permanent foam-polyurethane sheets.
(q) Suppressing the droplets above the surface by a single layer screen with 100–200 mesh, fibre beds, foam or granular bed, multiple layer or knitted mesh screen, water spray to beat the wastewater droplets, flat plate/slat over the tank, rotating brush.
(r) Consider disinfecting the airborne microbes by ultraviolet light.

(B) Diseases
(a) Substitution of substances.
(b) Engineering controls.
(c) Standard practice.
(d) Frequent health checkup.
(e) Personal protective equipment.

(C) Noise
(a) Reduction at source by design/maintenance.
(b) Reduction of noise transmission.
(c) Maintenance of machine.
(d) Personal protection equipment.

(D) Vibration
(a) Source reduction.
(b) Isolation.
(c) Maintenance of machine.
(d) Dampening.

(E) Cold
(a) Clothing.
(b) Temperature control of system.

(F) Heat
(a) Clothing.
(b) Ventilation.
(c) Equipment/process change.
(d) Training.
(e) Personal protective equipment.

(G) Radiation
(a) Training.
(b) Monitoring.

   (c) Personal protective equipment.
   (d) Proper job procedure.
   (e) Proper maintenance.
(H) Illumination
   (a) Proper lighting.
   (b) Proper colouring.
   (c) Maintenance.
(I) Physical hazards
   (a) Housekeeping.
   (b) Load handling equipment.
   (c) Proper fencing/handrail.
(J) Chemical hazards
   (a) Isolation of incompatible substances.
   (b) Monitoring of toxic chemicals.
(K) Biological hazards
   (a) Training/education.
   (b) Monitoring.
   (c) Job procedure.
   (d) Housekeeping.
   (e) Personal hygiene.
   (f) Personal protective equipment.
(L) Electrical hazard
   (a) Installation of proper electrical device.
   (b) Training.
   (c) Frequent inspection.
   (d) Installation of safety device.
   (e) High-voltage equipment is closed and secured.
   (f) Control panels are closed and secured.
   (g) Wiring, insulation and fixtures are in good condition.
   (h) Earthing is tested and is in good condition.
   (i) Explosion proof fixtures in flammable dust or vapour areas.
   (j) Flexible chords are free of joins.
   (k) Lockout provisions are made where necessary.
   (l) Electrical equipment is protected from fluids.
   (m) Provisions for safe restarting after power failure.
(M) Ergonomics
   (a) Design/engineering.
   (b) Training/education.
   (c) Administrative control.
   (d) Labour-saving devices.
(N) Physiological hazards.
   (a) Supervisor training.
   (b) Counselling/employee-assistance programme.
   (c) Positive behaviour reinforcement.
(O) Procedural issues
   (a) Proper task procedure.

(b) Proper task instructions.
(c) Task observation.
(d) Compliance with legislation.
(e) Proper labelling/signposting/warning posting/instruction posting.
(f) Proper disaster management plan and emergency-preparedness plan.
(g) Frequent mock drills.

(P) Emergency instructions
(a) Operational placards on emergency control.
(b) Emergency instruction at each work area.
(c) Hazard symbols at hazardous locations.
(d) Backup communication system.

(Q) Emergency rescue equipment
(a) Adequate equipment available and well maintained.
(b) Personnel know how to use the equipment.

(R) Exit
(a) Sufficient exits.
(b) Routes and exits are clearly marked as illuminated.
(c) More than one exit each work area.
(d) Approaches to exits unobstructed.
(e) Flammables are kept out of exit.
(f) Clear of snow/ice.

(S) Eyebath and showers
(a) Readily available where corrosive/irritant chemicals are used.
(b) Sufficient clean water supply is ensured.
(c) Proper signs and instruction are provided.
(d) Flushed frequently to clear clogging and contaminants.

(T) Fire protection.
(a) Portable extinguishers are readily available.
(b) Extinguishers are inspected frequently.
(c) Fire hoses/equipment is properly maintained and marked.
(d) Adequate personnel protective equipment is available for use during firefighting.

(U) First aid
(a) Adequate material is available.
(b) Properly located.

(V) Floors
(a) Clean and free of slip, trip and fall hazards.
(b) Free of protrusions, refuse.
(c) Load limits posted on upper floors.

(W) Hand and portable tools.
(a) Proper storage when in use.
(b) Maintained in good condition.

(X) Ladders
(a) Free of grease and oil.
(b) Properly positioned.

(Y) Snakes, scorpions and insects
(a) Check presence of any dangerous living creatures.
(b) Periodically check for snakes using sniffer dogs.

Working in slippery places will lead to injuries and may expose people to pathogens. Corrosive, oxidizing and reactive chemicals create hazards and hence need precautions. Inadvertent intermixing may cause adverse reactions, which can release flammable/toxic materials. Hence corrosive, oxidizing and reactive chemicals should be separated from flammable and other incompatible chemicals to minimize intermixing during spills. Workers handling corrosive, oxidizing, or reactive material should be provided with specialized training. People handling corrosive, oxidizing and reactive waste should wear appropriate PPEs, like gloves, aprons, splash suits, face shields or goggles and so on. Where corrosive, oxidizing, or reactive substances are handled or stored, qualified first-aid staff need to be available at all times.

Organizations should ensure that all personnel are properly trained before beginning work. All employees must receive job-specific training along with general environment health and safety (EH and S) training on the following topics: (i) the location and content of the safety manual, (ii) physical, biological, chemical, laser and radiation hazards in the workplace, (iii) the location of references explaining hazards and safety practices associated with chemicals (iv) protective measures that employees have to take to avoid injury, (v) procedures for responding to emergencies, (vi) procedures for receiving medical care in the event of injury/exposure and (vii) proper record keeping. Departments and/or supervisors should maintain safety-training records of all personnel, which include site-specific training forms, training certificates, safety training history. Employee training records should be retained for at least a year after the end of employment.

Contaminated clothing should be removed after completion of a job; avoid washing work clothes at home. Showering at work and changing into clean clothes and shoes is good practice and it can substantially reduce the risk to the worker and the general public. Washing hands with water and soap after work or before eating/smoking/drinking will protect the health of the worker.

Confined-space entry issues should be closely monitored care should be taken to ensure that employees are trained properly and follow proper procedure. A confined space is large enough so that an employee can enter but has restricted means of exit or entry and is not intended for continuous occupancy. Confined spaces lacks ventilation and lighting and are bound to have the following dangers: (i) atmospheric dangers due to toxic gases, oxygen deficiency and flammable/explosive gases, (ii) engulfment agents like grain, sand and sludge can envelope and suffocate entrants, (iii) converging configurations can trap and squeeze entrants.

In the wastewater-treatment facility, confined-space entry can result in death. Potential confined spaces in a wastewater-treatment facility include (i) digesters, (ii) aeration basins, (iii) primary tanks, (iv) applicator machines, (v) vaulted sampling pits and (vi) manholes.

Dangers associated with confined-space entry include (i) falls and slips, (ii) germs and diseases, (iii) insects and animals, (iv) traffic and (v) chemical hazards. The written procedures should include (i) identification of confined spaces, (ii) the location and dangers of the confined spaces, (iii) identification of 'entry supervisors' and their duties, (iv) duties of confined space entry team members.

Many treatment tanks are below ground level. Hence, these tanks are provided with stairs for access for routine maintenance, inspection, sampling, testing and repairs. The fall-protection requirement depends on the activities, facilities and the job tasks performed. Ladder-safety systems, full-body harnesses, hoists and tripods are important fall-protection products.

An example of a standard operating procedure for a 'permit-required confined space entry' might be:

1. Ensure wireless/mobile communication to central base.
2. Ensure zero gas detector is present.
3. Make detector functional test.
4. Test space using detector from outside the workspace at every metre at an interval of half a minute at each level.
5. Ensure steps are sound and use a ladder if steps are not sound.
6. Set up tripod over opening.
7. Put on PPE (minimum requirement: safety boots, hardhat and gloves).
8. Entrants should don a full body strap along with a 'D' ring used with mechanical retrieval systems.
9. Hook up strap of entrant to tripod cable.
10. Check functioning and attachment of fall protection aspect of tripod.
11. Make sure supervisor completes the permit (Box 8.2) and sign it.
12. Enter confined space while atmosphere is continuously monitored.
13. An attendant outside the confined space should monitor the entrant continuously and remain alert for alarm and signs of danger.
14. The attendant should be ready to evacuate entrant in case of danger.

Routine maintenance, repairs, inspections and testing may require lockout/tagout. Activities that need this often require de-energizing of the electrical source that provides electricity to equipment like pumps, electrical motors, valves and mixing systems.

Employees need PPE when they perform daily duties. This includes hard hats, safety glasses, foot protection, gloves, face shields, chemical-protective clothing, respirators and fall protection. The usual PPE in water/wastewater works includes (i) goggles, (ii) jackets, (iii) safety helmets, (iv) gloves, (v) overalls, (vi) safety harnesses, (vii) boots, (viii) safety lines, (ix) lanyards, (x) resuscitators, (xi) gas/oxygen detectors, (xii) ventilators, (xiii) respirators, (xiv) self-contained breathing apparatus and (xv) a first-aid kit.

Stacking materials will be dangerous as falling objects and collapsing material can pin or crush people, causing injuries or death. To avoid such incidents, material handlers should take the following precautions:

- Observe height limitations while stacking objects.
- Stack material only up to more than 4 m high if handled manually.
- Stack up to 6 m if using a forklift.
- Stack and level material on properly supported bracing.
- Store baled material more than 0.5 m from walls/partitions inside a building.
- Remove all sharp objects while stacking.
- Stack bags in interlocking rows.
- Make sure that stacks are stable and self-supporting.
- Stack drums/kegs/barrels symmetrically.
- Place plywood sheets/planks/pallets between every tier of barrels/drums/kegs.
- Stack/block cylindrical objects to prevent tilting/spreading.
- Colour the walls or posts with stripes in order to indicate stacking heights.
- Chock the bottom tier of barrels/drums/kegs.

Box 8.2 Sample Confined Space Entry Permit

Confined Space Entry Permit

1. Date and time of Issue of permit: ______________________________
2. Date and time of expiry of permit: ______________________________
3. SOP No. to be followed: ______________________________
4. Job supervisor: ______________________________
5. Stand by personnel: ____________________
   ____
   ____
   ____
6. Site location and description:____________________
7. Purpose of entry:____________________
8. Atmosperic Checks:

Oxygen: __________ mg/m^3 (permissible limit:______)
Toxins: __________ mg/m^3 (permissible limit:______)
Time: __________ mg/m^3 (permissible limit:______)

9. Communication Procedure: ____________________

10. Resque Procedure: ____________________

11. Checklist

Gas monitoring equipment	☐ Yes	☐ No	☐ Not Available
Personal Protective Equipment	☐ Yes	☐ No	☐ Not Available
Fire Extinguisher	☐ Yes	☐ No	☐ Not Available
Full Boady Harness	☐ Yes	☐ No	☐ Not Available
First Aid Kit	☐ Yes	☐ No	☐ Not Available
Respiratos	☐ Yes	☐ No	☐ Not Available
Body strap along with "D" ring	☐ Yes	☐ No	☐ Not Available
Stand by Safety Personnel	☐ Yes	☐ No	☐ Not Available
All personnel are trained	☐ Yes	☐ No	

Emergency Numbers 1.Ambulance:________
2. Fire:____________
3. Police:____________
4. Others:____________
(Please specify name and designation)

Copy of persmit should be available at the time of entry until completion of operation.

Signature of permitting person/authority

Experience has shown that desalination is a reliable source of water but the safety of aquatic organisms is often questionable. The Gulf region produces more than half of the world's desalinated water but environmental risks have not been considered in depth (Khordagui, 2006). The desalination intake has two types of effects: impingement and the entrainment effect. As the seawater is screened and filtered, aquatic organisms are filtered (impingement effect). The smaller organisms that pass through the filter are exposed to high temperature, pressure and chemicals, which endanger their lives (entrainment effect). Cooling sea water for the desalination plant has resulted in the impingement of fishes and macrovertebrates in the Gulf region (Khordagui, 2006). Many cases of fish kills were reported around power-desalination plants. The discharge of residual chlorine poses a real risk to the marine environment. Chlorine reacts with natural organic substances in seawater resulting in the formation of triholomethanes (THMs). Some of the THMs species can enter in seafood and are mutagenic to humans. Antiscalants and metals such as Ni, Cu, Fe and Zn (due to corrosion in the desalination process) find their way into the marine environment, posing a threat to it. In addition to the pollutants already discussed, trace volatile liquid hydrocarbons (VLHs) due to leakage of oil and grease from operating power-desalination plants might also affect the environment.

8.6.1 Work-Permit System

The work-permit system within an organization enables the identification, control and review of hazards within work environment. Safe work permits must be required (i) at entries to confined spaces, (ii) at excavations, (iii) for hot work, (iv) for work in or around confined spaces and (v) when working at heights.

The advantages of the work-permit system are: (i) it ensures entry of authorized people, (ii) it provides clarity about the hazard, (iii) it specifies the precautions, (iv) it ensures the person in charge of the work knows about the work in progress, (v) it provides a system of control and (vi) it provides handover and hand-back procedure.

Before issuing a work permit, the issuing person and recipient should consider all hazards like material hazards, pressure, fumes, temperature, electrical power, hazardous areas, mechanical energy, height, radioactive sources, restricted space field vision, explosive materials, and so on. The work permit should have a checklist of precautions.

The procedure to be followed for obtaining a permit should include a written request followed by the issue of duly filled safety permit by the designated authority. Permits should be in triplicate, serially numbered. Colour codes may be adopted for different permits.

8.7 Disaster Management

Natural and manmade disasters have devastating effects on the environment and humankind. They erode wealth built by individuals and nations and are a serious risk to poverty reduction. Despite all the precautions taken and emergency preparedness, disasters do occur all over the world. Hence the organizations should have a disaster management plan (DMP) in place. The DMP should clearly elaborate probable failures and action to be taken. It should clearly indicate mock drills and frequent trainings. Such practices, which are usually adopted in industry, are absent with respect to dams, irrigation structures, and

Table 8.13 Water treatment method during disaster.

Sl. no.	Method	Description
1	Solar disinfection	In this method contaminated water is exposed to solar light in closed PET bottles for a few hours. The UV rays and heat is sufficient to disinfect the water. The bottle can be kept on roof tops. To increase the efficiency the water bottle can be placed on black surface.
2	Sodium dichloroisocyanurate (NaDCC) tablets	33 mg tablets can disinfect around 20 l of nonturbid water
3	Pot-based filter	Filtration candle can be fixed to earthen pots
4	Saree method	Local method of purifying water in Bangladesh, where people fold a saree into eight sections and pour water through it after placing it on a vessel.
5	Shock chlorination	Additon of 5 to 10 mg/l of bleach, chlorine powder, or liquid bleach to water in a well. The water is kept unused for a few hours and drawn from the well. The term 'shock' is used as the chemical is added a little at a time rather than continuously.
6	Pot chlorination	Bleach or chlorine powder and gravel mixture are put in a chlorination pot/container, with a few holes and placed inside a larger vessel.
7	Sand-filtration technique	Can be used where sand is available. The sand can be kept in a container with a hole for filtration.
8	Ceramic filters	Filters made up of ceramic material.
9	Fuel wood ash	Water and ash are mixed and filtered after keeping the mixture for two hours.
10	Biosanitizer ecochips	Chips made up of biocatalyst containing plant enzymes that degrade the organic matter and neutralize the pH of the medium.
11	Electrochlorinators	This is electrical equipment that can be made to work on solar energy. It can convert ordinary salt into sodium hypochlorite.

water/wastewater treatment/distribution/collection systems. Neither the staff nor citizens will have a clue about how to act during failure of dam, water contamination due to cross connection, or the failure of a canal. Warning systems are usually absent in the developing countries. They should alert people near rivers, canal and other water bodies to show that water levels are rising due to a disaster. Table 8.13 shows some of the common water treatment methods used during disasters in developing countries.

Figure 8.24 shows a schematic diagram of a disaster management plan. The objectives of the DMP are:

1. Rescue and treat casualties.
2. Isolate and cordon off the affected area for a proper rescue operation.
3. Safeguard the uninjured people.

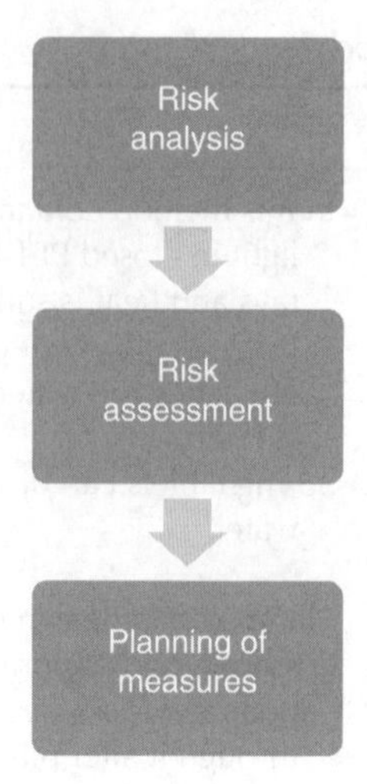

Figure 8.24 *Schematic diagram of disaster management plan.*

4. Contain the situation and bring it under control.
5. Minimize damage to property, people and surroundings.
6. Provide the required information to statutory agencies.
7. Safeguard and secure rehabilitation of the affected area.
8. Ward off prying onlookers and unsocial elements.
9. Provide information to the news media.
10. Counter rumours by providing relevant, accurate information to avoid panic amongst citizens.

The DMP should include both onsite and offsite plans and clearly explain the following:

1. Possible failures and disasters.
2. Geographical area affected due to possible disasters.
3. Responsibilities of each person.
4. Contact information of all the authorities/media to whom information is to be disseminated.
5. Frequency and methodology for mock drills.
6. Escape routes.
7. Location of shelters during disasters.
8. Sources of emergency medicine/food supplies.
9. First aid.
10. Location and requirement of control room.
11. Required infrastructure for communication.

References

Agard, L., Alexander, C., Green, S. *et al.* (2002) Microbial quality of water supply to an urban community in Trinidad. *Journal of Food Protection* **65**(8), 1297–1303.

Alan, J. and Tucci, M.P.A. (2010) *Bloodborne Pathogens*, training material, www.waterworldce.com/courses/23/PDF/bloodborne.pdf (accessed on 19 January 2014).

Anumba, C. and Bishop, G. (1997) Importance of safety considerations in site layout and organization. *Canadian Journal of Civil Engineering* **24** (2), 229–236.

Asfahl, C.R. (1990) *Industrial Safety and Health Management*, 2nd edn, Prentice-Hall, Englewood Cliffs, NJ, pp. 363–385.

Bailey, I.W. and Thompson, P. (1995) Monitoring water quality after disinfection. *Water Supply* **13** (2), 35–48.

Barna, Z. and Kádár, M. (2012) The risk of contracting infectious diseases in public swimming pools. A review. *Annali dell' Istituto Superiore di Sanità* **48** (4), 374–386, doi: 10.4415/ANN_12_04_05

Brown, N.J. (1997) Health Hazard Manual: Wastewater Treatment Plant and Sewer Workers, Cornell University ILR School.

Chen, B.-C., Chou, W.-C., Chen, W.-Y. and Liao, C.-M. (2010) Assessing the cancer risk associated with arsenic-contaminated seafood. *Journal of Hazardous Materials* **181**, 161–169.

Chen, C.J., Hsu, L.I., Wang, C.H. *et al.* (2005) Biomarkers of exposure, effect and susceptibility of arsenic-induced health hazards in Taiwan. *Toxicology and Applied Pharmacology* **206**, 198–206.

Dany, V., Visvanathan, C. and Thanh, N.C. (2000) Evaluation of water supply systems in Phnom Penh City: a review of the present status and future prospects. *International Journal of Water Resources Development* **16** (4), 677–689.

Das, D.B. (2002). Hydrodynamic modelling for groundwater flow through permeable reactive barriers. *Hydrological Processes* **16**, 3393–3418.

Das, D.B. (2005) Hydrodynamic modelling for coupled free and porous flow while designing permeable reactive barriers, in *Proceedings of the International Symposium on Permeable Reactive Barriers,* vol. **298**, IAHS, Wallingford, pp. 136–143.

Das, D.B. and Nassehi, V. (2002) A finite volume model for the hydrodynamics of flow in combined groundwater zone and permeable reactive barriers, in *Advanced Groundwater Remediation* (eds. T. Meggyes, F.-G. Simon and C. McDonald), Thomas Telford, London, pp. 251–263.

Egorov, A., Ford, T., Tereschenko, A. *et al.* (2002) Deterioration of drinking water quality in the distribution system and gastrointestinal morbidity in a Russian city. *International Journal of Environmental Health Research* **12** (3), 221–233.

El-Rayes, M. and Khalafallah, A. (2005) Trade-off between safety and cost in planning construction site layouts. *Journal of Construction Engineering and Management* **131**, 1186–1195.

Fang, H.Y. (1991) *Foundation Engineering Handbook*, 2nd edn, Van Nostrand Reinhold, New York, pp. 379–415.

Gadgil, A. (1998) Drinking water in developing countries. *Annual Review of Energy and the Environment* **23**, 253–286.

Gallagher, T.R. (2010) *Introduction to Distribution System Piping and Valving,* training material, www.waterworldce.com/courses/29/PDF/Intro%20Distrib%20sys.pdf (accessed 19 January 2014).

Geranio, L. (2007) Review of Zero Valent Iron and Apatite as Reactive Materials for Permeable Reactive Barrier. Term Paper SS 07/08, major in Biogeochemistry and Pollutant Dynamics, Department of Environmental Sciences, ETH Zurich, June 2007.

Gleick, P.H. (2006) Water and terrorism. *Water Policy* **8**, 481–503.

Hammer, W. (1981) *Occupational Safety Management and Engineering*, 2nd edn, Prentice-Hall, Englewood Cliffs, NJ 07632, pp. 62–63, 125–129.

Hata, A., Endo, Y., Nakajima, Y. *et al.* (2007) HPLCICP-MS speciation analysis of arsenic in urine of Japanese subjects without occupational exposure. *Journal of Occupational Health* **49**, 217–223.

Hegazy, T. and Elbeltagi, E. (1999) EvoSite: Evolution-based model for site layout planning. *Journal of Computing in Civil Engineering* **13** (3), 198–206.

Heinrich, H.W., Peterson, D. and Roos, N. (1980) *Industrial Accident Prevention*, McGraw-Hill, New York.

Helander, M. (ed.) 1981: *Human Factors/Ergonomics for Building and Construction*, John Wiley & Sons, Inc., New York, p. 32–36.

Hiller, K. and Schneider, C. (2001) OSHA raises the bar for steel construction safety. *Structural Engineering* **2**, 36–38.

Howell, G.A., Ballard, G., Abdelhamid, T. and Mitropoulos, P. (2002) Working near the edge: a new approach to construction safety. Paper presented at the Annual Conference on Lean Construction, 6–8 August 2002, Brazil.

ICLEI (International Council for Local Environmental Initiatives) (1996) *Local Agenda 21 Planning Guide: An Introduction to Sustainable Development Planning*, ICLEI, Toronto.

IFC (2007) *Environmental, Health and Safety General Guidelines*, http://www.ifc.org/wps/wcm/connect/554e8d80488658e4b76af76a6515bb18/Final%2B-%2BGeneral%2BEHS%2BGuidelines.pdf?MOD=AJPERES (accessed 19 January 2014).

IUCN, UNEP, WWF (1991) *Caring for the Earth. A Strategy for Sustainable Living*, Earthscan, London.

Jannadi, O.A. (2008) Risks associated with trenching works in Saudi Arabia. *Building and Environment* **43**, 776–781.

Kamei, D., Kuno, T., Sato, S. *et al.* (2012) Impact of the Fukushima Daiichi Nuclear Power Plant accident on hemodialysis facilities: an evaluation of radioactive contaminants in water used for hemodialysis. *Therapeutic Apheresis and Dialysis* **16** (1), 87–90.

Kinoshita, N., Suekia, K., Sasa, K. *et al.* (2011) Assessment of individual radionuclide distributions from the Fukushima nuclear accident covering central-east Japan. *Proceedings of the National Academy of Sciences of the United States of America* **108**, 19526–19529.

Khordagui, H. (2006) Desalination as a potential technological hazard to the environment in the arid Western Asia region, in *Real Risk* (eds. J. Griffiths and T. Ingelton), Tudor Rose, Leicester, pp. 148–150.

Kouwenhoven, W.B. (1998) Human Safety and Electric Shock. Electrical Safety Practices, Monograph, 112, Instrument Society of America, p. 93. November 1968.

Lee, E.J. and Schwab, K.J. (2005) Deficiency in drinking water distribution system in developing countries. *Journal of Water and Health* **3** (2), 109–126.

Li, L. (2013) *Shanghai Water Supply Affected after Chemical Spill Accident*, http://fmnnow.com/2013/01/11/shanghai-water-supply-affected-after-chemical-spill-accident (accessed 14 January 2013).

Lindberg, A.L., Rahman, M., Persson, L.A. and Vahter, M. (2008) Gender and age differences in the metabolism of inorganic arsenic in a highly exposed population in Bangladesh. *Toxicology and Applied Pharmacology* **230**, 9–16.

Lotze, H.K., Lenihan, H.S., Bourque, B.J. *et al.* (2006) Depletion, degradation and recovery potential of estuaries and coastal seas. *Science* **312** (5781), 1806–1809.

Merritt, F.S. (1986) *Standard Handbook for Civil Engineers*, McGraw-Hill, Singapore.

Moe, C.L., Sobsey, M.D., Samsa, G.P. and Mesolo, V. (1991) Bacterial indicators of risk of diarrhoeal disease from drinking-water in the Philippines. *Bulletin of the World Health Organization* **69** (3), 305–317.

National Bureau of Statistics of China (2008) *China Statistics Yearbook 2007*, China Statistics Press, Beijing.

National Bureau of Statistics of China (2009) *China Statistics Yearbook 2008*, China Statistics Press, Beijing.

Nellemann, C., Hain, S. and Alder, J. (eds) (2008) In Dead Water – Merging of Climate Change with Pollution, Over-harvest and Infestations in the World's Fishing Grounds, GRID, Arendal.

NIOSH (National Institute for Occupational Safety and Health) (2000) Worker Health Chartbook. DHHS (NIOSH), Atlanta, GA.

NTUA (National Technical University of Athens) (2000) *Literature Review: Reactive Materials and Attenuation Processes for Permeable Reactive Barriers*, www.perebar.bam.de/PereOpen/pdfFiles/Review_Reactive_Materials.pdf (accessed on 19 January 2014).

OSHA (2003a) *A Dangerous Worksite: The World Trade Center*, Occupational Safety and Health Administration Publications, Washington, D.C.

OSHA (2003b) *Occupational Safety and Health Administration*, http://www.osha.gov/archive/oshinfo/priorities/crane.html (accessed on 14 January 2014).

Rajendran, S. (2006) Sustainable Construction Safety and Health Rating System, Doctoral thesis. Oregon State University, http://ir.library.oregonstate.edu/xmlui/bitstream/handle/1957/3805/Rajendran%20Dissertation%20Report.pdf?sequence=1 (accessed 19 January 2014).

Rautenstrauch, K.R. and Krausman, P.R. (1989) Preventing mule deer drowning in the Mohawk canal, Arizona. *Wildlife Society Bulletin* **17**, 281–286.

Reilly, J.P. (1998) *Applied Bioelectricity: From Electrical Stimulation to Electropathology*, 2nd edn, Springer, New York. ISBN 978-0-387-98407-0.LCCN 97048860.OCLC 38067651.

Reimann, C., Birke, M. and Filzmoser, P. (2010) Bottled drinking water: Water contamination from bottle materials (glass, hard PET, soft PET), the influence of colour and acidification. *Applied Geochemistry* **25**, 1030–1046.

Ridley, J.R. (1983) *Safety at Work*, Butterworth, London, pp. 167–169, 632–643.

Sasso, S., Laterza, E. and Valenzano, B. (2012) A study about explosion hazards in the presence of an uncontrolled anaerobic digestive process. *Chemical Engineering Transactions* **26** (1), 135–140.

Selman, M., Greenhalgh, S., Diaz, R. and Sugg, Z. (2008) Eutrophication and Hypoxia in Coastal Areas: A Global Assessment of the State of Knowledge. WRI Policy Note Water Quality: Eutrophication and Hypoxia No. 1, World Resources Institute, Washington, DC.

Semenza, J.C., Robert, S.L., Henderson, A. *et al.* (1998) Water distribution system and diarrheal disease transmission: a case study in Uzbekistan. *American Journal of Tropical Medicine and Hygiene* **59** (6), 941–946.

Shotyk, W. and Krachler, M. (2007) Contamination of bottled waters with antimony leaching from polyethylene terephthalate (PET) increases upon storage. *Environmental Science and Technology* **41**, 1560–1563.

Singh, P. (2013) Uttarakhand horror: stranded pilgrims raped, murdered, Hindustan Times (22 June), http://www.hindustantimes.com/India-news/NorthIndiaRainFury2013/Uttarakhand-horror-stranded-pilgrims-raped-murdered/Article1-1080734.aspx (accessed on 14 January 2014).

Thinkhamrop, B., Chiruwatkul, A., Dobson, A. *et al.* (1997) Working and living conditions of construction workers: a comparison between large and small construction sites in Northeastern Thailand. *Southeast Asian Journal of Tropical Medicine and Public Health* **28** (1), 46–54.

Thompson, J, Porras, I.T., Tumwine, J.K. *et al.* (2000) *Drawers of Water II: Thirty Years of Change in Domestic Water Use and Environmental Health in East Africa*, Russell Press, Nottingham.

Tommelein, I.D., Levitt, R.E. and Hayes-Roth, B. (1992) Sight Plan model for site layout. *Journal of Construction Engineering and Management* **118** (4), 749–766.

Ueda, S., Hasegawa, H., Kakiuchi, H. *et al.* (2013) Fluvial discharges of radio caesium from watersheds contaminated by the Fukushima Dai-ichi Nuclear Power Plant accident, Japan. *Journal of Environmental Radioactivity* **118**, 96–104.

UNDP (2007) Human Development Report 2007/2008, Fighting Climate Change: Human Solidarity in a Divided World, United Nations Development Programme, New York.

USBLS (US Bureau of Labor Statistics) (2011) *Fatal Injuries in Construction*, http://www.cdc.gov/niosh/topics/construction/ (accessed on 4 March, 2013).

US Department of Labor (2006) Best Practices Guide: Fundamentals of a Workplace First-Aid Program, US Department of Labor Occupational Safety and Health Administration, Washington DC.

USEPA (2003) Cross-Connection Control Manual, US Environmental Protection Agency, Washington, DC.

WHO and UNICEF (2000) Global Water Supply and Sanitation Assessment 2000 Report, World Health Organisation and United Nations Children's Fund, Washington DC.

World Bank (2007) Water Pollution Emergencies in China: Prevention and Response, World Bank, Washington, DC.

World Commission on Environment and Development (1987) *Our Common Future*, www.un-documents.net/our-common-future.pdf (accessed 19 January 2014).

Yeh, I.C. (1995) Construction-site layout using annealed neural network. *Journal of Computing in Civil Engineering* **9** (3), 201–208.

Xiao, J., Chen, P., Peng, F. *et al.* (2011) Emergency drinking water treatment during source water pollution accidents in China: origin analysis, framework and technologies. *Environmental Science and Technology* **45**, 161–167.

Index

Sustainable Water Engineering: Theory and Practice, First Edition. Ramesha Chandrappa and Diganta B. Das.